# Engineering Materials

For further volumes:
http://www.springer.com/series/4288

Alexei Nazarov · J.-P. Colinge ·
Prof. Francis Balestra · Prof. Jean-Pierre Raskin ·
Prof. Francisco Gamiz ·
Prof. V.S. Lysenko
Editors

# Semiconductor-On-Insulator Materials for Nanoelectronics Applications

*Editors*
Prof. Dr. Alexei Nazarov
Lashkaryov Institute of Semiconductor
 Physics
National Academy of Sciences of Ukraine
Prospect Nauki 41
Kyiv 03028, Ukraine
e-mail: nazarov@lab15.kiev.ua

Prof. Francis Balestra
Sinano Institute
IMEP-LAHC
rue Parvis Louis Néel 3
38016 Grenoble Cedex 1
France

Prof. Francisco Gamiz
Nanoelectronics Research Group
Departmento de Electrónica
Universidad de Granada
Granada
Spain

Prof. J.-P. Colinge
Tyndall National Institute "Lee Maltings"
Prospect Row
University College
Cork
Ireland
e-mail: jean-pierre.colinge@tyndall.ie

Prof. Jean-Pierre Raskin
Microwave Laboratory
Université Catholique de Louvain
Place du Levant 3 á
B-1348 Louvain
Belgium

Prof. V.S. Lysenko
Lashkaryov Institute of Semiconductor
 Physics
National Academy of Sciences of Ukraine
Prospect Nauki 41
Kyiv 03028, Ukraine

ISSN 1612-1317

ISBN 978-3-642-15867-4

DOI 10.1007/978-3-642-15868-1

e-ISSN 1868-1212

e-ISBN 978-3-642-15868-1

Springer Heidelberg Dordrecht London New York

© Springer-Verlag Berlin Heidelberg 2011

This work is subject to copyright. All rights are reserved, whether the whole or part of the material is concerned, specifically the rights of translation, reprinting, reuse of illustrations, recitation, broadcasting, reproduction on microfilm or in any other way, and storage in data banks. Duplication of this publication or parts thereof is permitted only under the provisions of the German Copyright Law of September 9, 1965, in its current version, and permission for use must always be obtained from Springer. Violations are liable to prosecution under the German Copyright Law.

The use of general descriptive names, registered names, trademarks, etc. in this publication does not imply, even in the absence of a specific statement, that such names are exempt from the relevant protective laws and regulations and therefore free for general use.

*Cover design:* deblik, Berlin

Printed on acid-free paper

Springer is part of Springer Science+Business Media (www.springer.com)

# Introduction

This Book is devoted to the fast evolving field of nanoelectronics, and more particularly to the physics and technology of nanoelectronic devices built on semiconductor-on-insulator (SemOI) substrates. It compiles the results of research work from leading companies and universities in Europe, Russia, Brazil and Ukraine. Main of the Authors are involved in the NANOSIL Network of Excellence and the Thematic Network on Silicon on Insulator Technology, Devices and Circuits EUROSOI+ , both of which are funded by the European Commission under the 7th Framework Program of the European Community.

This Book describes different technologies and approaches used to fabricate Semiconductor-On-Insulator materials, devices and systems. The contributed papers are articulated around four main themes:

1. Fabrication of new semiconductor-on-insulator materials
2. Physics of modern SemOI devices
3. Advanced characterization of SemOI devices
4  Sensors and MEMS on SOI.

The first chapter is focused on techniques for producing new SemOI materials. The formation of thin germanium-on-insulator films allows one to fabricate a wide variety of devices and constitutes a fundamental step towards the design of monolithic hybrid Si–Ge systems and integrated circuits. SOI wafers on high-resistivity substrates can be used for RF and mixed-signal system-on-a chip (SoC) applications. Special substrates employing porous silicon technology allows one to fabricate III–V and column-IV alloy semiconductor films on dielectric substrates. Confined and guided growth of silicon nanoribbons creates new technological opportunities for the fabrication of field-effect transistors and ICs.

The second Chapter is devoted to the physics and electrical properties of novel SemOI devices such as ultrathin-body, fully depleted SOIMOSFETS scaled down to 22 nm and beyond, multigateFinFET devices, nanowire transistors using either lightly doped or highly doped silicon (junctionlessMOSFETs), carbon nanotubes (CNTMOSFETs) and single-electron devices. Novel phenomena such as quantum effects and Coulomb blockade effect occurring in nanoscale devices are described as well.

The third part of the Book focuses on advanced electrical characterization techniques for nanoelectronic devices, such as novel methods for extracting mobility, transconductance and noise.

Finally, the fourth Chapter is devoted to application of SemOI materials for biosensors, chemical sensors and MEMS. The use of SemOI substrates allow for considerable increase of sensitivity of the sensors, as well as for the fabrications of MEMS compatible with CMOS technology.

This Book will be useful not only to specialists in nano- and microelectronics but also to students and to the wider audience of readers who are interested by new directions in modern electronics and optoelectronics.

The editors would like to express their gratitude to SergeyGordienko for his help collating and formatting the different manuscripts and, of course, to the Springer Editor Dr. Mayra Castro for her highly appreciated assistance throughout the editing process.

Alexei Nazarov<br>
J.-P. Colinge<br>
Francis Balestra<br>
Jean-Pierre Raskin<br>
Francisco Gamiz<br>
V.S. Lysenko

# Contents

**Part I New Semiconductor-On-Insulator Materials**

**Germanium Processing** .................................... 3
H. Gamble, B. M. Armstrong, P. T. Baine, Y. H. Low,
P. V. Rainey, S. J. N. Mitchell and D. W. McNeill

**Low-Temperature Fabrication of Germanium-on-Insulator Using
Remote Plasma Activation Bonding and Hydrogen Exfoliation** ...... 31
C. A. Colinge, K. Y. Byun, I. P. Ferain, R. Yu and M. Goorsky

**Engineering Pseudosubstrates with Porous Silicon Technology** ...... 47
N. P. Blanchard, A. Boucherif, Ph. Regreny, A. Danescu, H. Magoariec,
J. Penuelas, V. Lysenko, J.-M. Bluet, O. Marty, G. Guillot and G. Grenet

**Confined and Guided Vapor–Liquid–Solid Catalytic Growth
of Silicon Nanoribbons: From Nanowires to Structured
Silicon-on-Insulator Layers** ................................ 67
A. Lecestre, E. Dubois, A. Villaret, T. Skotnicki, P. Coronel,
G. Patriarche and C. Maurice

**SOI CMOS: A Mature and Still Improving Technology
for RF Applications**. ..................................... 91
Jean-Pierre Raskin

## Part II  Physics of Modern SemOI Devices

**Silicon-based Devices and Materials for Nanoscale FETs** .......... 123
Francis Balestra

**FinFETs and Their Futures**................................. 141
N. Horiguchi, B. Parvais, T Chiarella, N. Collaert, A. Veloso,
R. Rooyackers, P. Verheyen, L. Witters, A. Redolfi, A. De Keersgieter,
S. Brus, G. Zschaetzsch, M. Ercken, E. Altamirano, S. Locorotondo,
M. Demand, M. Jurczak, W. Vandervorst, T. Hoffmann and S. Biesemans

**Ultrathin Body Silicon on Insulator Transistors for 22 nm
Node and Beyond** ........................................ 155
T. Poiroux, F. Andrieu, O. Weber, C. Fenouillet-Béranger,
C. Buj-Dufournet, P. Perreau, L. Tosti, L. Brevard and O. Faynot

**Ultrathin n-Channel and p-Channel SOI MOSFETs**............... 169
F. Gámiz, L. Donetti, C. Sampedro, A. Godoy,
N. Rodríguez and F. Jiménez-Molinos

**Junctionless Transistors: Physics and Properties** ................ 187
J. P. Colinge, C. W. Lee, N. Dehdashti Akhavan, R. Yan, I. Ferain,
P. Razavi, A. Kranti and R. Yu

**Gate Modulated Resonant Tunneling Transistor (RT-FET):
Performance Investigation of a Steep Slope, High
On-Current Device Through 3D Non-Equilibrium
Green Function Simulations**................................ 201
Aryan Afzalian, Jean-Pierre Colinge and Denis Flandre

**Ohmic and Schottky Contact CNTFET: Transport Properties
and Device Performance Using Semi-classical
and Quantum Particle Simulation** ........................... 215
Huu-Nha Nguyen, Damien Querlioz, Arnaud Bournel,
Sylvie Retailleau and Philippe Dollfus

**Quantum Simulation of Silicon-Nanowire FETs** ................. 237
Marco Pala

**Single Dopant and Single Electron Effects in CMOS Devices** ....... 251
M. Sanquer, X. Jehl, M. Pierre, B. Roche, M. Vinet and R. Wacquez

## Part III  Diagnostics of the SOI Devices

**SOI MOSFET Transconductance Behavior from Micro to Nano Era** ............................... 267
J. A. Martino, P. G. D. Agopian, E. Simoen and C. Claeys

**Investigation of Tri-Gate FinFETs by Noise Methods** ............. 287
N. Lukyanchikova, N. Garbar, V. Kudina, A. Smolanka,
E. Simoen and C. Claeys

**Mobility Characterization in Advanced FD-SOI CMOS Devices** ..... 307
G. Ghibaudo

**Special Features of the Back-Gate Effects in Ultra-Thin Body SOI MOSFETs** .................................... 323
T. Rudenko, V. Kilchytska, J.-P. Raskin, A. Nazarov and D. Flandre

## Part IV  Sensors and MEMS on Memory SOI

**SOI Nanowire Transistors for Femtomole Electronic Detectors of Single Particles and Molecules in Bioliquids and Gases** .......... 343
V. P. Popov, O. V. Naumova and Yu. D. Ivanov

**Sensing and MEMS Devices in Thin-Film SOI MOS Technology** .... 355
J.-P. Raskin, L. Francis and D. Flandre

**Floating-Body SOI Memory: The Scaling Tournament** ............ 393
M. Bawedin, S. Cristoloveanu, A. Hubert, K. H. Park and F. Martinez

## Part V  Afterword

**A Selection of SOI Puzzles and Tentative Answers** ............... 425
S. Cristoloveanu, M. Bawedin, K.-I. Na, W. Van Den Daele,
K.-H. Park, L. Pham-Nguyen, J. Wan, K. Tachi, S.-J. Chang, I. Ionica,
A. Diab, Y.-H. Bae, J. A. Chroboczek, A. Ohata, C. Fenouillet-Beranger,
T. Ernst, E. Augendre, C. Le Royer, A. Zaslavsky and H. Iwai

**Index** ........................................................ 443

# Part I
# New Semiconductor-On-Insulator Materials

# Germanium Processing

H. Gamble, B. M. Armstrong, P. T. Baine, Y. H. Low, P. V. Rainey,
S. J. N. Mitchell and D. W. McNeill

**Abstract** This paper reviews the development of germanium technology for applications in high performance CMOS ICs, rf and MMICs. The paper covers the development of MOSFET technology with respect to source/drain doping and gate dielectrics. Germanium has higher junction leakage currents than silicon on account of its lower energy bandgap. It is a scarce material, expensive and the wafer size is limited. To minimize these disadvantages germanium will be employed as a thin layer on an insulator substrate. Various methods of producing germanium-on-insulator (GeOI) substrates are outlined. These include the Smart-cut process, the condensation process starting with SOI wafers and the epitaxial growth of germanium on lattice matched crystalline oxides grown on silicon substrates. Partial GeOI layer techniques reviewed are dislocation necking of solid phase epitaxial layers grown in narrow high aspect ratio trenches and liquid phase epitaxy from rapid melt germanium confined in micro-crucibles. The fabrication of germanium on dielectric substrates such as quartz, sapphire and alumina are also discussed.

## 1 Introduction

Germanium is a single element semiconductor from group IV of the Periodic Table like silicon and also has a diamond like crystal structure. While the first transistors

---

H. Gamble (✉), B. M. Armstrong, P. T. Baine, Y. H. Low, P. V. Rainey,
S. J. N. Mitchell and D. W. McNeill
School of Electronics, Electrical Engineering and Computer Science, The Queen's University of Belfast, Belfast, BT9 5AH N, Ireland
e-mail: h.gamble@qub.ac.uk

B. M. Armstrong
e-mail: b.armstrong@ee.qub.ac.uk

were fabricated in germanium, it was superceded by silicon largely on account of the latter having a stable native oxide. The silicon dioxide can simply be produced by heating the silicon in oxygen. The electrical properties of the silicon dioxide to silicon interface are very good and this has led to the all pervasive CMOS technology over the last 40–50 years. However, as the lateral and vertical dimensions of the MOS transistors were reduced to improve performance and packing density, the silicon dioxide layer thickness needs to be scaled below 2 nm. At this thickness, tunnelling currents through the silicon dioxide layer became intolerable. To meet the charge requirements of state-of-the-art devices it was necessary to replace the silicon dioxide for the gate dielectric with a material of higher permittivity.

Thus the relationship between a thermally grown silicon dioxide layer on silicon was broken and replaced by a deposited higher permittivity (high-k) material. If a deposited high k dielectric could be satisfactory for silicon devices then there was the possibility that it could provide a satisfactory interface with other semiconductors. This brought germanium back into focus as it has the highest mobility for holes and a higher mobility for electrons than silicon. Thus germanium offers the promise of high performance CMOS. Germanium is lattice matched to gallium arsenide (GaAs), and has been used for several years as a substrate for epitaxially grown GaAs solar cells. Germanium substrates with epitaxial layers of GaAs are very attractive for very high performance MOS with p-channel Ge MOSTs and n-channel GaAs MOSTs exploiting both the high mobility holes in germanium and the high electron mobility in GaAs. Epitaxially grown GaAs layers on germanium offers the possibility of integrating optical and electronic circuits. It is a very suitable substrate for circuits employing mixed semiconductor devices.

## 2 Gate Dielectrics for Ge MOSTs

In the 1960s MOS transistors were fabricated using chemical vapour deposited $SiO_2$ as the gate dielectric. LL Chang et al. fabricated enclosed geometry n-channel MOSTs with diffused antimony source/drain, a 450 nm $SiO_2$ gate dielectric and a 24 μm long gate which had an effective electron mobility of 540 $cm^2$/V-s. Since this date n-channel MOSTs have had disappointing carrier mobility values. A similar device was recently produced by the present authors with a 100 nm APCVD $SiO_2$ gate dielectric and phosphorus implanted source/drains, Fig 1. The peak low field mobility was 1,200 $cm^2$/V-s and shows that for thick gate dielectrics there is no fundamental obstacle in producing n-MOSTs with high carrier mobility. The repeatability of employing a deposited oxide was not universally accepted and considerable effort was expended in investigating the growth of a suitable native oxide of $GeO_2$. For thermal oxidation of Ge in oxygen Cresman [1] concluded that there was a strong dependence of oxidation rate on oxygen pressure and that germanium atom diffusion was not the rate controlling process. For sub-atmosphere pressure the reaction rate is governed by the sublimation rate of GeO and the conversion of GeO to $GeO_2$. Below 525°C an

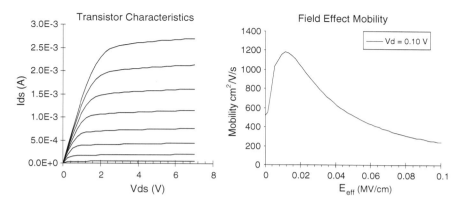

**Fig. 1** A non self aligned aluminium gate circular enclosed geometry n-channel Ge MOST with 100 nm LTO gate dielectric and phosphorus implanted source/drains. The gate length was 30 μm and the width 460 μm

orientation dependence was found with (111) orientation oxidising faster than (100). Above 600°C, layer growth rate was limited by sublimation of GeO from the surface before it could be oxidised to $GeO_2$. The refractive index of the oxidised layers increased with oxygen pressure, but the layers were soluble in deionised water.

To overcome the water solubility problem Hymes and Rosenburg [2] nitrided $GeO_2$ layers by annealing in ammonia. For long time anneals the layers were converted to $Ge_2N_2O$ with a refractive index of 1.8. However, shorter anneals produced better electrical interfaces with fast state densities below $10^{11}$ cm$^{-2}$ eV$^{-1}$. Employing a gate last process they fabricated n-MOSTs with 25 nm thermally grown $Ge_2N_2O$ gate dielectric, arsenic implanted source/drains and obtained low-field electron mobilities in the range 940–1,200 cm$^2$/V-s [3]. P-channel MOSTs were also produced with $BF_2$ implanted source/drains, which had peak low field hole mobility of 770 cm$^2$/V-s [4]. Jackson et al. [5] produced self-aligned aluminium gate p-MOSTs using only a 5 min 350°C anneal of boron implanted source/drains and obtained a hole mobility of 640 cm$^2$/V-s.

An alternative approach of obtaining a good electrical interface between a dielectric and germanium was taken by Vitkavage et al. [6]. A thin layer of silicon is deposited on the germanium surface prior to the deposition of a silicon dioxide layer by remote plasma enhanced chemical vapour deposition in an ultra high vacuum system. The 1 nm Si layer protects the germanium surface from oxidation and forms an excellent interface with both $SiO_2$ and germanium. The silicon layer deposited at 350°C is thick enough to contain the oxidation during the $SiO_2$ deposition at 300°C but thin enough to minimise its effect on channel transport. Optimising the technology, the midgap interface state density was reduced to $5 \times 10^{10}$ cm$^{-2}$ eV$^{-1}$ [7].

This work of employing a silicon interface has more recently been investigated by Reinking et al. [8] employing Ge layers with $2 \times 10^{16}$ cm$^{-2}$ Sb epitaxially grown on silicon substrates. Silicon layers 0–10 nm thick were deposited at 680°C before 30 nm of $SiO_2$ was deposited at 425°C by LPCVD. P-MOSTs were fabricated with self-aligned tungsten gates and $BF_2$ source/drain implants which were activated by rapid thermal annealing at 550°C. Hole mobility was found to increase as the silicon layer thickness decreased, giving an optimum value of 310 $cm^2$/V-s.

Shang et al. [9] employed a composite gate dielectric consisting of a 6 nm $Ge_2N_2O$ layer beneath a 3 nm low temperature deposited $SiO_2$ layer to give an effective oxide thickness of 8 nm. Employing a self-aligned tungsten gate process and phosphorus implanted source/drains in an enclosed geometry 100 μm gate length n-channel MOST on a $5 \times 10^{17}$ $cm^{-3}$ gallium doped substrate, they obtained $I_{on}/I_{off}$ ratio of $10^4$, a sub-threshold slope of 150 mV/dec and an electron mobility of 100 $cm^2$/V-s. The use of high permittivity hafnium based dielectrics with silicon, encouraged researchers to investigate its properties with germanium. To minimise the formation of $GeO_2$ during the deposition and annealing of sputtered $HfO_2$, Kita et al. [10] first deposited a 1–5 nm metallic Hf layer before reactive sputtering of $HfO_2$. The thin metallic hafnium layer was expected to oxidise during the reactive sputtering of $HfO_2$, but it would minimise the formation of $GeO_x$ at the germanium surface. While $GeO_2$ is not volatile up to 700°C in the absence of germanium they found volatilization of Hf–Ge–O during the reactive deposition of $HfO_2$.

The main difficulty in obtaining n-channel germanium MOSTs with electron mobility values greater than those in silicon has been attributed to the high density of interface states $D_{it}$ present near the conduction band edge in the energy band gap. These interface states are considered to arise due to desorption of GeO during thermal processing. Saito et al. [11] determined from first-principles total energy calculations that the defect density at the $Ge/GeO_2$ interface should be low. Thus suppression of Ge-atom emission during the oxidation process is crucial to the quality of the $GeO_2$/Ge electrical interface. Lee et al. [12] have confirmed the earlier work by Cresman [1] that the $GeO_2$ interface was improved by high pressure oxidation at 70 atm. Thermodynamic calculations for oxidation at 550°C shows that the interface pressure of GeO decreases as the oxygen pressure increases. Since GeO formed at the $GeO_2$/Ge interface diffuses through the $GeO_2$ layer to desorb at the surface, the desorption rate is decreased at higher oxidation pressure.

Germanium monoxide is formed at the dielectric/germanium interface by the reduction of $GeO_2$. Nitridation of the $GeO_2$/Ge interface after $GeO_2$ growth can be used to inhibit GeO formation for a limited time. Germanium nitride itself is not thermodynamically stable at temperatures above 550°C. The deposition of a thin silicon nitride layer on the germanium surface before $GeO_2$ deposition can retard the diffusion of Ge to the $GeO_2$, but eventually it will diffuse through the nitride layer. The deposition of an ultra-thin silicon layer on Ge surface before $GeO_2$ deposition also succeeds in retarding the formation of germanium monoxide until the germanium diffuses through the silicon barrier. Hafnium dioxide deposited on germanium does not prevent GeO desorption. Hafnium dioxide crystallises at temperatures above 500°C allowing the diffusion of GeO through the grain boundaries. However, a silicon cap layer on $GeO_2$ has been shown to be successful. Similarly a Ni$_3$Si cap layer suppresses GeO for temperatures up to 600°C, giving good MOSC characteristics.

Another approach to stop the desorption of GeO is to employ a high-k material that reacts with germanium to form an amorphous layer thus blocking GeO out-diffusion. Yttrium oxide (Y$_2$O$_3$) with a dielectric constant of 14 was found to react with germanium to form an amorphous layer YGeO$_x$ [13]. While germanium does diffuse in Y$_2$O$_3$ it does not degrade the C–V characteristics. Lanthanum oxide La$_2$O$_3$ has a higher dielectric constant, but is hygroscopic [14]. This can be overcome by combining the Lanthanum oxide with Yttrium oxide in a 1:1 ratio, which gives a dielectric constant $k = 29$. To minimise GeO desorption a silicon cap can be added.

## 3 Source/Drains

A key issue in germanium MOST technology is the reduction of junction leakage while minimizing the specific contact resistance to the metal interconnects. The diffusivity of acceptor dopants in germanium is very slow, so p$^+$–n junctions are formed either by ion implantation or in-situ doped epitaxial layers. Spann et al. [15] determined the specific contact resistivity of NiGe films to boron doped germanium as a function of the boron dopant concentration. They found that for in-situ doped germanium epitaxial layers the relationship between specific contact resistance and dopant concentration could be represented by $\rho_c \approx 6 \times 10^6 \times N_A^{-0.62}$ Ω cm$^2$. Thus for maximum boron concentration of $5 \times 10^{20}$ cm$^{-3}$ the specific contact resistivity is $9 \times 10^{-7}$ Ω cm.

However, the solid solubility of boron in germanium was determined by Hellberg et al. [16] to be about $1 \times 10^{18}$ cm$^{-3}$ at 800–900°C. Implanting boron into crystalline germanium and curve fitting with SIMS profiles, Uppal et al. [17] calculated the solid solubility to be $5.5 \times 10^{18}$ cm$^{-3}$. Satta et al. [18] achieved a maximum active boron concentration of $1.2 \times 10^{19}$ cm$^{-3}$ after furnace annealing. Impellizzeri et al. [19] found that a continuous amorphous layer is never formed with high dose boron implants into crystalline germanium. Instead two defective regions are observed one near the surface the other near projected range $R_P$ when the dose is greater than $7.6 \times 10^{15}$ cm$^{-2}$. The activation energy for recombination of point defects is small, thus the presence of implanted B mostly around $R_p$ enhances the recombination velocity favouring the damage annealing. The surface acts as a nucleation seed for the amorphous phase because the native oxide enhances the damage stability. They suggest that large inactive B–Ge complexes form during the room temperature implantation and once formed these complexes are stable at least up to 550°C. RBS shows residual damage in the lattice, which could be in the form of off-lattice Ge bonded into inactive B–Ge complexes. They plotted density of displaced Ge atoms to density of inactive boron and found a ratio of 8:1. They speculated that inactive boron B-Ge complexes with a 1:8 stoichiometry formed and are stable.

Chao et al. [20] found that the active concentration of B in Ge well above the solid solubility limit of $1 \times 10^{19}$ cm$^{-3}$ could be achieved by implantation into previously amorphized germanium. During the subsequent B implantation, dynamic annealing or ion beam induced epitaxial crystallisation occurred. This did not completely annihilate all of the implant induced defects and a rapid thermal anneal at 650°C was required to achieve a maximum activation of $4.7 \times 10^{20}$ cm$^{-3}$.

Mirabella et al. preamorphized Ge to a depth of 275 nm by implanting at 300 keV, a dose of $3 \times 10^{15}$ cm$^{-2}$ Ge ions at room temperature [21]. Boron was then implanted at 35 keV so that the boron profile was entirely contained within the a-layer. The boron was activated by SPE at 360°C for 1 h in $N_2$. The maximum active boron concentration $B_{max} = 5.7 \times 10^{20}$ cm$^{-3}$ is two to three times higher than in silicon. Thus as well as the higher hole mobility, the Ge can host more substitutional boron atoms than Si so p$^+$ germanium can have a lower sheet resistance than silicon. They found that the total hole carriers per surface area increases linearly with boron dose until $B_{max}$ is exceeded. If $B_{max}$ is exceeded, dopant precipitation occurs. Further annealing for 1 h at temperatures up to 550°C did not change the sheet resistance. Hole mobility for Ge was found to undergo a severe reduction with doping concentration, going from 300 cm$^2$V$^{-1}$s$^{-1}$ for $2.5 \times 10^{19}$ cm$^{-3}$ to 50 cm$^2$V$^{-1}$s$^{-1}$ for $5.8 \times 10^{20}$ cm$^{-3}$.

To prevent dynamic annealing during the implantation, Impellizzeri et al. [19] carried out the implantation of boron into crystalline germanium at liquid nitrogen temperature (LNT). The boron was implanted at 35 keV to give $R_P = 100$ nm with a current density of 0.1 µA/cm$^2$. The ion damage was accumulated in the form of an extended amorphous layer ranging from the surface to a depth of $\sim$185 nm for a dose of $2.8 \times 10^{15}$ cm$^{-2}$ and up to $\sim$205 nm for a dose of $7.6 \times 10^{15}$ cm$^{-2}$. After annealing at 360°C in $N_2$ for 1 h the residual damage was below the detection limit of RBS showing complete recrystallization of the Ge. The damage induced by the ion implantation of boron at LNT easily dissolves with 360°C thermal annealing unlike the case for RT implanted samples where more complex defects exist. Nuclear Reaction Analysis (NRA) showed that more than 85% of boron atoms are substitutionally positioned in the samples implanted with up to $7.6 \times 10^{15}$ cm$^{-2}$ giving similar maximum boron concentration of $5.8 \times 10^{20}$ cm$^{-3}$ as for pre-amorphized implants.

In summary implantation of boron into crystalline germanium does not yield a continuous amorphous layer even for high doses. Inactive B–Ge complexes with a 1:8 stoichiometry are formed during the implantation, which are stable with activation anneals up to 550°C. Thus the density of active boron is well below the solid solubility limit. High activation levels of boron of $4-6 \times 10^{20}$ cm$^{-3}$ can be facilitated by the presence of amorphous Ge produced either by pre-amorphization or during implantation at LN temperature.

Chui et al. [22] determined that during RTA of implanted phosphorus, diffusion takes place through double negatively charged vacancies. The maximum equilibrium solid solubility of phosphorus in germanium is approximately $2 \times 10^{20}$ cm$^{-3}$ [23]. The phosphorus diffusion is strongly concentration dependent and gives impurity profiles with an abrupt drop in concentration levels at $2 \times 10^{19}$ cm$^{-3}$. This is also the case for solid source Phosphorus diffusion. A significant out diffusion towards the surface takes place both with uncapped and capped germanium. Maximum phosphorus activation level was limited to $5-6 \times 10^{19}$ cm$^{-3}$. RBS analysis shows that 450–500°C anneals are required for full re-crystallisation.

Satta et al. [24] implanted a 15 keV phosphorus dose of $5 \times 10^{15}$ cm$^{-3}$ to give an amorphous layer of 37 nm and a maximum dopant concentration higher than the solid solubility limit. After 1 s anneal at 500°C the SIMS plot displays a chopped portion starting at the boundary of the original amorphous/crystalline interface. They concluded that initially during SPE, the moving amorphous/crystalline boundary pushes excess phosphorus in front of it and phosphorus is substitutionally incorporated into the crystalline Ge to a level of $4 \times 10^{20}$ cm$^{-3}$. However, as the moving boundary accelerates with increasing phosphorus concentration the diffusion rate in the a-Ge is exceeded at a concentration of $\sim 1 \times 10^{21}$ cm$^{-3}$. The phosphorus no longer can be pushed forward fast enough and phosphorus becomes trapped in non-substitutional sites. The phosphorus precipitates do not dissolve with long anneals at 500°C, but the substitutional phosphorus diffuses giving a BOX profile with a knee at $1-2 \times 10^{19}$ cm$^{-3}$.

Carroll and Koudelka developed a diffusion model incorporating an extrinsic diffusivity coefficient combined with a segregation component at the oxide germanium interface to account for the significant dopant loss at the surface [25]. For long anneals they found that a doubly charged diffusivity $D$ to be sufficient to match the experimental results. The diffusivity coefficient showed Arrhenius behaviour with an activation energy $E_a = 2.3$ eV which is slightly higher than the 2.07 eV determined by Chui. However the diffusivity values determined for these long diffusion times do not fit the results for implants subjected to RTA. Thus it is considered that the damage from the implant or possibly the rapid SPE could result in transient enhanced diffusion (TED).

Before implanting phosphorus, Posselt et al. [26] amorphized the germanium surface by implanting Ge ions. The phosphorus dose of $3 \times 10^{15}$ cm$^{-2}$ was then implanted into this pre-amorphized surface to give a peak concentration of $8 \times 10^{20}$ cm$^{-3}$. This peak concentration is less than that found by Satta et al. [24] to make the amorphous/crystalline boundary movement comparable to the phosphorus diffusion rate in the amorphous germanium at 500°C. The solid phase epitaxial growth was found to proceed at a rate of $V = 5.895 \times 10^{14} \exp - \left(\frac{1.926}{kT}\right)$ nms$^{-1}$. Thus 60 s at 400°C is capable of re-crystallising 135 nm of amorphous germanium. During the SPE a very fast redistribution of phosphorus takes place with the excess above $4 \times 10^{20}$ cm$^{-3}$ being pushed ahead of the amorphous/crystalline interface leading to phosphorus loss to the SiO$_2$ covering layer. This all happens in the time required for the SPE to reach the surface. Further annealing results in concentration dependent diffusion in crystalline germanium giving a box like profile with $D = D_O \left(\frac{n}{n_i}\right)^2$. Further annealing can lead to phosphorus deactivation either through phosphorus—vacancy centre acceptors or precipitate clusters reducing to 3–7 × $10^{19}$ cm$^{-3}$ after high temperature flash lamp annealing. Koffel et al. [27] suggest that with ion implantation the excess interstitials created also affects the diffusivity. They concluded that the fraction of phosphorus that moves by interstitials is 0.05. For extrinsic diffusion at 535°C Canneaux et al. [28] found the diffusivity $D = 1.2 \times 10^{-18} \left(\frac{n}{n_i}\right)^2$ suitable for up to 1 h.

Kim et al. [29] considered that part of the apparent non activation of the higher dose phosphorus implants was due to the incurable damage in the germanium after the implant anneal acting as acceptors. They reasoned that minimizing implantation damage could increase the effective solid solubility of n-type dopant atoms. The phosphorus atomic radius is 10% smaller than that of germanium so a localized tensile strain is created in the vicinity of a phosphorus atom. Since the atomic radius of antimony is 16% greater than that of germanium it would create a localized compressive strain. Thus co-implantation of phosphorus and antimony could minimize the local strain, reducing the defect density and increasing the effective activation of the dopants. A dopant activation of over 1 × $10^{20}$ cm$^{-3}$ was achieved after a 60 s RTA at 500°C of co-implanted phosphorus and antimony. Raman spectra show that with single type implants, damage remains in the germanium after annealing, but after co-implantation of phosphorus and antimony the spectra is as for un-implanted germanium.

To avoid ion implantation damage and the fast phosphorus diffusion during the recrystallisation step, researchers sought to avoid the implantation step. Yu [30] in-situ doped epitaxial germanium layers grown at 400–600°C. A 1 min growth at 600°C had a peak concentration of 2 × $10^{19}$ cm$^{-3}$ with 100% activation. On/off ratio of n$^+$p diodes at ±1 V was 1.1 × $10^4$ and forward current density was 120 A/cm$^2$. They concluded that the high forward current was achieved due to the absence of implant damage.

Scappucci et al. [31] adsorbed PH$_3$ molecules onto the surface of germanium before depositing 2 nm of undoped germanium. Peak concentration as given by SIMS was 1 × $10^{21}$ cm$^{-3}$ but only about 20% of the total phosphorus was activated.

Morii et al. [32] employed MOCVD based gas phase doping from tertialybutylarsine (TBA) at 600°C for 60 min to produce a surface doping concentration $C_s = 1 \times 10^{19}$ cm$^{-3}$ with nearly 100% activation for (110) Ge. $C_{max}$ for (100) Ge was 5 × $10^{18}$ cm$^{-3}$. The diffusivity of the phosphorus was slow compared to that obtained during post implant anneals. For 14 min 600°C diffusions the junction depth in 1 × $10^{16}$ cm$^{-3}$ p-Ge was 300 nm and the diode on/off ratio was 4 × $10^5$ at ± 1 V. The ideality factor was 1.2 compared to $n = 1.77$ for a reference implanted diode. The reverse current decreased with decreasing temperature with an activation energy of $E_a = 0.35$ eV ($E_g/2$) for the diffused diodes, suggesting that generation current dominates the reverse bias. For the implanted diode, $E_a$ is reduced suggesting defect-assisted tunneling current.

Posthuma et al. [33] diffused phosphorus from a spin-on–dopant (SOD) source to produce n$^+$-p junctions for solar cells. When the phosphorus concentration in the SOD was greater than the solid solubility of phosphorus in germanium of $2 \times 10^{20}$ cm$^{-3}$, concentration dependant diffusion took place. A 10 min diffusion at 580°C gave a junction depth of 0.5 μm. Due to the high phosphorus concentration in the SOD, germanium atoms are displaced, forming interstitial pairs with phosphorus atoms (PI)s. When the surface concentration of phosphorus falls below the solid solubility, the generation of interstitials is reduced and the diffusion rate falls. Thus box like profiles are formed with a flat concentration of $\sim 3 \times 10^{19}$ cm$^{-3}$ followed by a sharp drop. 100% activation of the Phosphorus is achieved.

## 4 Germanium on Insulator (GeOI)

There is a limited supply of the element germanium. It is expensive, heavy, fragile and susceptible to chemical attack. The low energy band gap results in higher junction leakage currents and the higher dielectric constant increases the short-channel effects. Employing a thin layer of germanium on an insulating (GeOI) substrate reduces these negatives. GeOI substrates can be produced by bonding either a germanium layer grown on silicon, or a bulk germanium wafer to an oxidised silicon wafer. The excess donor substrate material can then be removed by grinding and polishing to give a >2 μm thick GeOI layer. The use of a bulk germanium wafer gives better quality material, but limited to 200 mm diameter. The Smart-cut technology that has been so successful for silicon on insulator (SOI) wafers can also be applied to germanium. The technology developed by Bruel [34] consists of implanting hydrogen ions into the donor wafer before bonding it to a handle wafer. A post bond anneal at 400°C liberates bonded hydrogen which increases the hydrogen pressure in the platelets forcing them to increase laterally splitting the wafer near the project hydrogen range. It has been found that in the case of silicon the total implant dose required for splitting is reduced by factor of three when co-implantation of hydrogen and helium is used [35] If the donor wafer was not bonded to a handle wafer the hydrogen would cause blisters to form on the wafer surface. There is a strong relationship between the thermal budget required to form blisters on unbonded wafers with that required to cause splitting of a bonded wafer. In the case of germanium bonded to oxidised silicon substrates there is a mismatch in the thermal coefficients of expansion (TCE) of the germanium and silicon wafers. The mismatch in TCEs causes sufficient stress that the germanium wafer will craze if the bonded pair is heated above 300°C.

To determine the hydrogen implant dose required to split germanium at or below 300°C the time at temperature required to form optically detectable blisters in hydrogen and co-implanted hydrogen and helium was determined for different anneal temperatures. Blisters in the germanium surface after 350°C 30 min annealing of hydrogen implanted wafers as revealed by a Talysurf optical interference system are shown in Fig. 2. A comparison of blister formation for silicon and germanium under similar conditions is shown in the Arrhenius plot of

**Fig. 2** A Talysurf CCL 6000 image of blisters formed at 35°C 30 min on a 250 μm × 250 μm scanned area

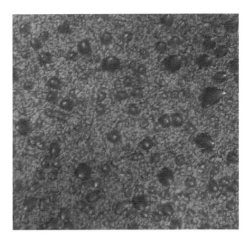

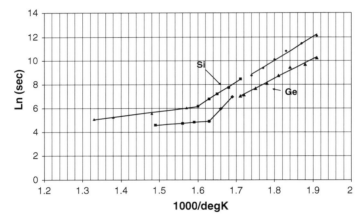

**Fig. 3** Time for blister detection at various annealing temperatures for Ge and for Si processed under similar conditions

ln (*t*) against temperature in Fig. 3 [36]. It can be seen that for a given implant dose the germanium will split at a lower temperature than silicon. The two plots follow the same three region profiles indicating that similar processes are involved. After splitting, the germanium surface has an RMS roughness of 11 nm [37]. The lower the implant dose employed for the splitting of the germanium the rougher the resulting surface after splitting. For transfer of germanium to silicon handle wafers ≤ at temperatures ≤300°C a hydrogen dose ≥6 × $10^{16}$ cm$^{-2}$ is required. It was shown by Sedgwick [38] that hydrogen in contact with germanium produces a high density of surface states. Ruddell et al. [39] found a significant shift along the voltage axis in capacitance–voltage of MOS capacitors after a forming gas anneal. Using density-functional theory calculations Van de Walle and Neugebauer [40] developed a universal alignment of hydrogen levels in a range of materials. They

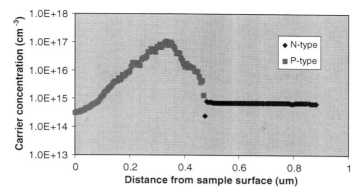

**Fig. 4** Carrier concentration of Hydrogen implanted Germanium as determined by spreading resistance probe. The H dose was $3 \times 10^{16}$ cm$^{-2}$ and the anneal was at 400°C for 2 h

showed that the transition level between the charge states of H$^+$ and H$^-$ was approximately 4.5 eV below the vacuum level. For most semiconductors the transition energy level lies within the energy bandgap. Thus for silicon the charge on hydrogen atoms will be positive when the Fermi level is in p-type material and negative in n-type material. The hydrogen in silicon is amphoteric acting as a donor in p-type material and an acceptor in n-type silicon reducing the conductivity in either case. However in germanium the transition energy level is just 0.04 eV below the valence band maximum. The hydrogen takes up a tetrahedral interstitial position and will be negatively charged independent of the dopant type and concentration. Thus hydrogen in germanium will always act as an acceptor state. Omachi et al. [41] found that epitaxial germanium layers grown on silicon by an evaporation technique were p-type. The acceptor concentration decreased and hole mobility increased with germanium layer thickness and with deposition temperature. The acceptor concentration was closely associated with the structural imperfections of the layer.

The Smart-cut process introduces both hydrogen and implant damage to the surface germanium layer being transferred. Low et al. [37] study the effect of the process on acceptor formation by the hydrogen implant and the residual acceptors in the transferred layer. An n-type germanium wafer containing $7 \times 10^{14}$ cm$^{-3}$ Sb atoms was subjected to a hydrogen implant dose $3 \times 10^{16}$ cm$^{-2}$ and was annealed at 400°C in $N_2$ for 2 h. The dose was insufficient to cause blistering and the anneal corresponds to that which might be used in a Smart-cut process. Spreading resistance analysis (SRP) gave an acceptor profile corresponding to the hydrogen implantation with a peak acceptor concentration of $1 \times 10^{17}$ cm$^{-3}$ (Fig. 4). Employing a hydrogen implant dose of $6 \times 10^{16}$ cm$^{-2}$, a germanium layer containing $2 \times 10^{16}$ cm$^{-3}$ Sb was transferred to an oxidised silicon wafer. After transfer portions of the GeOI wafer were annealed at 600 or 800°C for 2 h to restore the crystallinity of the transferred layer. SRP showed that the 600°C annealed layer had a net uniform donor concentration of $1-2 \times 10^{16}$ cm$^{-3}$. The

GeOI layer annealed at 800°C was p-type with a net uniform acceptor concentration of $7 \times 10^{15}$ cm$^{-3}$ at a depth below 0.2 μm. The acceptor concentration fell rapidly towards the surface to value of $\sim 3 \times 10^{11}$ cm$^{-3}$. Allowing for compensation by the Sb concentration this again indicates that the Smart-cut process has introduced an acceptor concentration of the order of $10^{16}$ cm$^{-3}$. For reasons of cost, and layer quality several alternative methods for producing GeOI are being investigated.

## 5 Ge on Crystalline Dielectric

Bojarczuk et al. [42] advanced the opinion that the most convenient way of producing ultra thin germanium layers on an insulating substrate was by epitaxially growing both the buried insulator and the germanium on a silicon substrate. On the <111> silicon substrate they initially grew an epitaxial layer of (La$_{0.27}$ Y$_{0.73}$)$_2$ O$_3$ which has a lattice constant twice that of silicon. An amorphous layer of Ge, 4–20 nm thick, was deposited to avoid 3D Volmer–Weber type growth. The layer was annealed at 450°C to achieve solid phase epitaxial growth seeded by the crystalline oxide. To avoid roughening of the germanium surface during the SPE, a flux of $10^{15}$ cm$^{-3}$ of antimony atoms was applied. After the crystallization the antimony was confined to the top 0.4 nm of the germanium layer. Employing the silicon substrate as the gate electrode, circular geometry p-MOS FETs were fabricated with a 38 nm (La$_{0.27}$ Y$_{0.73}$)$_2$O$_3$ gate dielectric and boron source/drain implants. These back gate devices yielded a square-law I–V characteristic and a gate leakage across the crystalline oxide in the order of $1 \times 10^{-8}$ A/cm$^2$.

To obtain (001) oriented germanium on silicon Seo et al. [43] employed MBE grown perovskite oxides of SrHf$_x$ Ti$_{x-1}$O$_3$ (SHTO) and SrHfO$_3$ (SHO). Strontium titanium which is nearly lattice matched to (001) Si is grown first followed by strontium hafnium titanium oxide to increase the lattice spacing. The hafnium fraction is gradually increased until Sr(Hf$_{0.5}$, Ti$_{0.5}$)O$_3$ is obtained. Germanium growth on the SHTO layer was 3-D Volmer–Weber type. Growth in the presence of a surfactant did not improve the situation. To overcome this 3-D growth problem, (001) oriented islands were seeded at 610°C. This was followed by germanium growth at the lower temperature of 350°C which promoted homogenous coverage of the oxide. The roughness of the layers was correlated with the coalescence of islands formed during the film growth. The layers were p-type with an acceptor concentration of $\sim 10^{17}$ cm$^{-3}$. Hall mobility was in the range 100–300 cm$^2$/V-s. Defects in the layer are the most likely cause of the p-type conductivity. Similar results were obtained with a crystalline oxide of (Ba$_{0.5}$Sr$_{0.5}$)TiO$_3$ in place of the Sr(Hf$_{0.5}$, Ti$_{0.5}$)O$_3$ layer.

Recently Guissani et al. [44] have investigated the use of praseodymium dioxide (P$_r$O$_2$) as a buried insulator/seed layer for the growth of germanium on insulator substrate. The type-B oriented P$_r$O$_2$ (111) layers were produced by MBE followed by ex-situ oxygen annealing. The cubic lattice structure of the oxide transfers the stacking information enabling twin free growth of the FCC germanium. During the initial germanium deposition the $PrO_2$ reacts with the germanium to form volatile GeO and cubic $Pr_2O_3$. Thus a clean surface is presented for germanium epitaxial growth. Since the cubic $Pr_2O_3$ has a lattice spacing between that of silicon and germanium it effectively reduces the lattice mismatch between the two semiconductors. No silicon or praseodymium was found in the germanium even after a post deposition anneal at 825°C for 30 min.

The growth of germanium on single crystal dielectrics, epitaxially grown on silicon, offers potential for high quality ultra thin germanium layers. The layers can be transferred to oxidised silicon substrates to give GeOI substrates. The technique is free from hydrogen implantation and selective etches avoids the need for polishing of transferred layers. GeOI layer thickness down to 15 nm with ±5% uniformity can be obtained [45]. It also promises a means of obtaining germanium layers on non-silicon substrates such as glass or quartz through layer transfer.

## 6 The Condensation Process

Tezuka et al. [46] proposed a technique for producing relaxed Ge rich SiGe layers on a buried oxide for (SGOI). A SiGe layer with a low Ge content is grown epitaxially on a SIMOX wafer Fig. 5. When the wafer is subjected to a high temperature oxidation, the Ge is rejected by the growing $SiO_2$ layer and condenses into the underlying silicon layer to form SGOI. As the oxidation proceeds the SGOI layer reduces in thickness and the germanium fraction increases. The total amount of germanium is conserved in the process so the final germanium fraction depends on the initial germanium incorporated and the final SGOI layer thickness. While the technique was originally developed as a virtual substrate for the epitaxial growth of tensile stressed silicon for MOSFETs, it has subsequently been used for the selective formation of very high germanium content regions for PMOSFETs [47]. The almost pure germanium in the recessed channels can be as thin as 5 nm. The layers have compressive strain which should boost the hole mobility to 10× that of silicon SOI-MOSFETs. A maximum hole mobility of

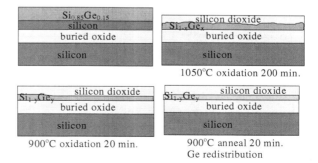

**Fig. 5** The condensation process for producing GeOI substrates

1,590 cm$^2$/Vs was obtained. For a blanket GeOI layer, 7 nm thick, prepared by the condensation process, Raman spectroscopy was used to determine the silicon fraction and the strain in the layer. The silicon fraction was less than 0.5% and the strain was 1.1% compressive. A cross-hatch pattern is observed in the Ge surface giving an RMS roughness of 0.4 nm in a 10 μm square [48]. However, MOSFETs fabricated on GeOI substrates produced by the condensation process exhibited a large off state current which could be attributed to crystal defects in the germanium layer. Vincent et al. [49] found that stacking faults were generated within the SGOI layers fabricated by the condensation process. Transmission Electron Microscopy analysis showed that for Ge enrichment above 82%, strain relaxation led to the formation of stacking faults. The highest critical enrichment was achieved with the highest initial Ge content and SiGe thickness.

Nakaharai et al [50] found that there was a rapid increase in the line density as the % Ge increased above 50%. For low Ge fractions, stacking faults were dominant but for high Ge fractions, microtwins were dominant. They found that the density of planar defects decreased significantly as the Ge fraction exceeded 96%. The threading dislocation density in GeOI layers was of the order of $10^8$ cm$^{-2}$. They concluded that the relaxation of the compressive strain in the GeOI layer was caused by glide of 60° perfect dislocation lines on (111) planes. They are not revealed in plan-view TEM-images, because the strain field associated with these dislocations is relaxed by deformation of the buried oxide layer. Misfit dislocations are generated during the initial stage of the condensation at the interface between the SiGe layer and the SOI substrate [51] due to relaxation of the SiGe layer. As the condensation process proceeds the misfit dislocations fragment and then disappear as threading dislocations appear. The density of the threading dislocations significantly increases during the temperature ramping up to 1,100°C. To minimise the generation of threading dislocations during the ramping up a two step oxidation/condensation method can be used [52]. Since the oxidation of the surface layer introduces stress and creates interstitials, a ramp up in an inert atmosphere would reduce these parameters. To protect the surface layer from roughening during the high temperature oxidation, a 15 nm protective oxide layer was formed at 900°C. The temperature is then ramped up to 1,200°C in an inert atmosphere before the condensation oxidation is commenced. After the 900°C initial oxidation, misfit dislocations are formed as the SiGe layer relaxed. After the 1,200°C oxidation misfit fragments remain around the initial SiGe/SOI interface. However, the rising of the misfit fragments and the threading dislocations observed for a ramp up in oxygen are absent. Seeco-etching shows that the etch pit density of a 20% Ge SGOI layer was as low as $10^3$ cm$^{-2}$ compared to $10^8$ cm$^{-2}$ for the conventional process.

To prevent loss of Ge during the initial oxidation stage a 2 nm silicon cap layer can be deposited in-situ after the epitaxial growth of the $Si_{1-x}Ge_x$ layer. The $Si_{1-x}Ge_x$ layer thickness and composition is chosen to be lower than the critical thickness for plastic relaxation. During the condensation processing, relaxation occurs through perfect dislocations (60° dislocations). These then dissociate into partial dislocations (30° and 90°). Nguyen obtained 10 nm GeOI layers with a threading dislocation density of less than $10^5$ cm$^{-2}$ [53]. The GeOI layers were p-type with a resistivity of $1.7 \times 10^{-2}$ Ω-cm. This low resistivity is attributed to defects induced from the condensation process. Maida et al. [54] found a Hall mobility for holes of 410 cm$^2$/V-s and an acceptor concentration of $1.3 \times 10^{17}$ cm$^{-3}$. Hole and electron mobilities in the 10 nm GeOI layer obtained using the Pseudo-MOSFET technique were 180 and 20 cm$^2$/V-s, respectively. The density of interface states at the BOX–Si$_{1-x}$Ge$_x$ interface was found to increase as the germanium fraction increased, degrading the sub-threshold slope. The GeOI layer takes up the orientation of the original SOI layer and so is compatible with the fabrication of (110) GeOI substrates. Dissanazake et al. [55] found a 2.3 times improvement in hole mobility for (110) GeOI as compared to (100) GeOI, both produced by the condensation process. The condensation process is compatible with ultra thin GeOI layers required for fully depleted MOSFETs. It is also suitable for localised GeOI regions and for producing SGOI substrate for tensile strained silicon and compressively strained germanium for n-MOSFETs and p-MOSFETs, respectively. Only CMOS compatible processes such as SiGe epitaxial growth and oxidation are needed to form the GeOI substrates.

# 7 Liquid Phase Epitaxy

The potential of GeOI substrates and the low melting point of germanium had led researchers to investigate the growth of crystalline germanium from the liquid state. The basic idea was to pattern a deposited layer of germanium so that on one side it had a fine point. A line beam of sufficient energy to melt the germanium would scan across the wafer from the pointed side of the patterns. The solidifying grain at the point would then act as a seed for epitaxial growth as the molten germanium progressively solidified away from the point [56, 57]. The need for the energy beam scanner was removed by a new technique based on defect necking proposed by Liu et al. at Stanford [58, 59]. Seed windows are patterned through a silicon nitride layer deposited on a silicon substrate. A blanket layer of germanium is deposited by evaporation, sputtering, or PECVD and then patterned into rectangular islands a few microns wide and a few tens of microns long. These islands are then encapsulated in a thick layer of PECVD SiO$_2$. The whole wafer is then subjected to an RTA $\geq$940°C for a few seconds and then allowed to cool naturally. Since the germanium melting point is 937°C, the germanium melts, but is held in the SiO$_2$ crucibles. The germanium in contact with the silicon solidifies first taking up the crystal structure of the silicon. As the solid germanium layer thickens in the seed window it relaxes through the formation of threading dislocations which propagate along (111) planes. The dislocations terminate on the covering SiO$_2$ layer and the lateral cooling germanium takes up the germanium crystal structure to give dislocation free strips (Fig. 6). Thus while there is a very high density of defects due to the lattice mismatch between germanium and silicon and intermixing in the seed window they terminate over a very short distance. The length

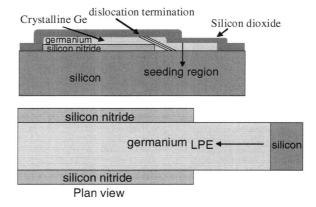

**Fig. 6** Micro-crucibles with seed windows for rapid melt liquid phase epitaxy

over which crystalline germanium can be sustained is a balance between the rate of lateral solidification and random nucleation. Single crystal germanium strips up to 80 μm long have been achieved. The orientation of the LPE germanium strips follows that of the silicon seed windows [58]. In the case of germanium deposition by LPCVD, an initial 4 nm of silicon was deposited in-situ over the seed windows and the insulator. This avoids the selective deposition of the germanium in the silicon seed windows with respect to the insulator. The LPE germanium layer will then contain a small percentage of silicon. Etch pit density was found to be lower than $5 \times 10^5$ cm$^{-2}$. The surface roughness of the LPE germanium is similar to that of the initial germanium deposit as the molten germanium is restrained by the oxide crucibles. Typical RMS values being 1.98 nm as deposited and 2.11 nm after crystallisation. GeOI layers down to 22 nm have been produced by this LPE technique.

P-channel MOSTs employing a LTO SiO$_2$ layer on a nitrided germanium surface were fabricated in the LPE GeOI layers [59]. Low field hole mobilities of 120 cm$^2$/V-s were obtained for 2 μm wide devices with gate lengths of 1.5 μm. The values are similar to those obtained on bulk germanium with the same gate stack. However, the LPE devices did have some source–drain leakage attributed to pinning at the germanium-buried insulator interface. GeOI photodetectors were also fabricated consisting of an array of 1 μm × 5 μm parallel P-i-n diodes. Photo responsivity values exceeded 160 mA/W and FWHM impulse response was less than 100 ps. Tweet et al. [60] found that the crystallised germanium had a tendency to exhibit twist about the long axis. In 2 μm wide strips they found that a twist of $\sim 10°$ could result if the germanium had been heated well above the germanium melting point. The twist, was minimised by heating just above the melting point and employing thicker layers. Despite the twist, the germanium on top of the buried Si$_3$N$_4$ layer was observed to be free of dislocations and stacking faults.

The technology allows the integration of P-channel MOSFETs in GeOI with n-channel MOSFETs in silicon as demonstrated by Feng et al. [61]. The n-channel devices are fabricated first since their source/drain implants are annealed at temperatures above the melting point of germanium. The GeOI layers were formed by crystallising a deposited germanium layer with an RTP at 945°C for 1 s. This, and the subsequent germanium MOSFET processing added little to the thermal budget of the silicon n-MOSFET. The P-channel germanium MOSFETs suffered from drain leakage which could be due to inversion or drain leakage at the back interface. Negative charges trapped in the buried oxide and/or interface states could result in a conducting back-channel.

The GeOI layers grown by rapid melt LPE were also used to implement P-channel FinFETs [62]. The FinFETs height was 0.95 μm and both partially and fully depleted devices were fabricated. The (110) channel enhanced the effective mobility by 60% at an effective field of 0.4 MV/cm as compared to the silicon universal hole mobility.

The integrity of the encapsulating dielectric during the melt period is critical. Balakumar et al. [63] found that if the $SiO_2$ layer delaminated the germanium segregated into balls. They employed an undercut of the bottom insulator to produce stable single-crystal growth of germanium without melting by annealing at a temperature of 925°C. They achieved single crystal germanium strips up to 60 μm long. Raman spectroscopy analysis showed that the strips had a tensile strain. Hashimota et al. [64] discovered that in 100 μm wide strips the germanium had aggregated into columns. They were employing a 1 min insertion into a furnace at 1,035°C for the LPE. The conclusion was that the thin germanium layer balled up due to the high interface energy at the germanium-silicon dioxide and pushed up the softened $SiO_2$ capping layer. In narrower strips the capping oxide area to germanium volume is increased thereby decreasing the pressure exerted by the molten germanium. In an effort to minimise the interface energy between liquid germanium and the buried dielectric they replaced the $SiO_2$ layer by $HfO_2$ or $La_2O_3$. The former produced results similar to those of $SiO_2$, but the $La_2O_3$ produced a significant improvement in the germanium layer quality. The wetability of $La_2O_3$ with molten germanium was found to be greater than that of $SiO_2$. There was thus a lower interface energy between germanium and $La_2O_3$ and a reduced drive to agglomerate. This also increases the nucleation time, decreasing random grain growth, resulting in longer and wider single crystal germanium strips.

The lateral diffusion of silicon atoms from the seed window along the germanium LPE strip was investigated by Miyao et al. [65] using Raman spectroscopy. For the case of LPE carried out at 1,000°C for 1 s they found that the silicon percentage varied from 4% at the seed window edge down to zero at 70 μm. The 3 μm wide stripes were single crystal germanium over a length of 400 μm. Employing (100), (110) and (111) silicon substrates they verified that the germanium orientation was governed by the silicon seed.

To determine whether the germanium solidification proceeded from the silicon seed window due to the thermal dissipation through the seed window, or from the spatial gradient of the solidification temperature of the $Si_xGe_{1-x}$ in the seed windows, they employed a quartz substrate and polycrystalline silicon seed islands.

The poor thermal conductivity of the quartz would suppress heat flow from the seed windows. After germanium deposition, patterning, encapsulation and rapid thermal annealing, as for LPE on silicon substrates, the germanium strips were found to be single crystal over the 400 μm length. The crystal orientations were either (100) or (110) consistent with the polysilicon seeds used. These results show that thermal dissipation from the seed window is not the crucial process, but that silicon–germanium mixing at the seeding window is a key initial driving force. The melting point decreasing as the silicon fraction decreases. However, far from the seed window the silicon fraction is zero, but the lateral crystallisation proceeds. The latent heat of solidification at the solid–liquid interface is thus a main driving force in the process. The direct growth of single crystal germanium stripes on quartz opens new avenues for germanium technology. The GeOI layers are p-type with an acceptor concentration of $6 \times 10^{16}$ cm$^{-3}$ and a hole mobility of 1,040 cm$^2$/V–S [66]. For the case of thin polycrystalline seeds the silicon concentration in the GeOI layer falls to zero within 5 μm of the seed window. The rest of the 400 μm stripe is pure single crystal germanium. GeOI stripes that nucleate with (101) or (111) orientation were found to propagate laterally by rotating the crystal orientation. As a result greater than 75% of the stripes more than 200 μm away from the seed edge were (100).

## 8 Epitaxial Necking/Aspect Ratio Trapping

To reduce threading dislocations in GaAs grown on silicon, Fitzgerald and Chand [67] investigated the use of patterned substrates. On 40 μm wide mesas, the GaAs layers had a cathodoluminescence equal to that of control GaAs/GaAs grown layers. The small diameter mesas allowed the mismatch generated dislocations to grow out to the mesa edge. Langdo et al. exploited this necking of dislocations to produce defect free germanium surfaces on silicon [68]. They employed selective epitaxial growth of germanium grown on exposed (100) silicon areas at the bottom of narrow holes etched in a silicon dioxide layer, Fig. 7. Threading segments of dislocations, caused by plastic relaxation of the germanium on the lattice mismatched silicon, propagate in <100> directions on the (111) planes make a 45° angle to the underlying silicon (100) substrate. Thus if the aspect ratio of the holes in the oxide is greater than unity the threading dislocations will be blocked by the oxide sidewall and the top germanium surface will be dislocation free. The thickness of the germanium layer required before it becomes defect free can be quite small when submicron holes in the oxide are used.

Langdo et al. employed a hot-walled UHV CVD reactor to selectively deposit germanium at 650°C from a germane/hydrogen atmosphere. Epitaxial lateral overflow was continued until coalescence from neighbouring seed windows. Microtwins and stacking faults were found at half of the coalescence boundaries since the hole spacing is not necessarily an exact multiple of the germanium

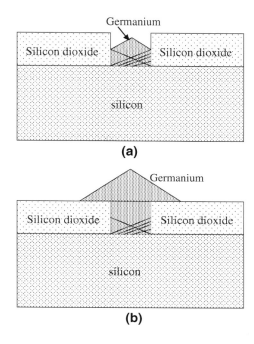

**Fig. 7 a** Necking within the oxide via, **b** after epitaxial layer overgrowth

lattice. The defect morphology for germanium grown in 100 nm holes through a120 nm oxide was confined to the initial 25 nm.

Selective growth of germanium through vias in $SiO_2$ was achieved using solid source MBE by Li et al. [69]. After germanium deposition at 0.7 monolayers (ML) per min at 650°C they observed that germanium had diffused into the silicon seed during the deposition. They reasoned that the intermixing of the germanium with the silicon resulted in a graded region which effectively reduced the lattice mismatch and aided the formation of a low defect density growth. Cross-section TEM showed that stacking faults were generated within the seed window, but terminated within 70 nm of the germanium–silicon interface. This is consistent with the theory of Lunyi and Suhir [70] that strain energy density decays exponentially from the heterojunction and that the characteristic length of decay is comparable to the dimension of the heterojunction which were 200 nm wide in this case.

Park et al. [71, 72] extended the necking technique from small area template windows to long narrow trenches. The trenches in a 500 nm $SiO_2$ layer were aligned along [110] and the two temperature process was used to selectively deposit the germanium. They found that the dislocations were trapped in a layer at the silicon seed of a thickness similar to trench width. Above this defect layer the germanium was defect free. Thus germanium stripes 400 nm wide and free from near-surface defects were obtained from 450 nm thick germanium deposits. The germanium within the trenches was (311) faceted but in epitaxial lateral overgrowth (ELO) the growth changed to (001) starting at the layer valley where two growths first met.

## 9 Germanium on Non-silicon Handles

Germanium on oxidised silicon substrates has advantages in that the equipment and procedures for handling of silicon substrates are well established. However, the use of other insulating substrates could be advantageous in optical, rf and sensor systems on a chip. Germanium on Quartz (GOQ) can be achieved by direct wafer bonding as can germanium to TCE matched glass. Miyao [73] has produced single crystal germanium stripes on quartz from sputtered germanium using polycrystalline silicon seeds. Quartz is a very good thermal insulator and so has heat transfer limitations. The thermal coefficient of expansion (TCE) of quartz is very much less than that for germanium. This can lead to processing problems and/or in the operation of the germanium devices at cryogenic temperatures. Gamble et al. [74] have proposed the use of germanium on sapphire (GeOS) to overcome these problems.

Sapphire (crystalline alumina) has long been used as the substrate in silicon-on-sapphire (SOS) wafers. Cut along certain crystallographic planes it has a lattice constant which is approximately matched to silicon. However, the mismatch can be up to 12% and gives rise to dislocations, twinning and stacking faults in epitaxially grown silicon layers. A double solid phase epitaxial process has been developed by Peregrine [75], which results in device quality layers. The technology was first used for providing radiation hardness for MOS transistors for military and aerospace applications. Sapphire is an ideal dielectric with a loss tangent tan $\delta$ less than $10^{-3}$ at 3 GHz. It is thus proven to be an ideal substrate for the integration of passive and active devices in rf applications. SOS substrates have found applications in patch clamped amplifiers, CMOS image sensors, energy harvesting devices and nanowatt ADCs. A disadvantage of SOS is that the silicon layer has a TCE much lower than that of sapphire. This means that SOS devices require special processing and are thus limited to specialist production laboratories.

Germanium devices are mainly targeted at operating frequencies higher than those achievable by silicon devices. Thus the use of a low loss substrate like sapphire should enhance performance. Sapphire has a thermal conductivity of 25–46 W/cm°K while being only three to six times less than that of silicon it is 25 to 46 times greater than that of quartz. The TCE of sapphire is similar to that of germanium and GaAs. Thus GeOS substrates could be suitable for the integration of optical and electronic circuits on a transparent substrate. With similar TCEs there should be no temperature processing difficulties with GeOS.

GeOS substrates can be fabricated by the bonding of germanium wafers directly to sapphire. Good bond strength has been achieved with low temperature post bond anneals. To minimise the generation of microvoids at the bonded interface during post bond thermal processing, Baine et al. [76] introduced a LTO $SiO_2$ layer on the sapphire surface before bonding Fig. 8. They found it necessary to anneal the LTO $SiO_2$ layer at the maximum subsequent process temperature to be used to avoid thermal voids. The interfacial oxide layer absorbs any annealing by-products Fig. 9a. However, a good electrical interface is required between the germanium

**Fig. 8** Germanium on Sapphire fabrication sequence

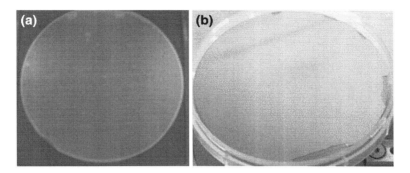

**Fig. 9** a Germanium wafer bonded to sapphire, b Germanium wafer bonded to an alumina wafer

layer and the interface (BOX) layer. Layers of $HfO_2$, $Ge_2N_2O$ and LTO $SiO_2$ were investigated. After bonding, this interface has to be stable during post bond anneal and all subsequent device processing. LTO $SiO_2$ was found to provide acceptable results. For GeOS layers greater than 2 µm the grind and polish back process produces the best electrical results.

For sub-micron thick germanium layers on sapphire the Smart-cut process can be employed. The process is the same as for GeOI with silicon handle, but the splitting can be carried out with a lower hydrogen dose and at a higher temperature if desired. There are no constraints due to mismatch of TCEs. Self-aligned polysilicon p-channel MOSTs fabricated on Smart-cut GeOS suffered from high source/drain resistance and back channel conduction. P-MOS TFTs fabricated on 0.5 µm GeOI produced by bond and polish back gave low field peak mobilities of $\sim 300$ $cm^2V^{-1}s^{-1}$. It was concluded that high source/drain resistance was again degrading the performance. However, enclosed geometry devices with large central drain contacts, yielded a low field maximum mobility of 500 $cm^2$/V-s. The main difference is attributed to the comparatively small contact windows used in the TFT design.

Since the GeOS wafers are prepared by wafer bonding and not by epitaxial growth there is no need for the sapphire substrate to be single crystalline. The requirements for the handle wafer are that it is smooth, flat, has a TCE similar to germanium, is a good dielectric, is chemically resistant, has a good thermal conductance, is low cost and is available in suitable wafer sizes. Polycrystalline alumina substrates have also been studied for this process. Corstek Superstrate 996 wafers, 4inches in diameter, have a specified roughness of less than 25 nm and are about half the cost of sapphire wafers. The surface is too rough for direct bonding so an 800 nm polycrystalline silicon layer was deposited by LPCVD. The polycrystalline silicon was polished down to 300 nm. These substrates were then bonded to the germanium wafers Fig. 9b [77]. After bonding the germanium was ground back to 80 μm. P-channel MOSTs were fabricated on these thick GeOI wafers employing a 100 μm gate enclosed geometry structure. The source/drains were boron implanted and the W/L ratio was 9. These tungsten gate MOS transistors have a sub-threshold slope of 140 mV/dec and a low field peak mobility of 650 cm$^2$/V-s. These results are close to those on bulk germanium, suggesting that the germanium layer is of device quality. Measurements were also carried out at low temperature to verify that because of the matched TCEs there would be no increased stress problems introduced by varying the temperature. The effect of germanium surface roughness on transistor performance was determined by using one GeOS wafer with an RMS roughness of 0.9 nm and another with RMS of 2.5 m. For the latter the effective hole mobility was found to decrease with decreasing temperature. The transistor hole mobility on the wafer with an RMS roughness of 0.9 nm was found to increase from 410 cm$^2$/V-s at 20°C to 510 cm$^2$/V-s at −150°C. The sub-threshold slope decreased with decreasing temperature for devices on both wafers, but the rougher surface had an order of magnitude greater sub-threshold slope.

## 10 Conclusions

Germanium continues to be an attractive semiconductor for future devices in a range of applications. However, much work still needs to be done to optimise nearly all aspects of the technology. GeO$_2$ gives the best electrical interface with both p-type ad n-type germanium. It has been shown recently that n-channel MOSTs with high electron mobility can be achieved if the thermal budget after the growth of the GeO$_2$ is minimised. In a gate first process the dielectric has to endure the source/drain activation implant anneals. The generation of a high density of states at the dielectric–germanium interface has been adjudged to be due to the generation of GeO by the reduction of GeO$_2$ by germanium at the oxide germanium interface, its diffusion through the oxide and desorption at the surface. High k dielectrics such as HfO$_2$ crystallise into polycrystalline material enabling the GeO to diffuse out through the grain boundaries during the implant anneals. The practice of employing a few monolayers of silicon to passivate the germanium surface works reasonably well on n-type germanium, but has not been so effective on p-type germanium. Some researchers have turned to Schottky barrier diodes in an effort to reduce the thermal budget and to minimise source/drain series resistance. The problem of low levels of activation of implanted dopant has only partially been solved with pre-amorphization implants. Since several major manufacturers of silicon ICs employing high-k gate dielectrics have gone over to a gate last process, this might also be a solution for germanium. The requirement for small equivalent oxide thickness (EOT) may not be satisfied if $GeO_2$ is used as an interface layer below a high-$k$ dielectric. To block the formation and diffusion of GeO, yttrium oxide which reacts with germanium to form an amorphous layer of $YGeO_x$ has been used. A composite dielectric of lanthanum oxide and yttrium oxide has a dielectric constant k = 29 and is fairly chemically stable.

It is accepted that for most applications germanium will be used in the form of GeOI. For optimum MOS performance, the GeOI layer will be sub −10 nm and the electrical properties of the back interface are practically as important as those of the front interface. Several ways of producing ultra thin germanium layers on oxidised silicon substrates are being investigated. For blanket GeOI layers the main technique is still Smart-cut, but the process is not easy and wafer supply is at a premium. For the large wafer market the starting material will be epitaxial germanium grown on either virtual substrates or by a low temperature, high temperature process. An alternative process favoured by many laboratories is the condensation process. The starting material is SOI substrates with a thin silicon layer. Epitaxial $Si_{1-x}Ge_x$ layers are grown on the SOI substrates. The germanium diffuses into the SOI layer as the wafer is oxidised. For high fraction germanium layers, strain relaxation leads to the formation of stacking faults. Defects in germanium act as acceptor states giving the layer an equivalent p-type dopant concentration $>10^{17}$ $cm^3$. This, together with interface states at the germanium–BOX interface, leads to a back gate conduction channel from source to drain. While this can be overcome with a deep arsenic implant it is not an ideal solution.

A third technique for producing blanket GeOI is the use of a single crystal oxide layer on silicon as the seed for the epitaxial growth of germanium. The germanium layer is then bonded to an oxidised silicon wafer or other appropriate insulator wafer. The silicon wafer is then removed to leave the GeOI substrate. This is a relatively new process and the quality of the germanium layers has still to be revealed. The crystalline oxide and germanium layers are grown by MBE.

The cost of GeOI substrates produced by the above processes will be greater than that of the equivalent SOI wafers. There is also likely to be a supply problem and preferred customers. This has caused other researchers to investigate alternatives to blanket transfer processes. If the industry does not insist on the total freedom of blanket GeOI layers then the technologies based on dislocation removal by necking could play a prominent role. Device quality stripes or islands can be achieved at a much lower cost. There are two different technologies employing dislocation necking (i) Solid Phase Epitaxy (SPE) within narrow cavities and (ii) Liquid Phase Epitaxy with rapid melt.

In the SPE technology germanium is selectively epitaxially grown in narrow cavities or trenches in a silicon dioxide layer on the silicon surface. The germanium grows with 311 facets until the oxide surface is reached. Epitaxial overgrowth of defect free germanium proceeds until adjacent layers coalesce where defects may occur. The misfit dislocations are confined to a layer whose thickness at the bottom of the trenches is approximately equal to the trench width. Such substrates could be purchased with an array of stripes or could be produced in-house. Selective epitaxial germanium deposition is the only non-standard process. The constraints imposed on circuit designers do not appear to be insurmountable.

For applications where a high areal density of transistors is not required the liquid phase epitaxial growth from rapid melt germanium offers exciting prospects of cheap GeOI substrates. The process starts with standard silicon wafers and requires only standard processing equipment. The germanium can be deposited by sputtering, PECVD or CVD in the amorphous form. It is very suitable for mixed semiconductor circuits including silicon, germanium and III-Vs to provide maximum flexibility in circuit integration.

Germanium on sapphire offers an ideal substrate for rf and mm integrated circuits with integrated passives, for the integration of optical and electronic circuits and for cryogenic application. For system on a chip (SOC) applications where continuous semiconductor layers are not required, the rapid thermal LPE of GeOS could be a very cost effective substrate. The rapid melt LPE process with subsequent low temperature processes could be a very cost effective way of producing integrated active electronic circuits not requiring state of the art lithographic processes.

The growth of germanium on a single crystal dielectric epitaxially grown on silicon offers potential for cheap ultra-thin germanium layers on silicon. It also promises a means of obtaining germanium layers on non-silicon substrates such as glass or quartz through layer transfer.

# References

1. Cresman, E.E.: The oxidation of germanium surfaces at pressures much greater than one atmosphere. J. Electrochem. Soc. **129**(8), 1845–1848 (1982)
2. Hymes, D.J., Rosenberg, J.J.: Growth and materials characterization of native germanium oxynitride thin films on germanium. J. Electrochem. Soc. **135**(4), 961–965 (1988)
3. Rosenberg, J.J., Martin, S.C.: Self-aligned germanium MOSFETs using a nitrided native oxide gate insulator. IEEE Electron Device Lett. **9**(12), 639–640 (1988)
4. Martin, S.C., Hitt, L.M., Rosenberg, J.J.: p-channel germanium MOSFETs with high channel mobility. IEEE Electron Device Lett. **10**(7), 325–326 (1989)
5. Jackson, T.N., Ransom, C.M., DeGelormo, J.F.: Gate-self-aligned p-channel germanium MISFET's. IEEE Electron Device Lett. **12**(11), 605–607 (1991)
6. Vitkavage, D.J., Fountain, G.G., Rudder, R.A., et al.: Gating of germanium surfaces using pseudomorphic silicon interlayers. Appl. Phys. Lett. **53**(8), 692–694 (1988)
7. Hattangady, S.V., Fountain, G.G., Rudder, R.A., et al.: Interface engineering with pseudormorphic interlayers: Ge metal insulator-semiconductor structures. Appl. Phys. Lett. **57**(6), 581–583 (1990)
8. Reinking, D., Kammler, M., Hoffmann, N., et al.: Ge p-MOSFETs Compatible with Si CMOS-Technology. ESSDERC, pp. 300–304 (1999)
9. Shang, H., Frank, M.M., Gusev, E.P., et al.: Germanium channel MOSFETs: opportunities and challenges. IBM J. Res. Dev. **50**(4/5), 377–386 (2006)
10. Kita, K., Kyuno, K., Toriumi, A.: Growth mechanism difference of sputtered $HfO_2$ on Ge and on Si. Appl. Phys. Lett. **85**(1), 52–54 (2004)
11. Saito, S., Hosoi, T., Watanabe, H., et al.: First-principles study to obtain evidence of low interface defect density at $Ge/GeO_2$ interfaces. Appl. Phys. Lett. **95**, 11908-3 (2009)

12. Lee, C.H., Tabata, T., Nishimura, T., et al.: Ge/GeO$_2$ interface control with high-pressure oxidation for improving electrical characteristics. Appl. Phys. Express **2**, 071404 (2009)
13. Li, C.X., Lai, P.T.: Wide-bandgap high-k Y$_2$O$_3$ as passivating interlayer for enhancing the electrical properties and high-field reliability of n-Ge metal-oxide-semiconductor capacitors with high-k HfTiO gate dielectric. Appl. Phys. Lett. **95**, 22910-3 (2009)
14. Zhao, Y., Kita, K., Kyuno, K., Toriumi, A.: Higher-$k$ LaYO$_x$ films with strong moisture resistance. Appl. Phys. Lett. **89**, 252905 (2006)
15. Spann, J.Y., Anderson, A., Thornton, T.J., et al.: Characterization of nickel germanide thin films for use as contacts to p-channel germanium MOSFETs. IEEE Electron Device Lett. **26**(3), 151–153 (2005)
16. Hellberg, P.E., Gagnor, A., Zhang, S.L., Petersson, C.S.: Polycrystalline Si$_x$Ge$_{1-x}$ Films. J. Electrochem. Soc. **144**(11), 3968–3973 (1997)
17. Uppal, S., Willoughby, A.F.W.: Diffusion of ion implanted boron in germanium. J. Appl. Phys. **90**(8), 4293–4295 (2001)
18. Satta, A., Simoen, E., Clarysse, T., et al.: Diffusion, activation, and recrystallization of boron implanted in preamorphized and crystalline germanium. Appl. Phys. Lett. **87**, 172109-3 (2005)
19. Impellizeri, G., Mirabella, S., Bruno, E., et al.: B activation and clustering in ion-implanted Ge. J. Appl. Phys. **105**, 63533-6 (2009)
20. Chao, Y.-L., Prussin, S., Woo, J.C.S.: Preamorphization implantation-assisted boron activation in bulk germanium and germanium-on-insulator. Appl. Phys. Letts. **87**, 142102-3 (2005)
21. Mirabella, S., Impellizzeri, G., Piro, A.M., et al.: Activation and carrier mobility in high fluence B implanted germanium. Appl. Phys. Lett. **92**, 251909-3 (2008)
22. Chui, C.O., Gopalakrishnan, K., Griffin, P.B., Plummer, J.D.: Activation and diffusion studies of ion-implanted p and n dopants in germanium. Appl. Phys. Lett. **83**(16), 32753277 (2003)
23. Satta, A., Janssens, T., Clarysse, T., et al.: P implantation doping of Ge: diffusion, activation, and recrystallization. Vac. Sci. Technol. B **24**(1), 494–498 (2006)
24. Satta, A., Simeon, E., Duffy, R., et al.: Diffusion, activation, and regrowth behavior of high dose P implants in Ge. Appl. Phys. Lett. **88**, 162118-3 (2006)
25. Carroll, M.S. Koudelka, R.: Accurate modelling of average phosphorus diffusivities in germanium after long thermal anneals: evidence of implant damage enhanced diffusivities. Semicond. Sci. Technol. **22**, S164–S167 (2007)
26. Posselt, M., Schmidt, B., Anwand, W., et al.: P implantation into preamorphized germanium and subsequent annealing: Solid phase epitaxial regrowth, P diffusion, and activation. J. Vac. Sci. Technol. B **26**(1), 430–434 (2008)
27. Koffel, S., Scheiblin, P., Claverie, A., Mazzocchi, V.: Doping of germanium by phosphorus implantation: Prediction of diffused profiles with simulation. Mater. Sci. Eng. B **154–155**, 60–63 (2008)
28. Canneaux, T., Mathiot, D., Ponpon, P., et al.: Diffusion of phosphorus implanted in germanium. Mater. Sci. Eng. B **154–199**, 68–71 (2008)
29. Kim, J., Bedell, S.W., Maurer, L., et al.: Activation of implanted n-type dopants in Ge over the active concentration of $1 \times 10^{20}$ cm$^{-3}$ using coimplantation of Sb and P. Electrochem. Solid State Lett. **13**(1), H12–H15 (2010)
30. Yu, H.-Y., Nishi, Y., Saraswat, K.C., Cheng, S.-L., Griffin, P.B.: Germanium in situ doped epitaxial growth on Si for high-performance n+/p junction diode. IEEE Electron Device Lett. **30**(9), 1002–1004 (2009)
31. Scappucci, G., Capellini, G., Lee, W.C.T., Simmons, M.Y.: Ultradense phosphorus in germanium delta-doped layers. Appl. Phys. Lett. **94** 162106-3 (2009)
32. Morii, K., Iwasaki, T., Nakane, R., et al.: High performance GeO2/Ge nMOSFETs with source/drain junctions formed by gas phase doping. Electron Devices Meeting (IEDM), IEEE International, pp. 1–4 (2009)

33. Posthuma, N.E., van der Heide, J., Flamand, G., Poortmans, J.: Emitter formation and contact realization by diffusion for germanium photovoltaic devices. IEEE Trans. Electron Devices **54**(5), 1210–1215 (2007)
34. Bruel, M.: Silicon on insulator material technology. Electron Lett. **31**(14), 1201–1202 (1995)
35. Agarwal, A., Haynes, T.E., Holland, O.W., Eaglesham, D.J.: Efficient production of silicon-on-insulator films by co-implantation of $He^+$ with $H^+$. Appl. Phys. Lett. **72**, 1086–1088 (1998)
36. Hurley, R.E., Wadsworth, H., Montgomery, J.H., et al.: Surface blistering of low temperature annealed hydrogen and helium co-implanted germanium and its application to splitting of bonded wafer substrates. Vacuum **83**, S29–S32 (2009)
37. Low, Y.W., Rainey, P., Hurley, R., et al.: Hydrogen implantation in Germanium. ECS Trans. **28**(1), 375–383 (2010)
38. Sedgewick, T.O.: Dominant surface electronic properties of $SiO_2$ passivated Ge surfaces as a function of various annealing treatments. J. Appl. Phys. **39**(11), 5066–5077 (1968)
39. Ruddell, F.H., Montgomery, J.H., Gamble, H.S., Denvir, D.: Germanium MOS technology for infra-red detectors. Nucl. Inst. Method Phys. Res. A **573**, 65–67 (2007)
40. Van de Walle, C.G., Neugebauer, J.: Universal alignment of hydrogen levels in semiconductors, insulators and solutions. Nature **423**, 625–628 (2003)
41. Omachi, Y., Nishioka, T., Shinoda, Y.: The heteroepitaxy of Ge on Si(100) by vacuum evaporation. J. Appl. Phys. **54**(9), 5466–5469 (1983)
42. Bojarczuk, N.A., Copel, M., Guha, S., et al.: Epitaxial silicon and germanium on buried insulator heterostructures and devices. Appl. Phys. Lett. **83**(26), 5443–5445 (2003)
43. Seo, J.W., Dieker, Ch., Tapponnier, A., et al.: Epitaxial germanium-on-insulator grown on (001) Si. Microelectron. Eng. **84**, 2328–2331 (2007)
44. Guissani, A., Rodenbach, P., Zaumseil, P., et al.: Atomically smooth and single crystalline $Ge(111)/cubic-Pr_2O_3(111)/Si(111)$ heterostructures: Structural and chemical composition study. J. Appl. Phys. **105**, 33512-6 (2009)
45. http://www.iqesilicon.com
46. Tezuka, T., Sugiyama, N., Mizuno, Y., et al.: A novel fabrication technique of ultrathin and relaxed SiGe buffer layers with high Ge fraction for sub-100 nm strained silicon-on-insulator MOSFETs. Jpn. J. Appl. Phys. **40**, 2866–2874 (2001)
47. Tezuka, T., Sugiyama, N., Takagi, S.: Fabrication of strained Si on an ultrathin SiGe-on-insulator virtual substrate with a high-Ge fraction. Appl. Phys. Lett. **79**(12), 1798–1800 (2001)
48. Tezuka, T., Nakaharai, S., Moriyama, Y., et al.: Selectively-formed high mobility SiGe-on-insulator pMOSFETs with Ge-rich strained surface channels using local condensation technique. In: Symposium on VLSI Technology. Digest of Technical Papers, pp. 198–199 (2004)
49. Vincent, B., Damlencourt, J.-F., Delaye, V., et al.: Stacking fault generation during relaxation of silicon germanium on insulator layers obtained by the Ge condensation technique. Appl. Phys. Lett. **90**, 074101 (2007)
50. Nakaharai, S., Tezuka, T., Hirashita, N., et al.: The generation of crystal defects in Ge-on-insulator (GOI) layers in the Ge-condensation process. Semicond. Sci. Technol. **22**, S103–S106 (2007)
51. Tezuka, T., Moriyama, Y., Nakaharai, S., et al.: Lattice relaxation and dislocation generation/annihilation in SiGe-on-insulator layers during Ge condensation process. Thin Solid Films **508**, 251–255 (2006)
52. Sugiyama, N., Nakaharai, S., Hirashita, N., et al.: The formation of SGOI structures with low dislocation density by a two-step oxidation and condensation method. Semicond. Sci. Technol. **22**, S59–S62 (2007)
53. Nguyen, Q.T., Damlencourt, J.F., Vincent, B., et al.: High quality germanium-on-insulator wafers with excellent mobility. Solid State Electron. **51**, 1172–1179 (2007)
54. Maeda, T., Ikeda, K., Nakaharai, S., et al.: Thin-body Ge-on-insulator p-channel MOSFETs with Pt germanide metal source/drain. Thid Solid Film **508**, 346–350 (2006)

55. Dissanayake, S., Tomiyama, K., Sugahara, S., et al.: High performance ultrathin (110)-oriented Ge-on-insulator p-channel metal–oxide–semiconductor field-effect transistors fabricated by Ge condensation technique. Appl. Phys. Express **3**, 041302 (2010)
56. Namra, S.: Crystallization of vacuum-evaporated germanium films by the electron beam zone-melting process. J. Appl. Phys. **37**, 1929–1930 (1966)
57. Takai, M., Tanigawa, T., Gamo, K., Namba, S.: Single Crystal germanium island on insulator by zone melting recrystallization. Jpn. J. Appl. Phys. **22**(10), L624–L626 (1983)
58. Liu, Y., Deal, M.D., Plummer, J.D.: High-quality single-crystal Ge on insulator by liquid-phase epitaxy on Si substrates. Appl. Phys. Lett. **84**(14), 2563–2565 (2004)
59. Liu, Y., Gopalafishan, K., Griffin, P.B., et al.: MOSFETs and high-speed photodetectors on Ge-on-insulator substrates fabricated using rapid melt growth. IEDM 40.4, pp. 1–4 (2004)
60. Tweet, D.J., Lee, J.J., Maa, J.-S., et al.: Characterization and reduction of twist in Ge on insulator produced by localized liquid phase epitaxy. Appl. Phys. Lett. **87**,141908 (2005)
61. Feng, J., Liu, Y., Griffin, P.B., Plummer, J.D.: Integration of germanium-on-insulator and silicon MOSFETs on a silicon substrate. IEEE Electron Device Lett. **27**(11), 911–913 (2006)
62. Feng, J., Woo, R., Chen, S., et al.: P-channel germanium FinFET based on rapid melt growth. IEEE Electron Device Lett. **28**, 637–639 (2007)
63. Balakumar, S., Roy, M.M., Ramamurthy, B., et al.: Fabrication aspects of germanium on insulator from sputtered Ge on Si-substrates. Electrochem. Solid State Lett. **9**, G158–G160 (2006)
64. Hashimoto, T., Yoshimoto, C., Hosoi, T., et al.: Fabrication of local Ge-on-insulator structures by lateral liquid-phase epitaxy: effect of controlling interface energy between Ge and insulators on lateral epitaxial growth. Appl. Phys. Express **2**, 066502 (2009)
65. Miyao, M., Tanaka, T., Toko, K., et al.: Giant Ge-on-insulator formation by Si–Ge mixing-triggered liquid-phase epitaxy. Appl. Phys. Express **2**, 045503 (2009)
66. Miyao, M., Tanaka, T., Toko, K., et al.: High-quality single-crystal Ge stripes on quartz substrate by rapid-melting-growth. Appl. Phys. Lett. **95**, 022115 (2009)
67. Fitzgerald, E.A., Chand, N.: Epitaxial necking in GaAs grown on pre-patterned Si substrates. J. Electr. Mater. **20**, 839–844 (1991)
68. Langdo, T.A., Leitz, C.W., Currie, M.T.: High quality Ge on Si by epitaxial necking. Appl. Phys. Lett. **76**(25), 3700–3702 (2000)
69. Li, Q., Han, S.M., Brueck, S.R.J., et al.: Selective growth of Ge on Si(100) through vias of $SiO_2$ nanotemplate using solid source molecular beam epitaxy. Appl. Phys. Lett. **83**(24), 5032–5034 (2003)
70. Luryi, S., Suhir, E.: New approach to the high quality epitaxial growth of lattice mismatched materials. Appl. Phys. Lett. **49**, 140–142 (1986)
71. Park, J.-S., Bai, J., Curtin, M., et al.: Defect reduction of selective Ge epitaxy in trenches on Si(100) substrates using aspect ration trapping. Appl. Phys. Lett. **90**, 52113-3 (2007)
72. Park, J.-S., Curtin, M., Hydrick, J.M., et al.: Low-defect-density Ge epitaxy on Si(100) using aspect ratio trapping and epitaxial lateral overgrowth. Electrochem. Solid State Lett. **12**(4), H142–H144 (2009)
73. Miyao, M., Toko, K., Tanaka, T., et al.: High-quality single-crystal Ge stripes on quartz substrate by rapid-melting-growth. Appl. Phys. Lett. **90**, 22115-3 (2009)
74. Gamble, H.S., Baine, P.T., Wadsworth, H., et al.: Germanium on sapphire. Int. J. High Speed Electr. Syst. **18**(4), 805–814 (2008)
75. http://www.psemi.com
76. Baine, P.T., Gamble, H.S., Armstrong, B.M., et al.: Germanium on sapphire by wafer bonding. Solid State Electr. **52**, 1840–1844 (2008)
77. Baine, P.T., Gamble, H.S., Armstrong, B.M., et al.: Germanium bonding to $Al_2O_3$. ECS Trans. **16**(8), 407–414 (2008)

# Low-Temperature Fabrication of Germanium-on-Insulator Using Remote Plasma Activation Bonding and Hydrogen Exfoliation

C. A. Colinge, K. Y. Byun, I. P. Ferain, R. Yu and M. Goorsky

**Abstract** Low-temperature germanium to silicon wafer bonding was demonstrated by in situ radical activation and bonding in vacuum. After low temperature direct bonding of Ge to Si followed by annealing at 200 and 300°C, advanced imaging techniques were used to characterize the bonded interface. The feasibility of transferring hydrogen-implanted germanium to silicon with a reduced thermal budget is also demonstrated. Germanium samples were implanted with hydrogen and a two-step anneal was performed. The first anneal performed at low temperature ($\leq$150°C for 22 h) to enhance the nucleation of hydrogen platelets. The second anneal is performed at 300°C for 5 min and is shown to complete the exfoliation process by triggering the formation of extended platelets.

## 1 Introduction

Germanium is gaining interest as a semiconductor material because bulk germanium has the highest hole mobility of all semiconductors (maximum $\mu_p = 1,900$ cm$^2$V$^{-1}$s$^{-1}$ and an electron mobility that is potentially twice that of silicon (maximum $\mu_p = 3,900$ cm$^2$V$^{-1}$s$^{-1}$). Since germanium is a much rarer element than silicon, it is impractical (unpractical means impossible—remember that one form aa picky HS teacher...) to use bulk germanium wafers. More suitable is the formation of thin germanium layers directly deposited on silicon or on oxidized

C. A. Colinge (✉), K. Y. Byun, I. P. Ferain and R. Yu
Tyndall National Institute, University College Cork, Cork, Republic of Ireland
e-mail: cindy.colinge@tyndall.ie

M. Goorsky
Department of Material Science and Engineering, UCLA, Los Angeles, CA, USA

silicon. It is also possible to produce mixed substrates in which n-channel devices are made in a thin silicon-on-insulator (SOI) layer and germanium p-channel transistors are made in a thin germanium-on-insulator (GeOI) film [1]. GeOI can be produced by the germanium condensation technique. In this technique a $Si_{0.9}Ge_{0.1}$ layer is first epitaxially grown on SOI. Oxidation is performed to produce a $SiO_2$ layer and a Ge-enriched layer ($Si_{0.25}Ge_{0.75}$). Pure germanium is subsequently grown on the $Si_{0.25}Ge_{0.75}$ layer [2]. The epitaxial approach has the drawback of being carried out at a relatively high temperature (>650°C). Since silicon and germanium have very different thermal expansion coefficients, use (or 'the use') of any high-temperature processing step results in the formation a high density of thermal mismatch defects in the germanium epilayer [3, 4].

## 2 Low-Temperature Bonding Mechanisms

Low-temperature direct wafer bonding eliminates the aforementioned problems associated with the epitaxial and condensation techniques. Low defect density is of prime importance for heterogeneous integration of dissimilar materials such as III–V on silicon. Low-temperature bonding eliminates the severe thermal stress that can be induced by high-temperature annealing [5]. Ge to Si direct wafer bonding has been studied for use in high-performance photodetectors as well as high-quality epitaxial templates for GaAs growth [6, 7].

Radical activation of the wafers is a key factor for the success of low-temperature bonding. The effects of free radical activation for Si to Si bonding have been previously reported in the literature [8]. In that study, a comparison of different surface treatments for direct Si to Si wafer bonding was made. Hydrophilic and hydrophobic Si wafers were exposed to a range of pretreatments, involving oxygen and nitrogen radical activation before in situ wafer bonding in a vacuum. After low-temperature annealing at 200 and 300°C, the formation of voids was observed by using scanning acoustic microscope inspection. A comparison of the bonding energy was conducted and analyzed as a function of the surface treatments, which demonstrated that the remote plasma pretreatment is a very suitable process for surface modification of hydrophilic and hydrophobic Si to Si direct wafer bonding.

Here, we will focus on the characterization of activated Ge surfaces using oxygen and nitrogen radicals and show successful low temperature Ge to Si direct bonding using radical activation and degassing channels. The chemical species on the activated Ge surfaces were investigated using angle-resolved X-ray photoelectron spectroscopy (ARXPS), while structural analysis was performed by scanning acoustic microscopy (SAM) and high-resolution transmission electron microscopy (HR-TEM).

In a first experiment, 100-mm, <100> oriented p-type Ge wafers (Ga doped with a resistivity of 0.016 $\Omega$ cm) were selected. The oxygen and nitrogen radical activated Ge surfaces were studied in a Vacuum Science Workshop Atomtech

ESCA system using Al $K\alpha$ radiation ($h\nu = 1{,}486.6$ eV). The Ge wafers were cleaned in an SC1-equivalent solution without ozone using a Semitool Spray Acid Tool (SAT) prior to surface activation. Wafers were then loaded into an Applied Microengineering Limited (AML) AW04 aligner bonder and vacuum was applied. The wafers were then exposed for 10 min to either oxygen or nitrogen free radicals at a pressure of 1 mbar and a plasma power of 100 W. The radicals were generated by a remote plasma ring. A reference sample of Ge cleaned and bonded without exposure was also prepared. Wafers were bonded under a pressure of 1 kN applied for 5 min and immediately transferred for XPS analysis. The three bonded Ge to Ge samples were then de-bonded using a razor blade, cleaved into $2 \times 2$ cm size samples and loaded into a high-vacuum sample holder of the XPS tool. The total exposure to ambient air was kept <1 min. The photoelectron peaks and chemical composition of the Ge surface were analyzed at various take-off angles.

In addition, blank Ge wafers were bonded directly to Si using the same recipe and bonder used for the Ge bonded to Ge, i.e. 10 min oxygen or nitrogen radical activation at low temperature. (100)-oriented n-type Czochralski grown prime grade bare Si wafers with a diameter of 100 mm were bonded with the Ge wafers. Prior to bonding, the Ge and Si wafers were cleaned in an SC1-equivalent solution with ozone for Si and without ozone for Ge. After loading into the AML wafers were activated and bonded in situ under a pressure of 1kN applied for 5 min at a chamber pressure ranging from 5 to 10 m bars. The wafers were annealed in situ at 100°C for 1 h with an applied pressure of 500 N in vacuum followed by an ex-situ anneal at 200°C for 24 h in order to increase bond strength. The bonded pairs were then annealed again at 300°C for 24 h. The ramp-up rate was set to 0.5°C/min in both cases. After the anneal, Ge-Si bonded pairs remained intact despite their coefficient of thermal expansion (CTE) mismatch, owing to the slow ramp-up rate. Bonded interfaces were imaged by SAM. Then Ge was directly bonded to a patterned Si wafer with 2-μm deep channels at a pitch of 400 μm using oxygen radical activation. The bond strength of the patterned Ge-Si wafer pair was

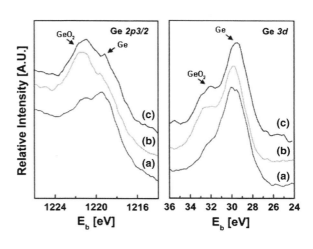

**Fig. 1** Ge 2p3/2 and Ge 3d photoelectron features (take-off angle = 75°): **a** reference cleaned sample **b** oxygen radical activated sample **c** nitrogen radical activated sample

measured using the crack opening method. Structural analysis of buried interfaces was studied by SAM and HR-TEM inspection.

Binding state configurations of germanium atoms at the activated surfaces were evaluated by ARXPS using the de-bonded cleaved samples. Figure 1 shows the Ge 2p3/2, Ge 3d photoelectron features (take-off angle = 75°). As expected the Ge 2p3/2 signal shows two binding energy contributions at 1,218.5 and 1,221.0 ($\pm 0.2$) eV [9]. These can be assigned to zero- and quadra-valent germanium, respectively. Zero-valent corresponds to bulk Ge while quadra-valent indicates the presence of a $GeO_2$ layer. The Ge 3d peaks exhibit similar information. Two features can be resolved at 29.8 and 32.5 eV that once again are attributed to zero- and quadra-valent Ge respectively [10–12]. There was no evidence in this work of the formation of detectable quantities of sub-stoichiometric oxides, which agrees with the Ge 3d signal (which is bulk sensitive) and Ge 2p signal (which is surface sensitive).

Quantitative relative contributions of the oxygen and germanium features are found by curve-fitting the XPS data. These data are reported in Table 1 as peak area ratios and are used to derive stoichiometric amounts using known sensitivity factors. The data collected at 75° are the most accurate representation of the actual stoichiometry close to the Ge surfaces. Curve fitting reveals a peak indicating the presence of OH– which is located at a binding energy which is higher than the binding energy of $GeO_2$ [17]. This data is shown in the first column of Table 1 where the ratio of OH– to $GeO_2$ (labeled $O^-/O^{2-}$) is represented. In the second column we have derived the O/Ge ratio which is the total O 1 s peak area divided by total Ge 2p3/2 peak area. From the second column of Table 1, it is clear that exposure of hydrophilic cleaned Ge substrates to free radicals produces some oxidation.

For samples exposed to either nitrogen or oxygen radicals, the quantification shown in Table 1 shows higher values of O/Ge ratio, which is more likely due to a

**Table 1** ARXPS peak area ratio

| Sample/take-off angle | | $O^-/O^{2-}$ | O/Ge total |
|---|---|---|---|
| Reference substrate | 0° | 0.34 | 0.04 |
| | 25° | 0.24 | 0.04 |
| | 50° | 0.09 | 0.08 |
| | 75° | 0.00 | 0.25 |
| Post oxygen radical | 0° | 0.11 | 0.06 |
| | 25° | 0.11 | 0.07 |
| | 50° | 0.02 | 0.12 |
| | 75° | 0.00 | 0.40 |
| Post nitrogen radical | 0° | 0.07 | 0.05 |
| | 25° | 0.09 | 0.06 |
| | 50° | 0.06 | 0.11 |
| | 75° | 0.00 | 0.35 |

$O^-/O^{2-}$ is the $O^-$ signal area divided by $O^{2-}$ signal area in O 1 s spectra. O/Ge is the total O 1 s peak area divided by total Ge 2p3/2 peak area

thicker oxide film compared to the reference Ge sample. The nitrogen radical exposed Ge also shows a significant GeO$_2$ formation; this is similar to previous studies where Ge exposed to nitrogen plasma resulted in a more hydrophilic surface signifying formation of an oxide [13]. These quantifiable XPS data were used to estimate the film thickness as 0.39, 0.58 and 0.72 nm for the reference, nitrogen radical activated and oxygen radical activated samples, respectively.

Most extensive hydroxylation, which is indicated in the first column (O$^-$/O$^{2-}$) of Table 1, was observed for the reference samples. Nitrogen radical activated samples gave the lowest –OH concentration and oxygen radical activated samples an intermediate value. The presence of hydroxyl species is important because increased hydroxyl groups can give rise to more hydrophilic reactions at bonded interfaces, which can then induce intrinsic void generation due to trapped reaction by-product. The signal due to O$^-$ (OH$^-$) decreases towards grazing emission for all samples, which is due to a sub-surface hydroxide or sub-stoichimetric GeO$_x$.

The buried interfaces of bonded pairs were inspected using SAM after annealing at 200°C 24 h and additionally at 300°C for 24 h. Figure 2 shows scanning acoustic microscope images of bonded pairs: (a) pre-cleaned by SC1-equivalent solution, (b) post oxygen radical exposure, and (c) post nitrogen radical exposure. The hydrophilic reaction at the interfaces can be described the following equations:

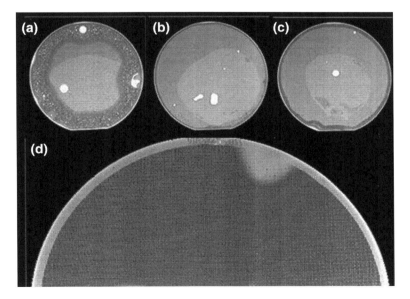

**Fig. 2** Scanning acoustic images of Ge/Si bonded wafer pairs after annealing at 200°C for 24 h, and additionally at 300°C for 24 h: **a** cleaned in an SC1-equivalent solution, followed by **b** O$_2$ radical 10 min exposure, **c** N$_2$ radical 10 min exposure, and **d** formed 2 μm depth channel in Si wafer prior to oxygen radical activated (10 min) bonding creates an exhaust path for trapped gas

$$Ge-OH + OH-Si = Ge-O-Si + H_2O \qquad (1)$$

$$Ge(Si) + 2H_2O = GeO_2(SiO_2) + 2H_2 \qquad (2)$$

Figure 2 shows that hydrophilic reactions (Eqs. 1 and 2) at the interfaces may create trapped gas like water vapor or hydrogen following the hydrophilic chemical reaction, which appears as a different contrast in the SAM images. Some of the water molecules resulting in covalent bonding of Ge–O–Si of oxygen and nitrogen exposed samples can diffuse through the nanometer range oxide layer and react with the bulk germanium and silicon to form dioxide and hydrogen. In Fig. 2a, however, the buried oxide is so thin that the reactants and by-products cannot diffuse through the interface. Radical activated samples in Fig. 2b, c, shows minimal intrinsic voids formed due to ability of the relatively thick oxide interface, which enhances by-product diffusion. For nitrogen radical activated sample in Fig. 2c, which has a thinner oxide than Fig. 2b, we can clearly see the initiation of void generation of by-products near the wafer edge area. The formation of interface voids during low temperature anneal depends on the thickness of the stoichiometric oxide film, which is coincident with silicon wafer bonding [14]. Thus, the SAM result agrees with the oxide thickness extracted from ARXPS.

In Fig. 2d, we can clearly see that the 2 μm-deep channels formed at the surface of the Si wafer prior to oxygen radical activated bonding and anneal creates an escape path for trapped gas or moisture allowing high bond strength to be achieved. Additionally, the SAM image does show a uniformly colored SAM image indicative of a well bonded sample. The SAM image suggests that hydrophilic reaction at interface generate gas phase by-product, which can be diffused out by the channel which is similar to the result of Ge bonded to sapphire with patterned channels [15].

In Fig. 3, after low temperature anneal at 200 and 300°C, the bonding energy is so high that insertion of a razor blade results in bulk fracture of the germanium, which is consistent with the respective fracture stress of single crystal Si (62 MPa)

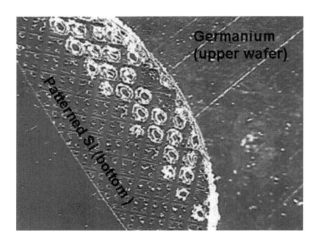

**Fig. 3** Optical microscope observation of razor blade tested sample after annealing at 200°C for 24 h and additional 300°C for 24 h. After crack opening test, bulk of the germanium crack (no de-bonding), revealing pieces of germanium strongly bonded to the thin buried oxide

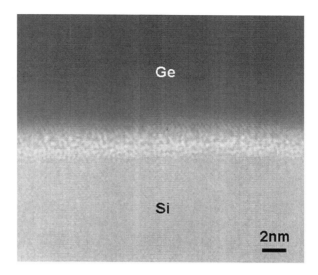

**Fig. 4** HR-TEM observation of buried interfaces after annealing at 200°C for 24 h and additional 300°C for 24 h

and Ge (49 MPa) [16]. The full bond strength is attributed to the oxygen radical activation.

Figure 4 shows cross-sectional HR-TEM image of buried interfaces using oxygen radical activation after 200°C and an additional 300°C 24 h anneal. The 2 nm-thick buried interface appears defect-free and a smooth bonded interface is visible. It corresponds with the calculated thickness (0.7 nm) by ARXPS of Ge exposed to oxygen and our previous TEM result which showed 1 nm-thick $SiO_2$ on oxygen radical activated Si. Additionally, compared to the Ge condensation technique [2], the TEM image shows no over oxidation or stacking faults generated during the low thermal budget ($\leq$300°C) anneal. This low temperature process using oxygen radical activation is the way to achieve high quality interfaces without generating stacking fault or inducing surface damage by direct plasma activation.

Chemical and structural analyses show that remote plasma treatment using oxygen and nitrogen radical facilitates oxidation on Ge surface to increase chemical reaction. Thus radical activation can enhance the bonding reaction allowing strong bond strength at low temperature while preserving the crystalline quality of the Ge. The radical activated bonding process is very suitable for a low temperature process as needed for thermally mismatched materials such as those used for photonic devices.

## 3 Low-Temperature Exfoliation of Germanium Layers

Transfer of thin semiconductor layers by exfoliation has received a lot of attention since its first use for the fabrication of Silicon-on-Insulator (SOI) substrates [17].

The implantation of hydrogen or inert gas into single crystalline semiconductor substrates leads to formation of a defective region below the surface. Under high-temperature treatment, usually in the range of 400–500°C, hydrogen molecules tend to be trapped in these defects and form pockets of gas at the projected range, commonly referred to as 'bubbles'. As temperature and/or anneal time increase, the internal pressure inside the bubbles increases and results in the formation of micro-cracks which triggers the splitting of a thin semiconductor layer [18, 19]. The mechanisms which govern defect formation in H-implanted semiconductors and the creation of micro-cracks has already been extensively characterized in silicon, germanium [20] and III–V compounds [21]. It is often addressed from a wafer bonding perspective and targets a range of applications varying from the fabrication of low defective substrates for CMOS compatible applications (SOI and Germanium on Insulator substrates 'GeOI'. The thermal budget required to generate micro-cracks is a sensitive matter for direct wafer bonding. The temperature range that is commonly considered lies above 400°C. However, such a high temperature is expected to induce significant modification of the bonded interface in heterogeneous substrates due to thermal expansion mismatch. This may result in poor bond strength and in degraded quality of the bonded interface [6, 22].

Here, we investigate the feasibility of transferring hydrogen-implanted germanium to silicon with a reduced thermal budget. Recently, co-implantation of hydrogen and helium for low-temperature (300°C) exfoliation of germanium has been successfully demonstrated [23]. That approach presents a relatively long time-to-blister anneal (40 min) at a temperature of 300°C. In this paper, an exfoliation process which does not require any helium co-implant is investigated which significantly reduces the time required for exfoliation at temperatures near 300°C. This process is based on a long defect nucleation step at low temperature ($\leq$150°C), followed by a very short time anneal (STA) at higher temperature (300°C). With this technique, complete exfoliation has been successfully demonstrated for hydrogen-implanted III-V materials such as InP and InAs [24, 25]. The benefits expected from this two-step process for direct wafer bonding are twofold: the low defect nucleation anneal enhances bond strength without degrading the bonded interface morphology; and the STA minimizes the impact of the dissimilar thermal expansion coefficients between the bonded pair.

<100>-orientated n-type germanium wafers with a diameter of 100 mm (Sb doped, 0.03 $\Omega$ cm) were used for this experiment. Prior to hydrogen implant, a 100 nm thick layer of PECVD silicon dioxide was deposited and densified at 600°C. Germanium substrates were implanted at room temperature with $H_2^+$ at a dose of $5 \times 10^{16}$ cm$^{-2}$ at 180 keV without active chuck cooling. The projected ion and germanium vacancy ranges that are expected are 650 and 590 nm below the germanium surface, respectively. Following the implant, the silicon dioxide layer was completely removed in a dilute HF solution. One of the implanted substrate was diced in small samples (1 cm $\times$ 1 cm) prior to anneal. Strain profile within as-implanted germanium samples has been assessed by high resolution X-ray Diffraction (XRD) measurements. The $\omega/2\theta$ diffraction patterns provide

information about strain introduced by the $H_2^+$ implant [26, 27]. Negligible strain profile variation was observed across the wafer, which suggests excellent implant uniformity among the 1 cm$^2$ samples. In addition, XRD measurements suggest that nucleation is already initialized during hydrogen implantation due to lack of wafer cooling during hydrogen implantation [27].

Following XRD measurements, a set of different anneal conditions was then considered in order to estimate the optimum thermal process to induce exfoliation. Implanted samples were encapsulated and sequentially annealed according to conditions described in Table 2. The onset of blistering was determined by tapping mode atomic force microscopy (AFM) and by optical microscopy. Prior to surface morphology characterisation, samples were cleaned in de-ionized (DI) water. Cross-sectional Transmission Electron Microscopy (X-TEM) was performed to characterise the evolution of cracks created by hydrogen implant, as a function of thermal budget.

In addition to blister tests, germanium exfoliation was tested after direct bonding of an $H_2^+$ implanted germanium wafer to a single side polished 100-mm p-type (100) silicon wafer. Prior to direct bonding, 100 nm of PECVD $SiO_2$ was deposited on the silicon wafer. After oxide densification, 2 μm-deep channels were patterned through the oxide and the silicon in order to facilitate the release of by-products generated during wafer bonding and annealing. The Si wafer was cleaned in a Standard Clean 1-equivalent solution. The germanium wafer involved in the direct bonding experiment was implanted with implant conditions comparable to those used for blister tests. The implanted germanium wafer was cleaned in a 1:1 $NH_4$:$H_2O$ solution dispended in a Spray Acid tool followed by four cycles alternating 1,200:1 HF:$H_2O$ and DI water cleans. Both wafers were then subjected to a Megasonic clean with DI water. Their surfaces are hydrophilic prior to the bonding. Wafers were loaded in an AML bonding chamber which was pumped down to $10^{-5}$ mbars. The wafers were exposed for 10 min to free oxygen radicals generated by a plasma ring [8]. Wafers were bonded under a pressure of 1,000 N applied for 5 min. The wafers were annealed in situ at 100°C for 1 h with an applied pressure of 500 N followed by an ex situ anneal at 130°C for 24 h in order to enhance bond strength and induce hydrogen platelet nucleation. The ramp-up rate was set to 0.5°C/min in both cases in order to minimize the formation of thermally generated voids at the bonded interface [28]. The exfoliation was triggered by a 5 min STA at 300°C.

Table 2 Germanium surface roughness, as measured by Atomic Force Microscopy, following long-time anneals at low temperature ($\leq 200$°C)

|  | RMS roughness (nm) | Scan area |
|---|---|---|
| As implanted | 1.5 | 50 μm × 50 μm |
| After 100°C anneal—22 h | 0.4 | 10 μm × 10 μm |
| After 130°C anneal—22 h | 1.7 | 50 μm × 50 μm |
| After 150°C anneal—22 h | 0.4 | 10 μm × 10 μm |

## 3.1 Defect Nucleation and Oswald Ripening Mechanism

Defect nucleation and hydrogen coalescence in implant-generated defects lead to the formation of hydrogen platelets and cracks in the bulk of the implanted semiconductor. These hydrogen-filled cavities have been reported to be located at a depth between the hydrogen projected range and the germanium vacancy range below the germanium surface [20]. Lattice deformation induced by these cavities generates surface blisters which are clearly visible (Fig. 5). The evolution of the nucleation process can thus be monitored using Atomic Force Microscopy or optical microscopy (image shows an optical micrograph).

Significant hydrogen coalescence in III-V materials such as InP and GaAs has already been reported after long anneals at 150°C [24, 25]. In addition, dependence between the lowest temperature required to trigger the nucleation process and the melting point of the implanted material has also been highlighted in previous work [21]. Since the melting point of germanium (937°C) is close to the melting point of InP (1,060°C), defect nucleation would be expected to occur in hydrogen-implanted germanium at 150°C or below the nucleation temperature in InP.

To study this effect, surface roughness measurements after long (22 h) anneal at 100, 130 or 150°C have been performed. RMS roughness values are detailed in Table 2. As compared to as implanted germanium, these long anneals at 100, 130 and 150°C do not modify surface roughness significantly. However these anneals promote some migration of the hydrogen, as shown by cross-sectional TEM and XRD. X-TEM micrographs confirm that low temperature annealing promotes the nucleation of small platelet defects. These nano-cracks are parallel to the substrate surface and are located close to the implant projected range (Fig. 6). The formation of these nano-cracks is known to be limited by the breaking of Ge–H lattice bonds and by hydrogen diffusion. The length of most of these cracks does not exceed 50 nm and their propagation causes minor lattice deformation.

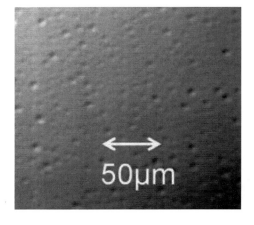

**Fig. 5** Optical micrograph showing the formation of blisters

Consequently, these nano-cracks do not modify the surface morphology significantly, as compared to as implanted germanium. It must be noted that this result does not hold for implanted germanium annealed at 200°C for 19 h suggesting that the minimum thermal budget needed to trigger germanium exfoliation lies between 150 and 200°C. This result is in agreement with previously reported time-to-blister for $H_2^+$-implanted germanium samples (implant energy and dose are 160 keV and $5 \times 10^{16}$ cm$^{-2}$, respectively): at a temperature equal to 200°C, the time-to-blister is estimated at 12 h [18].

The sample subjected to the long temperature anneal at 100°C was subsequently annealed at 200°C for 5 min. RMS roughness of this sample is not impacted by this STA. Consistently, $\omega - 2\theta$ diffraction patterns measured on this sample indicates very little relaxation of the strain created by the hydrogen implant (minor reduction of diffraction fringes), as compared to as-implanted germanium (Fig. 7a). This result points towards a limited hydrogen diffusion in the implanted region and insufficiently high temperature to trigger the blistering.

Additional samples annealed at 100, 130 and 150°C for 22 h were subjected to a STA at 300°C for 5 min. Surface roughness measurements suggest the formation of large hydrogen-filled cavities along the cracks and subsequent germanium exfoliation (Table 3).

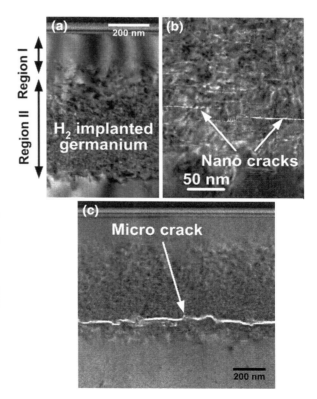

**Fig. 6** a TEM picture of the defective region created by $H_2^+$ implant, in as-implanted germanium. A $5 \times 10^{16}$ cm$^{-2}$ $H_2^+$ dose implanted at 180 keV generates a 600 nm thick implant-damaged region (region II) below a damage-free 150 nm thick region under the germanium surface (region I). **b** TEM picture of hydrogen-implanted germanium following a 22 h-anneal at 150°C 0°C showing the formation of nano-cracks at a depth close to the projected range, parallel to the substrate surface. **c** TEM picture of hydrogen-implanted germanium following a 22 h-anneal at 150°C 0°C and a 5 min anneal at 300°C 0°C showing the transformation of nano-cracks in an almost continuous and thick micro-crack line at 645 nm below the germanium surface

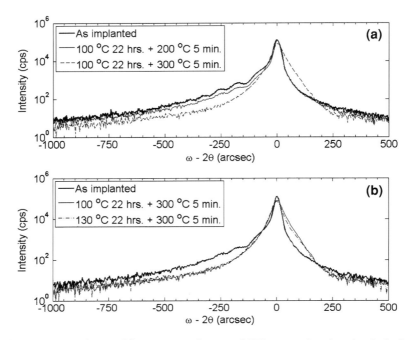

**Fig. 7** X-ray diffraction ($\omega$-$2\theta$) patterns: **a** impact of STA on strain relaxation in hydrogen implanted germanium after long anneals at 100°C; **b** impact of thermal budget during the nucleation process on implant-induced strain after STA at 300°C

**Table 3** Germanium surface roughness, as measured by Atomic Force Microscopy, following long time anneals at low temperature ($\leq$200°C) and STA at 200 or 300°C

| Anneal sequence | RMS roughness (nm) | Scan area |
| --- | --- | --- |
| After 100°C anneal—22 h, followed by 200°C anneal—5 min | 0.4 | |
| After 100°C anneal—22 h, followed by 300°C anneal—5 min | 7.6 | 10 μm × 10 μm |
| After 130°C anneal—22 h, followed by 300°C anneal—5 min | 14.6 | 50 μm × 50 μm |
| After 150°C anneal—22 h, followed by 300°C anneal—5 min | 28.1 | 50 μm × 50 μm |

Scan area is 50 μm × 50 μm unless specified

It should be noted that the height of surface blisters correlates well with thermal budget of the defect nucleation process: the higher the nucleation temperature, the larger the blisters. $\omega - 2\theta$ diffraction patterns for these samples confirm this enhanced hydrogen diffusion after completion of such two-step anneals, as most of the diffraction fringes induced by the hydrogen implant are strongly reduced as compared to $\omega - 2\theta$ diffraction patterns prior to STA at 300°C (Fig. 7b). Such modification of the diffraction pattern suggests that significant hydrogen diffusion occurs after a combined long time anneal at a temperature as low as 100°C and a short time anneal at 300°C.

In addition, the broadening of the germanium feature on triple-axis $\omega$ diffraction patterns suggests an increase of local deformation (Fig. 8a). The Full Width at 0.001 Height (FW0.001 M) increases from 240 arcsec. in as-implanted germanium to 600 arcsec. After a long time anneal at 100°C and a short time anneal at 300°C. This increase is attributed to local lattice deformations due to hydrogen Oswald ripening [27]. The latter is observed after the STA at 300°C, irrespective of the anneal temperature considered for completing the defect nucleation (Fig. 8b). This is a key result since it shows that a long time anneal at a temperature as low as 100°C reduces significantly the time-to-blister at 300°C, as compared to state-of-the-art data [18, 23].

Roughness measurements show evidence of blistering, which is confirmed by X-TEM analysis and suggested by X-ray diffraction patterns. On the sample annealed at 150°C for 22 h and subsequently annealed at 300°C for 5 min, the formation of micro-cracks is observed. The latter result from the merger of nano-cracks created at low temperature, which form longer cracks and cause germanium exfoliation (Fig. 6c). The formation of these micro-cracks proceeds in a similar way to silicon: during STA, hydrogen diffuses along the defect lines and forms large gas pockets (diameter > 5 nm) at the expense of smaller ones [28]. The internal pressure in large gas pockets increases and ultimately leads to the extension of small cracks into micro-cracks.

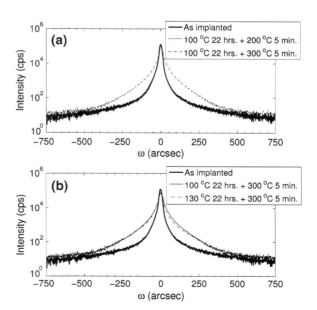

**Fig. 8** X-ray diffraction ($\omega$) patterns: **a** impact of STA on lattice deformation in hydrogen implanted germanium after long time anneals at 100°C; **b** impact of thermal budget during the nucleation process on the lattice deformation caused by the STA at 300°C

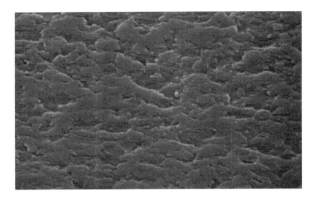

**Fig. 9** Scanning electron microscopy *graphs* of the Germanium-on-insulator sample resulting from an exfoliation carried out after a 24 h-long anneal at 130°C and a short time anneal at 300°C for 5 min: *top-down* tilted view of the germanium surface exposed after complete exfoliation

## 3.2 Low-Temperature Formation of Ge-on-Insulator

A bonded sample made of a hydrogen-implanted germanium wafer directly bonded to a silicon wafer was processed in order to demonstrate the feasibility of transferring a thin germanium layer at low temperature. The thermal treatment which was considered starts with a long (24 h) anneal at 130°C and ends with a 5-min anneal at 300°C. The purpose of the initial long time, low temperature anneal is twofold: it strengthens the bonds created at the germanium/oxide interface during the bonding operation; and promotes hydrogen platelet nucleation within the germanium substrate without modifying its morphology at the bonded interface. A 680 nm-thick layer of germanium was transferred onto 100 nm of $SiO_2$ deposited on the host silicon wafer. High bond strength was achieved, as suggested by the fact that the transferred germanium layer follows closely the pattern printed in the oxide layer prior to bonding. The germanium surface roughness as measured by AFM is 15 nm (germanium surface roughness after exfoliation is illustrated in Fig. 9). This value is consistent with the surface roughness value (14.6 nm) measured on bare implanted germanium after 22 h long anneal at 130°C (Table 3).

## 4 Conclusion

A low-temperature process for bonding and exfoliation of germanium has been developed using hydrogen-implanted germanium layers and surface plasma activation prior to bonding. It has been demonstrated proper engineering of the annealing sequence can promote hydrogen platelet formation and allow for bonding and complete germanium exfoliation after annealing at at temperatures no higher than 300°C. Our results demonstrates also that the lowest thermal budget required for defect nucleation is similar for germanium and III-V materials such as InP. Enhanced bond strength in directly bonded hetero-junctions -like GeOI or bonded III-V material for photonics- is the main benefit expected from such low temperature exfoliation process.

# References

1. Vinet, M., Le Royer, C., Batude, P., Damlencourt, J.F., et al.: Germanium on insulator and new 3D architectures opportunities for integration. Int. J. Nanotechnol. **7**, 204 (2010)
2. Le Royer, C., Damlencourt, J.F., Romanjek, K., Lecunff, Y., et al.: High mobility CMOS: first demonstration of planar GeOI pFETs and SOI nFETs. Proceedings of the sixth workshop of the thematic network on silicon-on-insulator technology, Devices and circuits (EUROSOI) 21 (2010)
3. Kim, M.J., Carpenter, R.W.: Heterogeneous silicon integration by ultra-high vacuum wafer bonding. J. Elec. Materi. **32**, 849 (2003)
4. Tezuka, T., Moriyama, Y., Nakahara, Si., Sugiyama, N., et al.: Lattice relaxation and dislocation generation/annihilation in SiGe-on-insulator layers during Ge condensation process. Thin Solid Films **508**, 251 (2006)
5. Tong, Q., Gan, Q., Hudson, G., Fountain, G., Enquist, P.: Low temperature InP/Si wafer bonding. Appl. Phys. Lett. **84**, 732 (2004)
6. Kanbe, H., Miyaji, M., Ito, T.: Ge/Si heterojunction photodiodes fabricated by low temperature wafer bonding. Appl. Phys. Express **1**, 072301 (2008)
7. Chen, L., Dong, P., Lipson, M.: High performance germanium photodetectors integrated on submicron silicon waveguides by low temperature wafer bonding. Opt. Express **16**, 11513 (2008)
8. Byun, K., Ferain, I., Colinge, C.: Effect of free radical activation for low temperature Si to Si wafer bonding. J. Electrochem. Soc. **157**, H109 (2010)
9. Pelissier, B., Kambara, H., Godot, E., Veran, E., Loup, V., Joubert, O.: XPS analysis with an ultra clean vacuum substrate carrier for oxidation and airborne molecular contamination prevention. Microelectron. Eng. **85**, 155 (2008)
10. Tabet, N., Faiz, M., Hamdan, N.M., Hussain, Z.: High resolution XPS study of oxide layers grown on Ge substrates. Surf. Sci. **523**, 68 (2003)
11. Molle, A., Bhuiyan, M.N.K., Tallarida, G., Fanciulli, M.: In situ chemical and structural investigations of the oxidation of Ge(001) substrates by atomic oxygen. Appl. Phys. Lett. **89**, 083504 (2006)
12. Signamarcheix, T., Allibert, F., Letertre, F., Chevolleau, T., et al.: Germanium oxynitride ($GeO_xN_y$) as a back interface passivation layer for Germanium-on-insulator substrates. Appl. Phys. Lett. **93**, 022109 (2008)
13. Ma, X., Chen, C., Liu, W., Liu, X., et al.: Study of the Ge wafer surface hydrophilicity after low-temperature plasma activation. J. Electrochem. Soc. **156**, H307 (2009)
14. Vincent, S., Radu, I., Landru, D., Leterte, F., Rieutord, F.: A model of interface defect formation in silicon wafer bonding. Appl. Phys. Lett. **94**, 101914 (2009)
15. Baine, P., Gamble, H., Armstrong, B., Mitchell, S., McNeill, D., Rainey, P., Low, Y., Bain, M.: Germanium bonding to $Al_2O_3$. ECS Trans. **16–8**, 407 (2008)
16. Komanduri, R., Chandrasekaran, N., Raff, L.: Molecular dynamic simulations of uniaxial tension at nanoscale of semiconductor materials for micro-electro-mechanical systems (MEMS) applications. Mater. Sci. Eng. A **340**, 58 (2003)
17. Bruel, M.: Silicon on insulator material technology. IEEE Electron Device Lett. **31**, 1201 (1995)
18. Tong, Q.Y., Gutjahr, K., Hopfe, S., Gösele, U.: Layer splitting process in hydrogen-implanted Si, Ge, SiC, and diamond substrates. Appl. Phys. Lett. **70**, 1390 (1997)
19. David, M.L., Pailloux, F., Babonneau, D., Drouet, M., et al.: The effect of the substrate temperature on extended defects created by hydrogen implantation in Germanium. J. Appl. Phys. **102**, 096101 (2007)
20. Zahler, J.M., Fontcuberta, A., Morral, I., Griggs, M.J., Atwater, H.A., Chabal, Y.J.: Role of hydrogen in hydrogen-induced layer exfoliation of Germanium. Phys. Rev. B **75**, 035309 (2007)

21. Hayashi, S., Goorsky, M., Noori, A., Bruno, D.: Materials issues for the heterogeneous integration of III-V compounds. J. Electrochem. Soc. **153**, G1011 (2006)
22. Yu, C.Y., Lee, C.Y., CLin, C.H., Liu, C.W.: Low-temperature fabrication and characterization of Ge-on-insulator structures. Appl. Phys. Lett. **89**, 101913 (2006)
23. Hurley, R.E., Wadsworth, H., Montgomery, J.H., Gamble, H.S.: Surface blistering of low temperature annealed hydrogen and helium co-implanted germanium and its application to splitting of bonded wafer substrates. Vacuum **83**, S29 (2009)
24. Hayashi, S., Bruno, D., Goorsky, M.S.: Temperature dependence of hydrogen-induced exfoliation of InP. Appl. Phys. Lett. **85**, 236 (2004)
25. Hayashi, S., Noori, A.M., Sandhu, R., Cavus Gutierrez, A., Aitken, A., Goorsky, M.S.: InAs on insulator by hydrogen implantation and exfoliation. ECS Trans. **3–6**, 129 (2006)
26. Christensen, D.H., Hill, J.R., Hickernell, R.K., Matney, K., Goorsky, M.S.: Evaluating epitaxial growth stability. Mater. Sci. Eng. B **44**, 113 (1997)
27. Miclaus, C., Goorsky, M.S.: Strain evolution in hydrogen-implanted silicon. J. Phys. D **36**, A177 (2003)
28. Fournel, F., Moriceau, H., Beneyton, R.: Low temperature void free hydrophilic or hydrophobic silicon direct bonding. ECS Trans. **3–6**, 139 (2006)

# Engineering Pseudosubstrates with Porous Silicon Technology

N. P. Blanchard, A. Boucherif, Ph. Regreny, A. Danescu, H. Magoariec, J. Penuelas, V. Lysenko, J.-M. Bluet, O. Marty, G. Guillot and G. Grenet

**Abstract** In this work, we use a controlled oxidation of a mesoporous silicon substrate as a tool for extending and adjusting the Si lattice parameter to other materials such as $Si_xGe_{1-x}$. Our approach involves four steps. First, a seed film is epitaxially grown on a single-crystal Si(100) wafer by Molecular Beam Epitaxy(MBE). Second, porosification is performed according to a standard electrochemical etching procedure but using a "two wafers technique". Third, the porous part of the sample is oxidized at mild temperatures (300–500°C) in a dry $O_2$ atmosphere, inducing a substantial in-plane expansion of the seed film. Fourth, an overgrowth by MBE of an epilayer is done to test the thus-obtained pseudosubstrate. The challenging task in this last step is to deoxidize the seed film surface at $\sim 900°C$ without losing the strain induced by oxidation of the porous part of the sample.

## 1 Introduction

Because of large lattice mismatches and big differences in thermal expansion coefficients between III–V materials and silicon, the integration of high-quality III–V based optical devices with low-cost Si based electronic devices remains a

---

N. P. Blanchard, A. Boucherif, Ph. Regreny, A. Danescu,
J. Penuelas, V. Lysenko, J.-M. Bluet, O. Marty, G. Guillot and G. Grenet (✉)
Institut des Nanotechnologies de Lyon (INL), CNRS UMR-5270, Université de Lyon, Lyon, France
e-mail: genevieve.grenet@ec-lyon.fr

H. Magoariec
Laboratoire de Tribologie et Dynamique des Systèmes (LTDS), CNRS UMR-5513, Université de Lyon, Lyon, France

difficult technological task. Whilst GaAs or InP can be grown on Si, threading dislocations, which cannot be completely avoided, deteriorate the electrical and optical performance of the optoelectronic devices. Therefore, engineering a "universal" substrate, i.e., "compliant" with any kind of epitaxial growth (even a strongly mismatched one), is presently a key requirement in materials research for optoelectronics [1, 2]. Moreover, if this "compliant" substrate could be made of silicon—the dominant material in the microelectronic industry—it would be a real breakthrough towards the monolithic integration of III–V optical devices with Si integrated circuits.

By way of general background, when a material is grown epitaxially on a substrate with a lattice parameter mismatch, it adapts to the in-plane substrate lattice at the interface, but as a result it must distort its lattice-cell in the perpendicular direction. During this so called "*pseudomorphic*" growth, elastic stress gradually increases with layer thickness, up to a point at which it has to be released. This relaxation process can be either:

- elastic (buckling, curving, islanding,...): This process has been addressed in many works dealing with either the Grinfeld effect [3–6] or self-autoorganised quantum dots [7–11] depending on the chosen continuous or discrete point of view.
- plastic (dislocations, cracks,...): the relaxation process [12, 13] is performed via extended crystallographic defects such as misfit dislocations and associated threading dislocations. These defects are responsible for most of the deterioration of the optoelectronic device properties.

In the past, many attempts have been made to overcome the problem. One of them—known as "*metamorphic growth*"—consists of growing a composition-graded buffer or a super lattice between the substrate and the layer in order to smooth-out the misfit effect or to confine dislocations within it. In both cases, the result is rather disappointing even though the threshold for the dislocation appearance is usually delayed [14–16].

In 1991, a new concept known as the "*compliance approach*" has been suggested by Lo [17] starting with the following idea: It is because the film is much thinner than the substrate that it bears most of the strain effect. This statement can be inverted, in other words if the substrate is thinner than the film, the substrate will be strained instead of the film and sustain elastic or plastic relaxation. Such a thin substrate can be a thin layer freed from its substrate, for example, by the chemical etching of a sacrificial layer. This "*free standing*" layer is then converted into a seed layer for a subsequent lattice-mismatched overgrowth [18, 19]. If this layer is a pseudomorphic layer which has relaxed its elastic stress energy completely before being used as a seed layer the technique is known as the "*paramorphic approach*" [20]. However, in both cases, the intrinsic weak point of these techniques lies specifically on the mechanical handling of such ultra-thin membranes. The lateral dimensions can barely be more than 300 μm.

This limitation could be managed by sticking the seed layer weakly on a thick host substrate but the bonding ought to allow in-plane gliding and no curvature.

Numerous solutions [1, 21–23] have been proposed to weaken this interface. Among them, let us mention Carter-Coman et al. [24] who have tried patterning in order to reduce the contact between the film and the host substrate. Lo et al. [25] have suggested to weaken the interface by a misorientation of the film/substrate relative crystallographic axes [16, 26–29]. The major inconvenience of all these solutions is that such structured interfaces generate a non-uniform strain field extremely favorable to island formation [30, 31].

A variant is to insert a thick amorphous viscous layer such as silicon dioxide ($SiO_2$) or borophosphorosilicate (BPSG) between a thin seed film strained in compression and its host substrate [32, 33]. The objective in this case is to favor either irreversible plastic ($SiO_2$) or reversible elastic (BPSG) processes into this amorphous layer rather than in the seed layer and by consequence in the overgrown layer. The main drawback of the technique is a wrinkling of the seed layer when it relaxes [34–36].

In this chapter, we will present and discuss an alternative method which employs controlled porous silicon (PS) oxidation as a means of extending and adjusting the Si lattice parameter to other materials such as $Si_xGe_{1-x}$ [37]. This technique has alreadybeen used successfully by Kim and co-workers [38–39] in order to obtain strained Si films in a Silicon-on-insulator (SOI) context. We have subsequently improved the method for heteroepitaxy pseudosubstrate purposes [40–41].

## 2 The Approach

Rather than relaxing a thin film strained in compression as in the previous approaches the idea is to strain a thin film in tension by homogeneously extending a thick substrate. First, the thin film being in tension cannot wrinkle. If the strain is too great, the only possible way of relaxation is plastic. Second, the new aspect ratio and the new stress repartition, i.e., a thick straining layer versus a thin strained layer provide good strain uniformity into the heterostructure.

The approach, summarized in Fig. 1, is divided in four steps.

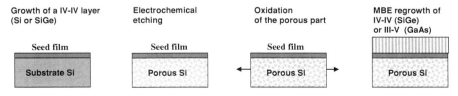

**Fig. 1** Controlled porous silicon oxidation as a means of extending and adjusting the Si lattice parameter to other materials

To be appropriate for engineering a pseudosubstrate, the seed layer must be an ultra-thin (50–100 nm) single crystal of large lateral dimension (at least 2″) free of any extended defects such as dislocations. Its thickness must be strictly constant (variation <1%) since after being elongated, the film must not present an inhomogeneous strain field which may induce preferential nucleation sites during overgrowth. Its surface must have a minimal roughness (RMS < 1 nm) and be perfectly clean without oxides, metals or any other contaminants.

## 2.1 Preliminaries

However, a challenge in meeting the requirements for engineering a pseudosubstrate is to avoid any significant curvature of the pseudosubstrate which can be detrimental during the overgrowth step. If a film with a finite thickness $h_f$ is coherently grown on a substrate with finite thickness $h_s$, the whole system will bend with a curvature $\rho$ given by

$$\rho = [6E_s E_f h_s h_f (h_s + h_f)\varepsilon]/[4E_s E_f h_s h_f (h_s^2 + h_f^2) + 6E_s E_f h_s^2 h_f^2 + E_s^2 h_s^4 + E_f^2 h_f^4]$$

where $\varepsilon$ is the mismatch, $E_f$ and $E_s$ are the biaxial Young modulus (given by $E = Y/(1 - v)$ where $Y$ and $v$ are the film Young modulus and Poisson ratio) for the film and the substrate, respectively.

In this respect, the knowledge of the Young modulus $E_s$, Poisson ratio $v_s$ and shear modulus $\mu_s$ of porous silicon is crucial. It has been shown [42] that the Young modulus as a function of porosity can vary over two orders of magnitude. In order to investigate these variations, we have implemented an extension to super-cells of a classical result of Keating [43]. The elastic energy of the super-cell is assumed to be the sum of all nearest neighbor interaction terms accounting for changes of the reference lengths and all next nearest neighbor interaction terms accounting for the distortions of reference angles (under periodic boundary conditions). The next step is the minimization of the total elastic energy of the super-cell with respect to the "internal" degrees of freedom and identification of the obtained result with the classical elastic energy (quadratic form with respect to strain components). Due to the minimization step in the above procedure the resulting macroscopic law may not have cubic symmetry, unless the pore and the super-cell also possess (geometric) cubic symmetry. In this later case we have explored several (parametric) geometries and we have computed the numerical values of the "macroscopic" Young modulus, Poisson ratio, and Coulomb coefficient. The obtained results agree very well with the available experimental data as shown in Fig. 2 and confirm that the empirical rule, $E_s(p) = A(1 - p)^2$, based on the model of Gibson and Ashby [44] holds for $A = 120$. In this equation p is the porosity, viz, the fraction of the volume of voids over the total volume.

For the Coulomb coefficient and the Poisson ratio, our numerical results [44] lead us to propose the following empirical rules: $G(p) = G_{bulk}(1 - p)^4$ and

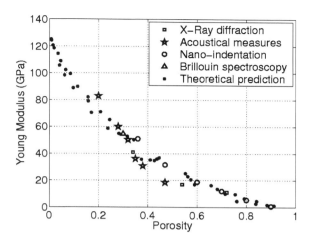

**Fig. 2** Numerical results for the PS Young modulus obtained for various porosities compared with experimental data from [41]

$v(p) = v_{bulk}(1 - p)$. The developed method also allows computing elasticities of cells with complex pore shapes exhibiting less than cubic symmetry, which is of particular interest to model the oriented electrochemical etching process.

In addition, we have extended the method to account for modification of structural changes at the free-surface of the pores during the oxidation process, i.e., to account for back-bond oxidation of porous silicon [45]. This consists in modifying physical characteristics of Si–Si bonds which contains at least one Si atom at the surface of the pore and introduces a pre-strain in the silicon network, which can be computed as well as the elasticities. Once more, the computed value of pre-strain is in good agreement with the experimental results based on Raman spectroscopy [46].

These theoretical preliminaries concerning the PS mechanical behavior are essential to determine the relative thickness of the monocrystalline and porous parts of pseudosubstrates. Let us now consider the sequence of steps to be addressed to engineer pseudosubstrates using PS silicon.

## 2.2 Step1: Growth of the IV–IV Seed Layer and Back Plate

The first step is to grow a nid-type IV–IV thin film by molecular beam epitaxy (MBE) on a p-type Si substrate. This thin film will be referred to as the seed layer in the following. In the anodization process, the interface between the p-type and nid Si acts as a barrier for the anodization front. Using this method, the thin film thickness and its surfaces can be controlled at the atomic scale right from the initial epitaxial stage. The quality of the initial wafers (highly doped p-type : 0.01 $\Omega$ cm, 2-inch, 200 μm thick) is of paramount importance: they must be highly crystalline with front and back sides strictly parallel to each other and perfectly polished. The n-type IV–IV semiconductor can be either Si (homoepitaxy) or SiGe (heteroepitaxy). In this latter case, the seed layer thickness must be less than the critical thickness for plastic relaxation, which means that at 500°C, less than 100 nm for $Si_{0.75}Ge_{0.25}$ (1% in compression) and less than 20 nm for $Si_{0.5}Ge_{0.5}$ (2% in compression). Our equipment is a Riber 2300 MBE system equipped with Si e-beam evaporator and a Ge effusion cell.

In fact, two IV–IV semiconductor films have to be grown because the porosification is done using "the two wafers technique": one (labeled wafer I): ∼50 nm thick will be the seed layer, the other (labeled wafer II): ∼150 nm thick will operate as a back plate. The role of this back plate is not only to increase the thickness of the n-type region artificially but also to protect the seed layer surface during the etching process [40]. It has been shown that when the thickness of the n-type barrier is less than 150 nm, it can be tunneled, resulting in a seed layer with pores and a rough surface. On the other hand, the thinner the seed layer is, the easier it is to strain.

## 2.3 Step 2: Anodization: "The Two Wafers Technique"

The wafers I and II are placed into contact, seed layer face to face with back plate layer, after the necessary HF-cleaning of their surfaces. These wafers must be stuck only by attractive van der Waals interactions without any extra reinforcement because they will have to be separated after the anodization process.

The anodization is carried out in a double bath cell as shown in Fig. 3 for 3.2 h with a current density of 90 mA/cm$^2$. The electrochemical solution is 1:1 volume of concentrated aqueous hydrofluoric acid (48%) and ethanol (100%). The anodization front progresses inside the wafer I from its back-side until it slows down considerably and stops at the interface between the thin intrinsic MBE-grown seed layer and the p-type wafer.

After the separation of the two wafers, wafer I with the seed layer is ready for a low temperature oxidation. This sticking/unsticking method does pose a serious

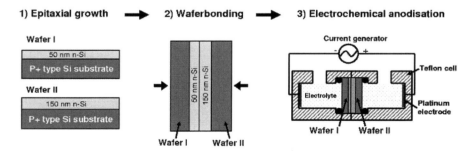

**Fig. 3** Schematic of the engineering of ultra-thin monocrystalline Si films by the two wafers technique [40]

Engineering Pseudosubstrates

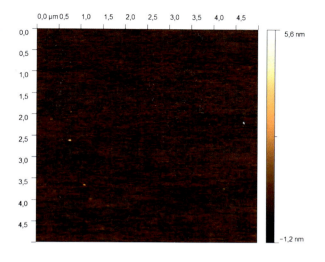

**Fig. 4** Typical 5 × 5 μm AFM image of the seed layer surface

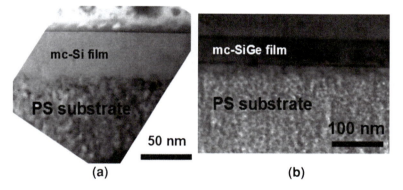

**Fig. 5** Cross-sectional TEM image of **a** 60 nm thick monocrystalline Si film on porous silicon substrate, **b** 50 nm monocrystalline SiGe film on porous silicon substrate

drawback: the wafer could break due to the mechanical contact between the two wafers and therefore thorough wafer bonding preparation is required. Note that a III–V semiconductor cap layer that can be removed by selective etching could circumvent the need for wafer bonding and should better protect the surface of the seed film.

At this point, the seed layer on its porous substrate is characterized by cross sectional transmission electron microscopy (X-TEM), Raman spectroscopy, X-ray diffraction (XRD), X-ray photoemission and atomic force microscopy (AFM). A typical 5 × 5 μm AFM image of the seed layer surface is shown in Fig. 4. The root mean square (RMS) is 0.2 nm whereas it is 1.3 nm if no back plate is used.

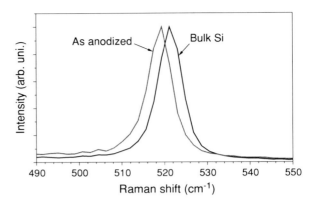

**Fig. 6** Raman spectra of 60 nm thick Si seed layer, measured at UV excitation wavelength 325 nm after substrate porosification

Representative TEM cross sections are shown in Fig. 5. For these samples, the seed layer thicknesses—60 nm in Si layer and 50 nm in SiGe layer—were in good agreement with what was expected from the MBE growth point of view and with spectroscopic ellipsometry measurements. No structural defects were detected.

Raman spectroscopy and X-ray measurements are systematically done and compared to check the seed layer single crystal quality and to evaluate its strain. Raman spectroscopy shows a slight extension of seed layer due to an increase of silicon substrate volume induced by its porosification [47]. Figure 6 exemplifies this effect: the Raman peak position, which corresponds to the Si seed layer, is shifted to lower frequencies with respect to that of a bulk Si wafer.

X-ray diffraction is a technique which allows precise measurements of the lattice parameters (in and out of plane), strain, and alloy composition [48–50]. However, there are few reported XRD studies focused on Si or SiGe thin films on PS heterostructures (see Bellet and Dolino [51], for a review). X-ray measurements shown in Fig. 7 were performed with a Rigaku four-circle diffractometer. The X-ray source is a rotating anode operating at 9 kW, monochromatized with a two-reflection Ge(400) crystal which selects the Cu $K\alpha_1$ radiation (wavelength = 0.15406 nm). The detection unit includes a Ge (400) analyzer.

Figure 7 compares the reciprocal space mapping around the Si and SiGe Bragg peaks (004) and (-2 -2 4) for a $Si_{0.72}Ge_{0.28}$ 50 nm thick film on PS substrate, before and after low temperature oxidation. After anodisation the mapping exhibits a broad component due to the formation of small Si crystallites and a narrow Bragg peak attributed to bulk-like PS. The PS in and out of plane lattice parameters were measured at 5.442 Å, which corresponds to a lattice expansion of $\Delta a/a = 2 \times 10^{-3}$ with respect to bulk silicon. This effect has already been observed by other groups [52]. The expansion was explained as due to the formation of a native oxide layer inside the pores due to aging in air [53]. An interesting feature in Fig. 7b is that the in-plane lattice parameter of the SiGe film is the same as that of the porous Si. At this stage the SiGe film is in compression.

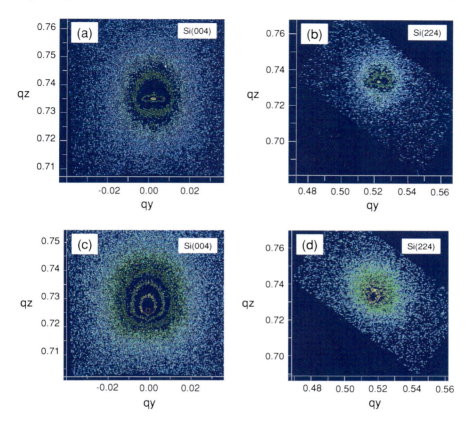

**Fig. 7** Bragg peaks (004) and (−2 −2 4) for the Si and SiGe in the case of a 60 nm thick $Si_{0.75}Ge_{0.25}$ film on Si before and after oxidation

## 2.4 Step 3: Low Temperature Oxidation

The oxidation of porous silicon is performed in a dry $O_2$ atmosphere at different low temperatures (400–500°C) and for different exposure times. The seed layer strain is evaluated by Raman backscattered spectroscopy (He-Cd lasers with a wavelength of 325 nm) at room temperature via the following formula $\varepsilon = 0.134 \, (\omega_s - \omega_0)$ where $\varepsilon$ is the lattice strain, $\omega_0$ the LO phonon wave number for a strain-free Si substrate and $\omega_s$ the wave number for the strained Si films. The result for a 60 nm thick Si seed layer before and after an oxidation at $T = 400°C$ and $T = 500°C$ for 5, 30, 55 and 80 min is shown in Fig. 8.

The seed film surface is monitored by optical microscopy. The images for the measurements labeled (a–d) in Fig. 8 are collected in Fig. 9. Clearly, for a 60 nm seed film, the achievable strain is only just over 1%, which is far below the 4% required for Ge or GaAs on Si. Above 1% strain, plastic relaxation by cracking becomes visible by optical microscopy.

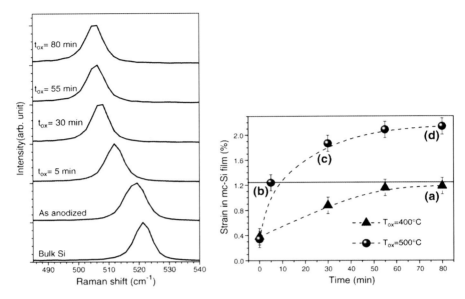

**Fig. 8** Raman spectra of the strained Si films measured at UV excitation wavelength 325 nm and deduced strain from Raman peak position as function of oxidation time and temperature [40]

To get round this difficulty, the starting point could be a 50 nm $Si_{0.72}Ge_{0.28}$ seed layer strained on Si(100) substrate in compression ($\varepsilon = -1.17\%$) according to Raman spectroscopy [54]. After porosification, the seed layer is still in compression even if slightly relaxed ($\varepsilon = -1\%$). When the sample is oxidized at 375°C, as expected, the seed layer has almost its relaxed lattice parameter due to the PS substrate expansion induce by oxidation. A further increase of the oxidation temperature implies a tensile strain of the $Si_{0.72}Ge_{0.28}$ thin film, which increases as temperature increases until reaching saturation at 600°C as shown in Fig. 10.

To be a proper pseudosubstrate, the seed film must support annealing at high temperature in vacuum in order to first desoxidize the surface ($\sim 900°C$) and then grow SiGe material (growth temperature $\sim 500$–$600°C$). Therefore, a 50 nm-thick film of $Si_{0.78}Ge_{0.22}$ on PS substrate with 0.43% tensile strain was fabricated with the previous procedure and annealed in vacuum (base pressure $\sim 10^{-8}$ Pa). Below 500°C, the annealing has no influence on the measured strain. However, for annealing temperatures >500°C the seed layer progressively returns to a relaxed state before recovering its original compressed state as one can see in Fig. 11. The reason is simple: when the porous silicon is heated above the oxidation temperature, the oxidation being reversible, the process reverses, first by a deoxidization of the porous silicon and then by its recrystallization.

After oxidation, a broadening of the Bragg peak in XRD mapping is observed and it becomes difficult to distinguish between SiGe and Si. The same lattice parameter is obtained in and out of plane, viz, 5.476 and 5.474 Å, respectively, which indicates that the film has recovered its natural lattice parameter.

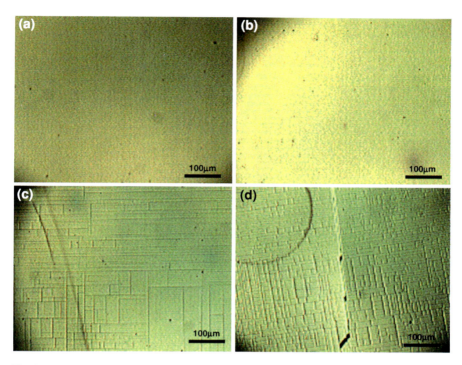

**Fig. 9** Top view of the Si films surface at locations labeled by (**a**), (**b**), (**c**), and (**d**) in Fig. 8 [40]

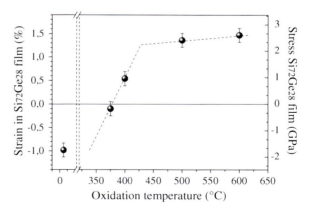

**Fig. 10** Strain deduced from Raman peak position as function of PS oxidation temperature for a 50 nm thick $Si_{0.72}Ge_{0.28}$ seed layer on a PS substrate

The expansion $\Delta a/a$ is $8 \times 10^{-3}$ and the PS substrate has a lattice parameter equivalent to that of the relaxed thin SiGe film. These results confirm the ability of the PS substrate to tune the lattice mismatch between the substrate and the epitaxially deposited material.

**Fig. 11** Raman spectra for a 50 nm thick $Si_{0.72}Ge_{0.28}$ seed layer on a PS substrate originally 0.43% strained by a PS oxidation at 375°C as a function of annealing temperature in vacuum

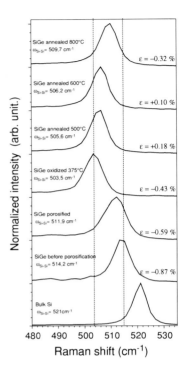

## 2.5 Step 4: Overgrowth of an Epilayer

As mentioned earlier in the text, the preparation of the initial Si(001) wafer is of utmost importance in order to avoid consequent strain inhomogeneity or surface roughness. This necessitates preparing a clean, flat and oxide free (2 × 1) reconstructed Si surface upon which the seed layer can been epitaxially grown. To accomplish this, an empirical study, using RHEED and *ex post facto* AFM, was carried out to find the optimum ex-situ and in situ preparation methods.

Typically, the sample surface is deoxidized ex-situ by HF etching followed by in situ annealing at 680°C. However, RHEED patterns revealed signs of carbon at the surface and the AFM images did not exhibit atomic steps (Fig. 12a). Increasing the in situ annealing temperature to 900°C resulted in an improved RHEED pattern, however, AFM revealed the presence of pitting. With the addition of a final ex-situ $UV/O_3$ treatment followed by an in situ annealing at 900°C, RHEED patterns were free of additional spots often attributed to the presence of carbon and atomic steps were resolved in AFM images (Fig. 12b). The conclusion of these studies is that a $UV/O_3$ controlled oxidation is required before introducing the wafer into the MBE chamber. Consequently, this method has been used to prepare the seed layer surface of the Si/PS and $Si_{1-x}Ge_x$/PS pseudo-substrates.

# Engineering Pseudosubstrates

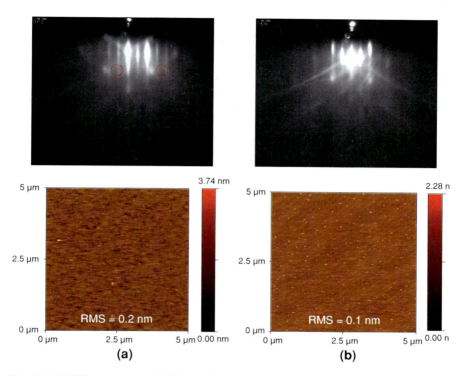

**Fig. 12** RHEED pattern along [110] and 5 × 5 µm AFM images for **a** HF oxide etching plus annealing at 680°C. **b** UV/O$_3$ controlled oxidation plus in situ annealing at 900°C

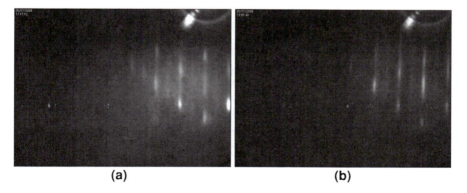

**Fig. 13** RHEED pattern along [110] for **a** Si/PS surface at 600°C, **b** SiGe/PS surface at 500°C

Homo and heteroepitaxial growths were performed on tensile strained Si/PS and relaxed Si$_{0.85}$Ge$_{0.15}$/PS pseudosubstrates at high growth temperatures (600 and 500°C, respectively). In all cases the in situ RHEED patterns show [2 × 1] reconstruction (Fig. 13) synonymous with two dimensional epitaxial growth both

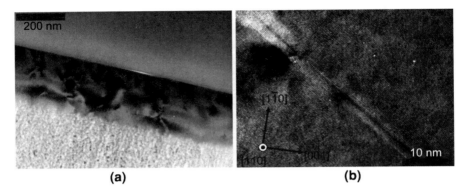

**Fig. 14** Cross-sectional TEM image of **a** a 160 nm SiGe regrown on Si/PS pseudosubstrate, **b** Twin originating from a defect on the seed film surface

after in situ annealing and during the overgrowth. However, the remaining spots on the pattern indicated some residual roughness induced by the processing and annealing.

Cross-sectional TEM images (Fig. 14) confirmed the quality of the grown layers but also revealed the presence of twins related to defects or particles coming from the anodization process. However, post growth Raman measurements performed on the films showed the same trend as the in situ annealing experiments presented in Fig. 10. This is clearly due to a deoxidation/recrystallization of the porous silicon during in situ heating. Evidently, conventional back-side heating of these particular pseudo-substrates (which are very poor thermal conductors) during the high temperature deoxidation of the seed layer surface is poorly suited. This particular reversible behavior of the porous substrate is a great limitation for epitaxy and special method using low temperature or rapid thermal process should be envisaged in order to get around this problem.

An in situ heating system permitting high temperature flashing of the seed layer surface to drive off the controlled $UV/O_3$ surface oxide could circumvent the problem of PS deoxidation/recrystallization, thus allowing 2D growth on the tensile strained (Si) or relaxed ($Si_{1-x}Ge_x$) seed film.

## 3 Conclusion and Miscellaneous

The aim of this work was to engineer a new type of pseudosubstrates for heteroepitaxial purposes with an adaptable lattice parameter. We used a thick porous silicon substrate as mechanical stressor: its bulk extension when oxidised is a very effective tool for extending and thus adjusting the lattice parameter of a thin standard Si layer.

Our approach involves four steps. The first step is the growth by MBE of an n-type IV–IV semiconductor film on a single-crystal highly doped p-type Si(100) wafer. The second step is the porosification of the p-type Si(100) wafer using "the two wafers technique". The role of the second wafer, the "back plate", is not only to increase the thickness of the n-type region artificially but also to protect the seed layer surface during the etching process. The third step, the oxidation of porous silicon, is performed in a dry $O_2$ atmosphere at low temperature (400–500°C) and for a given exposure time. At this point the thin film lattice parameter is tuned to the desired value. The pseudosubstrate is now ready for a lattice-matched or a mismatched overgrowth. Typically, it is deoxidized ex-situ by HF etching (with the addition of a final ex-situ UV/$O_3$ treatment) followed by an in situ annealing at 900°C. Homo and heteroepitaxial growths were performed on tensile strained Si/PS and relaxed $Si_{0.85}Ge_{0.15}$/PS pseudosubstrates at high growth temperatures (600 and 500°C, respectively). However, post growth Raman measurements performed on the films showed that a deoxidation/recrystallization of the porous silicon occurs during in situ heating. Evidently, the conventional back-side heating of our MBE reactor has induced the deoxidation/recrystallization of these poor thermal conductors. Low temperature or a rapid thermal process should be developed in order to get around this problem.

The above described technology can be used for other purposes than obtaining pseudosubstrates. In particular, it can be of interest for straining thin films in tension. Let us just mention a few examples of other possible applications of the technique.

## 3.1 Enhanced Mobility in Strained Si

Since the 1990's, Si is known to have an enhanced hole and electron mobility when strained in tension on SiGe [55] Therefore, a straightforward application could be to use the technique to obtain strained Si layers with large lateral dimensions

## 3.2 Indirect to Direct Band Gap Transition in Strained Ge

Bulk Ge is an indirect band gap semiconductor but a transition from indirect to direct band gap—and therefore the possibility of optical emission—has been predicted if Ge is strained by a biaxial tensile strain [56, 57]. This strain generation can be done by a micromechanical tool such as a cantilever [58] or a bulge device [59]. However, the use of a thick porous silicon layer as a mechanical stressor will give a better lateral homogeneity of the strain field (no curvature) and therefore of the emitted wave length. To test the idea, a Ge film has been grown in place of the Si thin film in step 1. As the mismatch with the Si bulk substrate is too high to allow the layer to be truly pseudomorphic, the number of dislocations must be restricted by lowering the growth temperature and by post annealing process. Ge dots or wells can also been included in the thin Si layer during its growth. In this case, the crystallinity of the IV–IV layer can be preserved and likewise the photo luminescence.

## 3.3 Adaptation to III–V Semiconductor Technology

The approach is not intrinsically restricted to Si; other materials like III–V semiconductors (InP or GaAs) or SiC can also be porosified. For example classical III–V heterostructures can be grown on standard InP substrates and strained by porosification and oxidation. The expected effect is a wavelength shift of the interband transition and the apparition of intra-band transitions. One can imagine adjusting III–V dot emission wavelength to photonic crystal design [60].

Note also that the approach is not limited to oxidation and that other chemicals can be used in order to obtain different results.

Some of these ideas have already been tried and the results are very promising.

**Acknowledgments** We would like to acknowledge the French Research Agency (ANR) for funding this work via a "Projet Blanc" N° BLAN06-1_144612.

## References

1. Kästner, G., Gösele, U.: Principles of strain relaxation in heteroepitaxial films growing on compliant substrates. J. Appl. Phys. **88**, 4048–4056 (2000)
2. Ayers, J.E.: Compliant substrates for heteroepitaxial semiconductor devices: theory, experiment, and current directions. J. Electron. Mater. **37**, 1511–1523 (2008)
3. Grinfeld, M.A.: Instability of the interface between a nonhydrostatically stressed elastic body and a melt. Dokl. Akad. Nauk SSSR **290**, 1358–1368 (1986)
4. Asaro, R.J., Tiller, W.A.: Interface morphology development during stress corrosion cracking, Part I: via surface diffusion. Metall. Trans. **3**, 1789–1796 (1972)
5. Srolovitz, D.J.: On the stability of surfaces of stressed solids. Acta Metal. **37**, 621–625 (1989)
6. Danescu, A.: The Asaro-Tiller-Grinfeld instability revisited. Int. J. Solids Struct. **38**, 4671–4684 (2001)
7. Moison, J.M., Houzay, F., Barthe, F., Leprince, L., André, E., Vatel, O.: Self organized growth of regular nanometerscale InAs dots on GaAs. Appl. Phys. Lett. **64**, 196–198 (1994)
8. Bimberg, D., Grundmann, M., Ledentsov, N.N.: Quantum Dot Heterostructures. Wiley, Chichester (1999)
9. Tersoff, J., LeGoues, F.: Competing relaxation mechanisms in strained layers. Phys. Rev. Lett. **72**, 3570–3573 (1994)
10. Müller, P., Kern, R.: The physical origin of the two-dimensional towards three-dimensional coherent epitaxial Stranski–Krastanov transition. Appl. Surf. Sci. **102**, 6–11 (1996)
11. Duport, C., Priester, C., Villain, J.: Equilibrium shape of a coherent epitaxial cluster. In: Zhang, Z., Lagally, M. (eds.) Morphological Organisation in Epitaxial Growth and Removal, Vol. 14, 73. World Scientific, Singapore (1998)
12. Matthews, J.W., Blakeslee, A.E.: Defects in epitaxial multilayers: I. Misfit dislocations. J. Cryst. Growth **27**, 118–125 (1974)

13. Freund, L.B., Nix, W.D.: A critical thickness condition for a strained compliant substrate/epitaxial film system. Appl. Phys. Lett. **69**, 173–175 (1996)
14. Inoue, K., Harmand, J.C., Matsuno, T.: High-quality InxGa1-xAs/InAlAs modulation-doped heterostructures grown lattice-mismatched on GaAs substrates. J. Cryst. Growth **111**, 313–317 (1991)
15. Behet, M., Van der Zanden, K., Borghs, G., Behres, A.: Metamorphic InGaAs/InAlAs quantum well structures grown on GaAs substrates for high electron mobility transistor applications. Appl. Phys. Lett. **73**, 2760–2762 (1998)
16. Fitzgerald, E.A., Xie, Y.H., Green, M.L., Brasen, D., Kortan, A.R., Michel, J., Mii, Y.J., Weir, B.E.: Totally relaxed $Ge_xSi_{1-x}$ layers with low threading dislocation densities grown on Si substrates. Appl. Phys. Lett. **59**, 811–813 (1991)
17. Lo, Y.H.: New approach to grow pseudomorphic structures over the critical thickness. Appl. Phys. Lett. **59**, 2311–2313 (1991)
18. Chua, C.L., Hsu, W.Y., Liu, C.H., Christenson, G., Lo, Y.H.: Overcoming the pseudomorphic critical thickness limit using compliant substrates. Appl. Phys. Lett. **64**, 3640–3642 (1994)
19. Jones, A.M., Jewell, J.L., Mabon, J.C., Reuter, E.E., Bishop, S.G., Roh, S.D., Coleman, J.J.: Long-wavelength InGaAs quantum wells grown without strain-induced warping on InGaAs compliant membranes above a GaAs substrate. Appl. Phys. Lett. **74**, 1000–1003 (1999)
20. Damlencourt, J.-F., Leclercq, J.-L., Gendry, M., Regreny, P., Hollinger, G.: High-quality fully relaxed $In_{0.65}Ga_{0.35}As$ layers grown on InP using the paramorphic approach. Appl. Phys. Lett. **75**, 3638–3690 (1999)
21. Brown, A.S., Doolittle, W.A.: The status and promise of compliant substrate technology. Appl. Surf. Sci. **166**, 392–398 (2000)
22. Bourret, A.: Compliant substrates: a review on the concept, techniques and mechanisms. Appl. Surf. Sci. **164**, 3–14 (2000)
23. Vanhollebeke, K., Moerman, I., Van Daele, P., Demeester, P.: Compliant substrate technology: integration of mismatched materials for opto-electronic applications. Prog. Cryst. Growth Charact. Mater. **41**, 1–55 (2000)
24. Carter Coman, C., Brown, A., Bicknell-Tassius, R., Marie-Jokerst, N., Fournier, F., Dawson, D.: Strain-modulated epitaxy: modification of growth kinetics via patterned compliant substrates. J. Vac. Sci. Technol. B **14**(3), 2170–2174 (1996)
25. Ejeckam, F.E., Lo, Y.H., Subramania, S., Hou, H.Q., Hammons, B.E.: Lattice engineered compliant substrate for defect-free heteroepitaxial growth. Appl. Phys. Lett. **70**, 1685–1687 (1997)
26. Ejeckam, F.E., Seaford, M.L., Lo, Y.H., Hou, H.Q., Hammons, B.E.: Dislocation-free InSb grown on GaAs compliant universal substrates. Appl. Phys. Lett. **71**, 776–778 (1997)
27. Zhu, Z.H., Zhou, R., Ejeckam, F.E., Zhang, Z., Zhang, J., Greenberg, J., Lo, Y.H., Hou, H.Q., Hammons, B.E.: Growth of InGaAs multi-quantum wells at 1.3 µm wavelength on GaAs compliant substrates. Appl. Phys. Lett. **72**, 2598–2600 (1998)
28. Tan, T.Y., Gösele, U.: Twist wafer bonded "fixed-film" versus "compliant" substrates: correlated misfit dislocation generation and contaminant gettering. Appl. Phys. A **64**, 631–633 (1997)
29. Kästner, G., Tan, T.Y., Gösele, U.: A model of strain relaxation in hetero-epitaxial films on compliant substrates. Appl. Phys. A **66**, 13–22 (1998)
30. Obayashi, Y., Shintani, K.: Critical thickness of a heteroepitaxial film on a twist-bonded compliant substrate. J. Appl. Phys. **88**, 105–115 (2000)
31. Bourret, A.: How to control the self-organization of nanoparticles by bonded thin layers. Surf. Sci. **432**, 37–53 (1999)
32. Leroy, F., Eymery, J., Gentile, P., Fournel, F.: Ordering of Ge quantum dots with buried Si dislocation networks. Appl. Phys. Lett. **80**, 3078–3080 (2002)
33. Hobart, K.D., Kub, F.J., Fatemi, M., Twigg, M.E., Thompson, P.E., Kuan, T.S., Inoki, C.K.: Compliant substrates: a comparative study of the relaxation mechanisms of strained films bonded to high and low viscosity oxides. J. Electron. Mater. **29**, 897–900 (2000)

34. Yin, H., Huang, R., Hobart, K.D., Suo, Z., Kuan, T.S., Inoki, C.K., Shieh, S.R., Duffy, T.S., Kub, F.J., Sturm, J.C.: Strain relaxation of SiGe islands on compliant oxide. J. Appl. Phys. **91**, 9716–9722 (2002)
35. Liang, J., Huang, R., Yin, H., Sturm, J.C., Hobart, K.D., Suo, Z.: Relaxation of compressed elastic islands on a viscous layer. Acta Mater. **50**, 2933–2944 (2002)
36. Sridhar, N., Srolovitz, D.J., Suo, Z.: Kinetics of buckling of a compressed film on a viscous substrate. Appl. Phys. Lett. **78**, 2482–2484 (2001)
37. Marty, O., Nychyporuk, T., de la Torre, J., Lysenko, V., Bremond, G., Barbier, D.: Straining of monocrystalline silicon thin films with the use of porous silicon as stress generating nanomaterial. Appl. Phys. Lett. **88**, 101909–101911 (2006)
38. Kim, J., Xie, Y.H.: Fabrication of dislocation-free tensile strained Si thin films using controllably oxidized porous Si substrates. Appl. Phys. Lett. **89**, 152117–152119 (2006)
39. Kim, J., Li, B., Xie, Y.H.: A method for fabricating dislocation-free tensile-strained SiGe films via the oxidation of porous Si substrates. Appl. Phys. Lett. **91**, 252108–252110 (2007)
40. Lysenko, V., Ostapenko, D., Bluet, J.M., Regreny, Ph., Mermoux, M., Marty, O., Boucherif, A., Grenet, G., Skryshevsky, V., Guillot, G.: Straining of thin Si films by partially oxidized mesoporous Si substrates. Phys. Stat. Solidi **206**, 1255–1258 (2009)
41. Boucherif, A., Blanchard, N.P., Regreny, P., Marty, O., Guillot, G., Grenet, G., Lysenko, V.: Tensile strain engineering of Si thin films using porous Si substrates. Thin Solid Films **518**, 2466–2469 (2010)
42. Bellet, D.: Mechanical and Thermal properties. In: Canham, L.T. (ed.) Properties of Porous Silicon. EMIS, Data Reviewbook Series 18, INSPEC, 38. Institute of Electrical Engineers, London (1997)
43. Keating, P.N.: Effect of invariance requirements on the elastic strain energy of crystals with application to the diamond structure. Phys. Rev. **145**, 637–645 (1966)
44. Gibson, L.J., Ashby, M.F.: Cellular Solids: Structure and Properties. Pergamon, New York (1988)
45. Magoariec, H., Danescu, A.: Modeling macroscopic elasticity of porous silicon. Phys. Stat. Solidi **6**, 1680–1684 (2009)
46. Magoariec, H., Danescu, A.: In: Steinmann, P. (ed) Macroscopic Elasticity of Nanoporous Silicon: Bulk and Surface Effects, IUTAM Book Series 17 (2009). doi:10.1007/978-90-481-3447-2_13
47. Dolino, G., Bellet, D.: In: Canham, L.T. (ed.) Properties of Porous Silicon. EMIS, Data Reviewbook Series 18, INSPEC, 118. Institute of Electrical Engineers, London (1997)
48. Chamard, V., Dolino, G.: X-ray diffraction investigation of n-type porous silicon. J. Appl. Phys. **89**, 174–181 (2001)
49. Erdtmann, M., Langdo, T.A.: The crystallographic properties of strained silicon measured by X-ray diffraction. J. Mater. Sci. **17**, 137–147 (2006)
50. Hartmann, J.M., Gallas, B., Zhang, J., Harris, J.J.: Gas-source molecular beam epitaxy of SiGe virtual substrates: II Strain relaxation and surface morphology. Semicond. Sci. Technol. **15**, 370–377 (2000)
51. Bellet, D., Dolino, G.: X-ray diffraction studies of porous silicon. Thin Solid Films **276**, 1–6 (1996)
52. Barla, K., Bomchil, G., Hérino, R., Pfister, J.C., Baruchel, J.: Determination of lattice parameter and elastic properties of porous silicon by X-ray diffraction. J. Cryst. Growth **68**, 727–732 (1984)
53. Kim, K.H., Bai, G., Nicolet, M.A., Venezia, A.: Strain in porous Si with and without capping layers. J. Appl. Phys. **69**, 2201–2205 (1991)
54. Tsang, J., Mooney, P., Dacol, F., Chu, J.: Measurements of alloy composition and strain in thin $Ge_xSi_{1-x}$ layers. J. Appl. Phys. **75**, 8098–8109 (1994)
55. Lee, M.L., Fitzgerald, E.A., Bulsara, M.T., Currie, M.T., Lochtefeld, A.: Strained Si, SiGe, and Ge channels for high-mobility metal-oxide-semiconductor field-effect transistors. J. Appl. Phys. **97**, 011101–011128 (2004)
56. Soref, R., Kouvetakis, J., Tolle, J., Menendez, J., D'Costa, V.: Advances in SiGeSn technology. J. Mater. Res. **22**, 3281–3291 (2007)

57. El Kurdi, M., Fishman, G., Sauvage, S., Boucaud, P.: Band structure and optical gain of tensile-strained germanium based on a 30 band k·p formalism. J. Appl. Phys. **107**, 013710–013717 (2010)
58. Lim, P.H., Park, S., Ishikawa, Y., Wada, K.: Enhanced direct bandgap emission in germanium by micromechanical strain engineering. Opt. Express. **17**, 16358–16365 (2009)
59. El Kurdi, M., Bertin, H., Martincic, E., de Kersauson, M., Fishman, G., Sauvage, S., Bosseboeuf, A., Boucaud, P.: Control of direct band gap emission of bulk germanium by mechanical tensile strain. Appl. Phys. Lett. **96**, 041909–041911 (2010)
60. Seidl, S., Kroner, M., Högele, A., Karraib, K., Warburton, R.J., Badolato, A., Petroff, P.M.: Effect of uniaxial stress on excitons in a self-assembled quantum dot. Appl. Phys. Lett. **88**, 203113–203115 (2006)

# Confined and Guided Vapor–Liquid–Solid Catalytic Growth of Silicon Nanoribbons: From Nanowires to Structured Silicon-on-Insulator Layers

A. Lecestre, E. Dubois, A. Villaret, T. Skotnicki, P. Coronel, G. Patriarche and C. Maurice

**Abstract** The stacking of crystal semiconductor thin films alternated with dielectric layers continuously arouses a sustained interest for its utility in three-dimensional (3D) integration of metal-oxide-semiconductor field-effect transistor (MOSFET). However, the growth of crystalline silicon without resorting to epitaxial growth from a crystal seed still constitutes an unresolved challenge. Although many different techniques ranging from solid-phase crystallization to thin-film bonding constitutes possible solutions with their respective advantages and weaknesses, little attention has been paid so far to the adaptation of a technique widely used for producing semiconductor nanowires, namely, the vapor–liquid–solid (VLS) catalytic growth. The basic idea developed in this chapter is to control VLS growth for synthesizing local silicon-on-insulator (SOI) layers at reduced thermal budget. Confined VLS growth is therefore proposed to produce single crystalline silicon (c-Si) film over an amorphous oxide layer, without

---

A. Lecestre and E. Dubois (✉)
Institut d'Electronique, de Microélectronique et de Nanotechnologie, IEMN-CNRS, Avenue Poincaré, BP 6006959652, Villeneuve d'Ascq, France
e-mail: emmanuel.dubois@isen.iemn.univ-lille1.fr

A. Lecestre, A. Villaret and T. Skotnicki
STMicroelectronics, 850 Rue Jean Monnet, 38926, Crolles Cedex, France

P. Coronel
CEA-LITEN, 17 Avenue des Martyrs, 38054, Grenoble, France

G.Patriarche
Laboratoire de Photonique et de Nanostructures, LPN-CNRS, Route de Nozay, 91460, Marcoussis, France

C. Maurice
Ecole des Mines, Centre SMS, PECM-UMR CNRS 5146, 158 Cours Fauriel, 42023, St Etienne, France

crystalline seeding. It is demonstrated that VLS growth in the spatial confinement of a cavity produces nanometer-thick c-Si ribbons over a micron area scale with a well controlled localization. The nature of grown silicon layers is characterized by SEM (Scanning Electron Microscopy), EBSD (Electron Backscattered Diffraction) and STEM (Scanning Transmission Electron Microscopy) to analyze its crystallinity and to check the impact of the confining cavity walls on the purity of grown silicon. Beyond the in-depth structural analysis of VLS grown nanoribbons, simple back-gated MOSFET structures have been fabricated and electrically characterized to extract transport properties. The obtained hole mobility of 53 $cm^2\ s^{-1}\ V^{-1}$ constitutes an excellent compromise for a processing temperature less or equal to 500°C.

# 1 Introduction

The increase of density, functionality and performance in deeply scaled integrated circuits (IC) is critically constrained by limited two-dimensional (2D) interconnection routing and signal delays. One possible solution to overcome this problem is the spreading of memory and logical functions across several stacked levels (3D integration) [1, 2]. However, 3D integration requires the stacking of active layers alternated with interlayer dielectrics (ILD). In that context, the difficult challenge consists in growing crystal-quality silicon (c-Si) ribbons starting from an amorphous substrate that can obviously not initiate the necessary crystal pattern. In other words, the primary condition to integrate CMOS transistors layers on top of each other is to grow c-Si without resorting to epitaxial growth from a crystal seed. Accounting for this hard constraint, many different techniques have been developed for producing silicon active layers that feature transport properties as close as possible to those of the crystal phase. Well before the emergence of 3D integration, research on flat panel displays has significantly contributed to the development of such techniques over the last three decades. The primary driver in that application domain was the integration of low temperature polycrystalline silicon thin-film transistors (TFTs) for pixel switching [3, 4]. For that sake, the most widely used method consists in amorphous silicon (a-Si) deposition followed by thermally-induced crystallization to obtain large grain polysilicon [5]. Solid-phase crystallization (SPC) by rapid thermal and/or furnace annealing at relatively low temperature (<700°C) constitutes the simplest techniques [6–9]. However, silicon films obtained using this approach suffer from the major drawback associated with any polycrystalline microstructure, i.e., a random position of grains and a limited carrier mobility due to grain-boundary energy barriers and traps. As an alternative to SPC, metal-induced lateral crystallization (MILC) [10–17], introduced in the early 90s, takes advantage of the reduced temperature of crystallization, typically 500–600°C, when some metals are brought in contact to a thin a-Si layer. However, one challenge associated to MILC remains the control of metal contamination [18]. A third approach for producing (poly) crystalline layers starting from an amorphous substrate is the excimer laser crystallization (ELC). It was first introduced in 1986 by Sameshima et al. [19], and considerably gained in popularity for its capabilities to melt and crystallize thin a-Si film while keeping the underlying substrate at low temperature. This distinctive advantage relies on the superficial confinement of the laser beam energy within the absorption depth of the exposed material at the considered laser wavelength. Because irradiation pulses typically lie in the nanosecond range, the thin top layer of silicon is melted and crystallizes upon sudden temperature decrease before heat diffusion towards the substrate takes place [20–24]. Intimately related to the highly transient lateral-thermal profile involved in laser processing, the phenomenon of super lateral growth (SLG) was observed to generate extremely large polysilicon film [25]. Im and Kim [26] provided a comprehensive analysis of this effect, showing in particular that solidification of melted silicon sensitively depends on the interface response and nucleation kinetics. They concluded that SLG was not suitable for industrial implementation due to the sensitive and critical nature of the phenomenon. To cope with the aforementioned drawback, Crowder et al. proposed a controlled manipulation of the SLG mechanism referred as the sequential lateral solidification (SLS) process. The improvement resides in the efficient production of either directionally solidified microstructures or location-controlled, single-crystal regions on $SiO_x$ [27, 28]. Among the recent implementation of ELC, it is worth noting that its application to pre-patterned a-Si stripes has been successfully demonstrated to serve as 3D fin-like channels in low temperature FinTFTs [29, 30]. Finally, the so-called $\mu$-Czochralski process has significantly advanced the field of thin a-Si layer crystallization by introducing a grain filter through the local modification of the substrate. Using conventional lithography, very small indentations are produced in $SiO_2$ and filled with a-Si. When this structure is irradiated by a laser beam, a small unmolten portion remains embedded in the indentation, seeding the subsequent lateral growth [31]. Following this strategy, the location control of large grains was obtained for a wide energy density window [32–36]. To conclude on laser-based techniques, ELC provides large grains free of randomly oriented grain boundaries but also leads to the formation of ridges coming from the impingement of grains laterally grown in opposite directions [37]. On the other hand, although the sophisticated $\mu$-Czochralski approach produces large grains at predetermined location, the formation of coincident site lattice (CSL) domain boundaries, also denoted as twin boundaries, has been highlighted using electron backscattered diffraction (EBSD) analysis [38]. Both types of defects can negatively impact the uniformity and transport properties of laser annealed polycrystalline silicon films. In recent years, self-assembling techniques including the vapor–liquid–solid (VLS) [39–43] catalytic growth of semiconductor nanowires (NWs) has spurred a sustained research effort for its inherent capabilities to produce crystal quality material. Although, semiconductor NWs aroused a considerable number of fundamental and application-driven studies [44], VLS growth has so far received little interest as a technique for structuring planar thin layers. To contribute to this unresolved problem, this chapter elaborates on the possibility to control VLS growth in such a way it can produce local silicon-on-insulator (SOI) layers at reduced thermal budget. In the following, the strategy of guided and confined VLS growth is first discussed and the fabrication sequence of cavities is described. The impact of growth conditions is analyzed for both the unconstrained and confined synthesis modes. Results of guided growth are subsequently discussed with a particular emphasis on the impact of the cavity size on the growth mechanism. Grown silicon layers are characterized using electron backscattered diffraction and transmission electron microscopy to analyze their crystal properties. Finally, transport properties of confined VLS silicon thin films are studied and compared to those provided by other growth techniques.

## 2 Guiding and Confining VLS Growth in the Plane

### 2.1 Strategy of Confinement

Conventional VLS growth of Si NWs generally uses a metal featuring a low eutectic temperature (e.g. 364°C for Au) that acts as an agent to catalyse whisker growth from an atmosphere containing a reactive gas such as $SiH_4$ or $SiCl_4$. Au and Si form a liquid alloy, which upon supersaturation, nucleates the growth of a Si NWs. Starting from a crystalline substrate covered by Au nanodots, it is widely recognized that the NW diameter plays a crucial role in the growth direction. For instance, Si NWs of diameter greater than 40 nm exhibit a preferential <111> direction while sub-20 nm diameter NWs feature a marked <110> orientation [45, 46]. Growth in the <112> direction has also been observed for the intermediate diameter range. Alternatively, <100> oriented NWs have rarely been observed for non epitaxial growth. It results that the fabrication of highly ordered arrays of NWs proves to be a difficult task that necessitates a suitable selection of the catalyst particle size, substrate and reaction conditions [47]. Without necessarily controlling their crystal orientation, some authors have proposed the synthesis of vertical NWs using self-assembled nanoporous materials like anodic aluminium oxide (AAO) [48, 49]. Although this approach combines the advantages of template-directed structuring and VLS synthesis, growth remains limited in a direction perpendicular to the initial substrate plane. As such, ordered AAO nanopore arrays are obviously not suitable for producing NWs or nanoribbons (NRs) parallel to the starting plane. To solve this problem, the idea developed in the present work is to couple the catalytic VLS growth to a confining guide leading to the formation of nanometer thick planar semiconductor blades over a micron area scale. Following this strategy, a confining cavity opened at one end to enable the transport of the reacting gas to a catalyst ingot positioned at the opposite closed extremity is expected to produce a crystal blade with known placement and orientation. One prerequisite for implementing the proposed guided growth technique is to devise an accurate cavity fabrication technique that ensures the tight control over the size and position of the catalyst ingot and the absence of any catalyst contamination on the inner surface of its walls. This method has been demonstrated for structuring in-plane NWs [50]. Whereas the crystalline nature of the grown semiconductor was identified by Raman spectroscopy, atomic-scale characterizations are still lacking [51]. Moreover, it is worth noting that the production of the cavity of confinement involved the etching of a gold sacrificial layer which potentially raises critical problems in terms of metallic contamination [52]. To alleviate this problem, a new cavity process scheme has been developed in which the inner volume of cavities is defined by using a sacrificial germanium layer and a highly selective etching step [53]. A schematic description of the cavity integration is given in Fig. 1. First, 50 nm thick Au ingots are defined on a thermally oxidized bulk Si substrate using a resist-based lift-off process (Fig. 1a). In the second step, sacrificial 50 nm thick Ge lines, deposited by e-gun evaporation, are aligned to the Au ingot using e-beam lithography and a second lift-off step. Note that the sacrificial Ge lines define the inner volume of the cavity (Fig. 1b). In the third step, a thin oxide liner (10 nm thick $Si_xO_y$) and a 100 nm thick nitride layer are conformally deposited by PECVD over the Au and Ge patterns (Fig. 1c). In step four, openings are defined in the oxide and nitride capping layers using dry etching (Fig. 1d). Finally, sacrificial Ge lines are selectively etched in $H_2O_2$ to reveal the cavity (Fig. 1e).

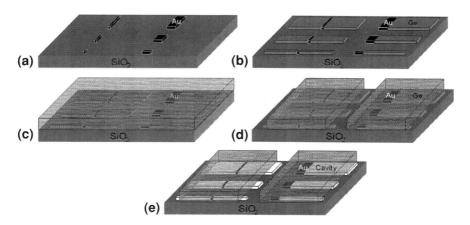

**Fig. 1** Schematic representation of confining cavities using for guided VLS growth. **a** Patterning of catalyst Au ingots by e-beam lithography and lift-off. **b** Formation of sacrificial Ge strips aligned to catalytic ingots by e-beam lithography and a second lift-off step. These strips define the cavity dimensions. **c** $SiO_x$ liner deposition and formation of a SiN conformal capping by PECVD. **d** Opening of the cavity extremity by lithography and RIE. **e** Selective wet etching of the sacrificial Ge lines

## 2.2 Experimental Conditions of VLS Growth

VLS growth was conducted starting from Au catalytic ingots patterned on a thermally grown oxide layer of 100 nm thick over a (100)-oriented Si substrate. A 50 nm thick Au layer was deposited by e-beam evaporation without any intermediate adhesion layer and was structured into ingots by lift-off after e-beam lithography over a positive-tone resist. The VLS process was performed in a LPCVD reactor at 500°C, using 5% silane ($SiH_4$) diluted in $H_2$/Ar as silicon precursor at flow rate of 100sccm, under a total pressure 0.5 mbar. Two growth configurations were subsequently considered and associated to the unconstrained growth and to the guided growth in a cavity, respectively. In order to properly observe NRs after confined growth, a de-processing step is necessary to selectively remove the top and lateral cavity walls without damaging the underlying materials. In this context, the role of the $Si_xO_y$ liner is to provide selectivity for the necessary etching steps. The nitride capping composing the cavity walls is etched by reactive ion etching and the oxide liner is finally eliminated by wet etching in hydrofluoridic acid.

## 2.3 Unconstrained Versus Confined Growth

The use of an amorphous substrate, the volume of the catalyst ingot, the dimensions of the cavities as well as the growth temperature, gas flow and pressure constitutes operational coupled constraints that can severely impact the final result. In order to check that these parameters are compatible with VLS in free condition, a preliminary study presents the main features associated to the unconstrained growth mode. For that sake, a plane amorphous substrate with Au catalyst ingots of identical dimensions than those used for guided growth (50 nm thick and from 50 nm to 2 μm wide) were considered (Fig. 2). Using the aforementioned growth conditions, Si NWs were obtained only for the smallest Au ingots (50 × 50, 100 × 100 and 200 × 200 $nm^2$). For larger dimensions (0.5 × 0.5, 1 × 1 and 2 × 2 $\mu m^2$), a mixture of Au and Si was observed without NWs growth. Figure 2 reveals that the diameter and the position of Si NWs can be controlled by the initial position of the catalyst. However, NWs present a random orientation and Au diffusion on sidewalls is clearly observed [54, 55]. One important outcome of this preliminary study is the observation of crystalline NWs growth providing that the initial 50 nm thick Au ingots have lateral dimensions smaller than 200 nm, for the specified growth conditions (500°C, 5% $SiH_4$ in $H_2$/Ar, 0.5 mbar).

A second major conclusion resulting from this experiment is that VLS growth produces an erratic distribution of NWs length and orientation when the starting surface is made of an amorphous material [56, 57]. This point constitutes a major obstacle to the synthesis of structured material that motivates the implementation a technique of confinement for producing crystalline ribbons with a controlled

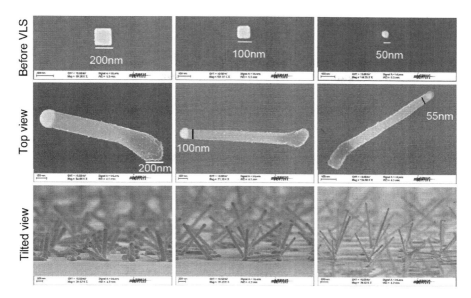

**Fig. 2** Unconstrained localized growth and influence of the catalyst ingots dimensions on the VLS growth: SEM images (top and tilted views) of Au slugs (50 × 50, 100 × 100 and 200 × 200 nm$^2$) on a plane amorphous substrate, before and after growth

orientation and placement. When constrained growth in the cavity is considered, silane penetrates into the cavity, reacts with the Au forming a liquid $Au_xSi_y$ alloy at the back of this cavity. Following the VLS mechanism, crystalline Si precipitates at the interface between the alloy and the cavity sidewall. Overall, the catalytic growth was successful for cavities featuring large dimensions: 50 nm thick, 10, 5 and 2 µm long, and for widths in the 200 nm to 2 µm range. Figure 3 shows one typical example of cavities (10 µm long and 1 µm wide) before and after the VLS growth.

Both long (2 µm) and short (50 nm) Au ingots successfully catalyzed Si growth. Moreover, the presence of an Au drop at the tip of every NRs tends to confirm that VLS growth took place. In order to control the dimensions and orientation of Si NRs with precision, it is worth noting that cavities with a width of 200 nm or less are the most suitable. For wider cavities, growth takes place under the form of several NWs in parallel without being guided by the sidewalls (Fig. 3b).

To be more specific, Fig. 4 shows the interplay between the length of the catalytic Au ingot ($L_{Au}$) and the width of the cavity ($W_{cav}$) that also corresponds to the width of the ingot. At first glance, two regimes can be readily distinguished: (i) for an ingot length less than approximately one-fifth of the cavity width, VLS systematically catalyzes the growth of several NRs in parallel, (ii) for a sufficiently long catalyst ingot ($L_{Au} > W_{cav}$), one single NR is synthesized but do not fully fill the cavity volume. A careful inspection reveals that a third regime exists when the

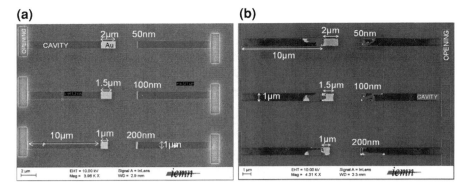

**Fig. 3 a** Cavities before VLS growth: SEM image (*top view*) of several cavities (50 nm thick, 10 μm long and 1 μm wide), containing Au catalyst ingots of different lengths at their closed extremity (from 50 nm to 2 μm). **b** SEM image (*top view*) of the same cavities after VLS growth (500°C, 5% SiH$_4$, 100sccm, 30 min). The SiN cavities walls (cap layer) were selectively etched to free the Si nanoblades. As in the case of unconstrained nanowires growth, the catalyst is pushed by the front of crystal growth. For long catalyst ingots, part of it remains at the initial position in the back of the cavity

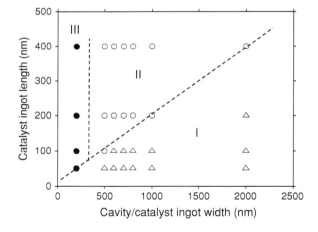

**Fig. 4** Impact of the length of the catalyst slug and cavity width on the morphology of VLS grown silicon (500°C, 5% SiH$_4$, 100sccm). In region I, VLS systematically catalyzes the growth of several NRs in parallel. In region II, one single NR is synthesized but do not fully fill the cavity width for a sufficiently long catalyst ingot ($L_{Au} > 0.2 \times W_{cav}$). In region III, a single NR fits the whole volume of the cavity and forms a regular domain with controlled width and thickness ($L_{Au} > 0.2 \times W_{cav}$ and $W_{cav} \leq 200$ nm)

second condition is fulfilled and the cavity width is reduced to 200 nm. In this latter case, a single NR fits the whole volume of the cavity and forms a regular domain with controlled width and thickness.

Beyond the catalytic dissociation of silane in the cavity, its thermally induced decomposition also produces the formation of an amorphous silicon layer that deposits over the entire surface of the sample. One consequence is the reduction of the gas flow at the cavity entrance. Another drawback is the concentration reduction of reactive gas because its decomposition generates a by-product ($H_2$) that must be evacuated through the cavity entrance which also feeds the reaction with silane. This point is consolidated by Fig. 5a which clearly outlines a growth rate reduction with time. AFM views in Fig. 5b confirm that amorphous Si is deposited inside the cavity with an average thickness of few nanometers and RMS roughness of 8 nm. Another element to take into account in the analysis of Fig. 5 is to consider that growth in confined environment is also strongly dependent on

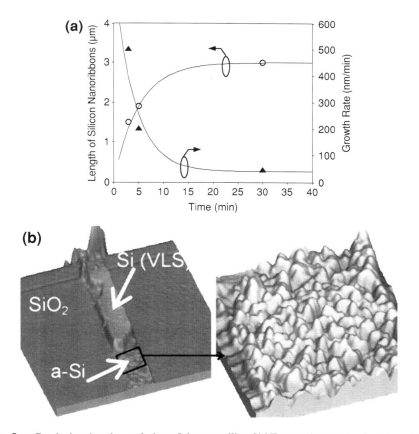

**Fig. 5** a Graph showing the evolution of the crystalline Si NR growth rate as a function of the growth time. Growth conditions: 500°C, 0.5 mbar, 5% $SiH_4$ in $H_2$/Ar, 100sccm. **b** AFM characterization of the Si NR after de-layering cavity walls: the *right image* corresponds to a zoom on the entrance of the cavity showing that amorphous Si also deposits during VLS growth

the dominant character of two competitive mechanisms: first, the chemical reaction rate on the cavity walls and catalyst surface, secondly, the rate of mass transport by diffusion. Neglecting the effect of dilution that results from the generation of by-products ($H_2$), it has been shown that the concentration of gas ($SiH_4$) at depth z with respect to the top of a cylindrical pore is exponentially dependent on the second Damköhler number defined by $D_{II} = k_w/D \times 4L^2/d$, where $k_w$ is the rate of reaction on the pore wall, D the diffusion coefficient of the reactive gas, L is the pore length and d its diameter. According to the magnitude of the $k_w/D$ ratio that quantifies the relative effect of surface reaction versus mass transport, the concentration of reactive gas deep inside the pore can be considerably lower when compared to its counterpart at the pore inlet. However, Lew et al. [49] showed that the Damköler number remains very low for the range of temperature considered in this work ($\leq 500^\circ C$) resulting in equal concentration distribution in the cavity.

## 2.4 Structural Analysis of Silicon Nanoribbons

### 2.4.1 Electron Backscattered Diffraction

Structural properties of Si NRs have been studied by electron backscattered diffraction (EBSD) and transmission electron microscopy (TEM). EBSD is an appropriate technique for the characterization of nanometer thick crystalline Si ribbons on an amorphous substrate over a large extent. In particular, spatially resolved crystal quality maps can be unambiguously deduced by analyzing diffraction patterns. Figure 6 shows a typical set of EBSD characterization results performed on a 500 nm wide Si NR obtained by confined VLS growth in a cavity. The growth conditions correspond to a temperature of 500°C, a pressure of 0.5 mbar and a gas flow of 100sccm (5% $SiH_4$ in $H_2$/Ar).

Consistent with the above discussion on cavity filling, the association of a 200 nm long Au catalytic ingot with a 500 nm wide cavity gives rise to the growth of two NRs in parallel. Figure 6b shows the band contrast mapping that reveals two separate single crystal nanoribbons, each of them composed of multi-twinned domains separated by Σ3 twin boundaries. To complete the analysis, Fig. 6c, d and e provide the crystal orientation maps corresponding to the longitudinal ($x$), transverse ($y$) and normal ($z$) directions. No dominant crystal orientation emerges among the different twinned sub-domains and none of them corresponds to a direction of high symmetry. The situation is markedly different for rectilinear nanoribbons grown in 200 nm wide cavity as exemplified in Fig. 7. In particular, the mapping of crystal orientation associated to the inverse pole figure in the longitudinal direction reveals a dominant growth direction close to [101]. However, it remains difficult to quantitatively ascertain this information because the epitaxial growth does not necessarily follows the alignment of the cavity.

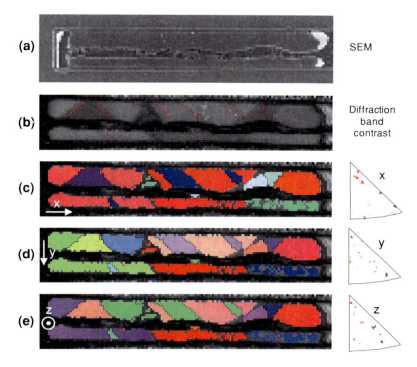

**Fig. 6** EBSD analysis showing the crystallinity of two parallel NRs grown under confined VLS conditions in a 500 nm wide, 50 nm thick cavity. Growth conditions are 500°C, 0.5 mbar, 5% SiH$_4$ in H$_2$/Ar, 100sccm. **a** SEM image of Si NRs after removing the cavity walls; **b** pattern quality image obtained from diffraction Kikuchi bands. High contrast is representative of crystal quality material. **c, d, e** Crystal orientation maps and associated inverse pole figures in the longitudinal, transverse and normal directions with respect to the initial substrate

### 2.4.2 Transmission Electron Microscopy

As outlined in the preceding section, VLS guided growth generates either several multi-twinned NRs with a scattered distribution of crystal orientations or a single rectilinear twinned NR occupying the entire volume of the cavity and featuring a more focused crystal orientation. To better understand the relation between the crystal orientation and the cavity width-dependence of the growth mechanism, TEM characterizations have been performed on selected samples featuring a cavity width larger than 200 nm. Figure 8a shows a detailed cross-section SEM view of a 2 μm wide cavity filled with VLS grown silicon except along the left sidewall where a reduced volume of gold is still present. Residual Au clusters can also be observed along the inner perimeter of the tunnel. TEM analyses shown in Fig. 8b, c and d confirm the presence of four crystalline domains evidenced under the [110] zone-axis diffraction conditions. Figure 8e provides the selective area electron diffraction (SAED) pattern corresponding to Fig. 8b which confirms that

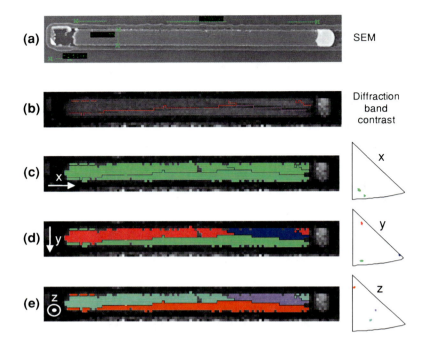

**Fig. 7** EBSD analysis showing the crystalline nature of a single rectilinear NR grown under confined VLS conditions in a 200 nm wide, 50 nm thick cavity. Growth conditions are 500°C, 0.5 mbar, 5% SiH$_4$ in H$_2$/Ar, 100sccm. **a** SEM image of Si NRs after removing the cavity walls; **b** pattern quality image obtained from diffraction Kikuchi bands. High contrast is representative of crystal quality material. **c, d, e** Crystal orientation maps and associated inverse pole figures in the longitudinal, transverse and normal directions with respect to the initial substrate. In particular, EBSD analysis reveals a dominant direction of growth close to [101] in the longitudinal direction

crystalline silicon is obtained. For this wide cavity, it is worth noting that twins also form over the thickness of the cavity as clearly shown by a {111} twin boundary that separates mirror-image domains (Fig. 8d). Reminding that VLS does not fill the entire volume of wide cavities (W$_{cav}$ > 200 nm), growth is likely to be initiated over a reduced area of the catalyst exposed surface without being guided by the cavity sidewall. As growth proceeds along a random direction, it is speculated that NRs are redirected when they come in contact with a cavity wall leading to lattice reflection and twin boundary formation in any of the space directions.

Complementary to cross-section observations, top view TEM characterizations have been performed to investigate how domain twinning develops along the cavity as growth proceeds. For that purpose, samples have been preliminary processed to remove the cavity sidewalls and to eliminate Au residues by chemical etching in potassium iodide (KI). Figure 9a shows a top view SEM image indicating that growth in the 500 nm wide cavity produced a NR occupying only a half of the cavity width. Figure 9b shows the same NR after it was peeled off with a micropipette and deposited on a TEM grid to enable top TEM observations. Figure 9c and d correspond to zoomed portions of the same NR: these bright field (BF) TEM characterizations clearly indicate the formation of multi-twinned domains with twin boundaries that presumably form at the edge of the growth front. From the lattice schematic model presented in Fig. 9e, it is easy to distinguish Si–Si dumbbells aligned along the <004> direction as well as twinned domains that appear as mirror reflection with respect to the (111) plane.

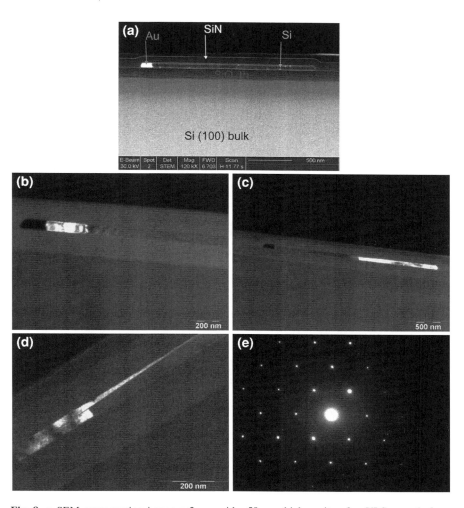

**Fig. 8** **a** SEM cross-section image a 2 μm wide, 50 nm thick cavity after VLS growth. **b, c, d** TEM images observed under the [110] zone-axis diffraction condition that highlights the formation of four twinned domains. **e** SAED diffraction pattern corresponding to TEM view shown in **b**

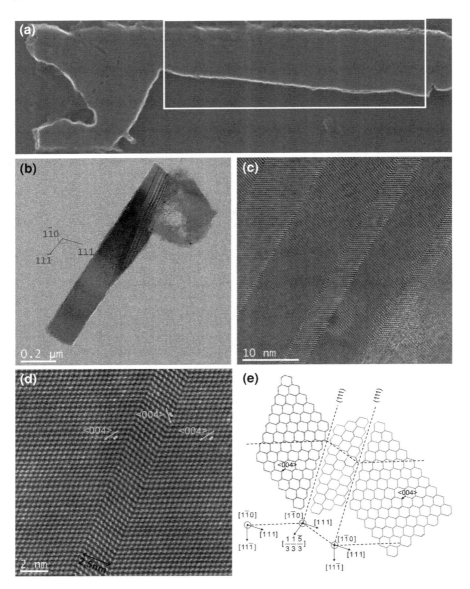

**Fig. 9 a** SEM top view image of a Si NR grown in a 500 nm wide cavity before preparation for TEM top-view observation. **b** Bright field top view STEM observation of the Si nanoribbon after it was peeled off from the substrate and deposited on a TEM grid. Note that the observation is made under the <110> zone-axis diffraction condition that reveals the multi-twinned structure of the Si blade. **c, d** High-resolution bright field (BF) image showing details of a twin boundaries taking place at {111} planes. Si–Si dumbbells are materialized by *yellow dots*. **e** Schematic model showing the coherent assembly of twinned domains from the viewpoint of coincident lattice site

One recurrent issue associated to VLS growth is the contamination resulting from catalyst diffusion along nanowires. This effect was unveiled by Hannon et al. [54] and further investigated by other groups [55, 56] to outline the influence of growth conditions like the partial pressure of silane, the growth temperature and even the role of oxygen incorporation. In the case of confined growth, the control of catalyst contamination takes on even greater importance because the diffusion of Au is further influenced by friction forces along the cavity walls. The formation of twin domains raises another question about the possible trapping or segregation of Au particles at the twin boundaries. To elucidate this problem, the same region of a NR has been characterized by bright field TEM to clearly delineate the twin boundaries and also under high angle annular dark field (HAADF) conditions to exacerbate the chemical contrast between Si and Au. Two important conclusions can be drawn from Fig. 10: first, no evidence of gold nanoclusters accumulation is observed at the twin boundaries and, secondly, most of contamination takes place along the outer envelope of the NR. Although Au residues standing on top and sidewalls were chemically etched before the placement of the NR on the TEM grid, it is believed that unintentional contamination remained at the back $SiO_2$/Si-NR interface. It is therefore difficult to ascertain whether contamination is a limited surface effect [57].

## 3 Electrical Transport in VLS Grown Nanoribbons

In order to characterize the transport in silicon nanoribbons grown by confined VLS, a simple integration scheme has been implemented to fabricate back-gated

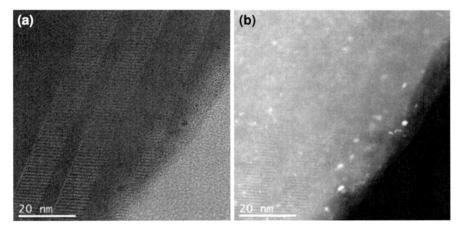

**Fig. 10 a** Bright field TEM top view of a NR grown using confined VLS in a 500 nm wide cavity: *dark spots* indicate the presence of Au clusters that essentially concentrate along the outer envelope of the NR. **b** Observation under high angle annular dark field (HAADF) conditions to exacerbate the chemical contrast between Si and Au

transistors. Using this strategy, electrical results can be rapidly obtained to characterize electrical transport without developing a complete MOS process. Among the most practical and fast characterization technique, the so-called pseudo-MOSFET (Ψ-MOSFET) method has widely served for in situ characterization and optimization of silicon-on-insulator (SOI) wafers [58]. The Ψ-MOSFET structure uses the silicon substrate as a gate electrode, the oxide layer underneath the silicon nanoribbon as a gate dielectric, and the silicon nanoribbon as the conduction channel. Usually, source/drain (S/D) probes are positioned directly onto the top silicon and are separated by a typical distance ranging for 100 μm up to 1–5 mm. Under this typical configuration, the probe-to-silicon series resistance that presents rectifying Schottky properties remains small with respect to the resistance of the silicon sheet. However, in the particular case of VLS grown NRs, the available silicon region extends over a typical surface of $1 \times 3$ μm$^2$, making the silicon channel resistance low and probe series resistance prohibitive. The direct consequence is an improper extraction of the mobility law and of the attenuation

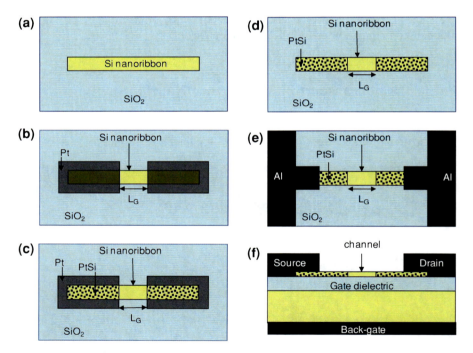

**Fig. 11** Schematic representation illustrating the fabrication steps of back-gated MOS transistors. **a** Initial crystalline silicon nanoribbon deterministically placed on a substrate covered by a silicon dioxide (SiO$_2$) layer. **b** Deposition of platinum (Pt) over the source/drain (S/D) region of the silicon nanoribbon. **c** Platinum silicide (PtSi) formation using a rapid thermal annealing at 400°C. **d** Selective etching of unreacted Pt. **e** Formation of aluminium pad contacts over the silicided S/D. **f** Cross section schematic representation of a back-gated MOS transistor using a VLS grown nanoribbon as the conducting channel

coefficient with gate bias [59]. To cope with this problem, a process using metal contacts has been developed to minimize series resistances, as shown in Fig. 11. Platinum silicide (PtSi) has been selected because it features a low Schottky barrier for holes leading to an acceptable contact resistance without resorting to degenerate doping [60, 61]. As silicon NRs are deterministically placed over the $SiO_2$ amorphous substrate, the deposition of a 5 nm thick platinum (Pt) layer on the S/D regions can be performed using the same alignment marks than those used for the cavity fabrication (Fig. 11b). Pt is subsequently subjected to a rapid thermal anneal at 400°C for 1 min to obtain the desired ~10 nm thick PtSi layer (Fig. 11c) and unreacted Pt over the adjacent $SiO_2$ region is selectively removed by aqua regia (HCl:$HNO_3$:$H_2O$, 3:1:2 at 50°C for 2 min) as schematized in Fig. 11d [62, 63]. Finally, the process is completed with 200 nm thick Al pads

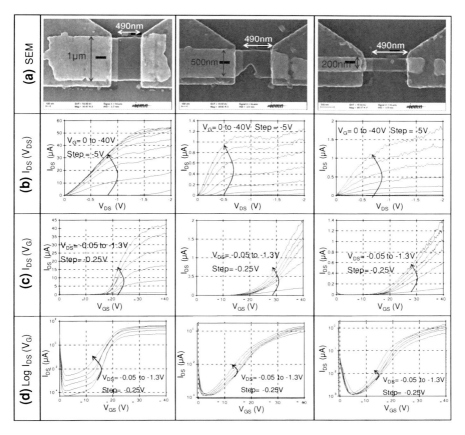

**Fig. 12** Electrical characterization of VLS grown Si nanoribbons of several widths (1 μm, 500 nm, 200 nm) integrated in a back-gated MOS transistor. **a** Top view SEM pictures of each Si NR showing the Schottky source/drain contacts. **b** $I_{ds} - V_{ds}$ characteristics. **c** and **d** $I_{gs} - V_{gs}$ represented in linear and log formats, respectively

to facilitate probing and further reduce series resistances. In that configuration, the distance separating the PtSi source from its drain counterpart defines the channel length over which the field effect developed by the back gate is exerted. It is worth noting that the thermal budget used to fabricate this back-gate transistor is kept under 500°C to keep in line with constraints imposed by 3D integration. SEM images presented Fig. 12a shows a typical implementation of back-gated transistors for 1 μm, 500 nm and 200 nm NRs grown using confined VLS. Typical $I_{ds} - V_{ds}$ characteristics with a back-gate voltage varied from 0 to $-40$ V by steps of $-5$ V reported in Fig. 12b confirm well behaved p-type MOSFET operation. In addition, Fig. 12c and d associated to the $I_{ds} - V_{gs}$ characteristics in both linear and logarithmic formats reflect an excellent gate field control over the current.

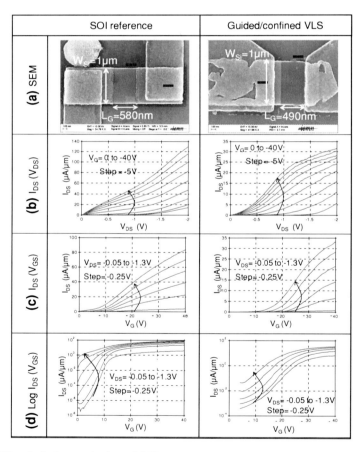

**Fig. 13** Electrical characterization of SOI and VLS nanoribbons (NRs): comparative analysis. **a** Top view SEM pictures of the SOI (*left*) and VLS (*right*) NRs. **b** $I_{ds} - V_{ds}$ characteristics. **c** and **d** $I_{gs} - V_{gs}$ represented in linear and log formats, respectively

For the sake of comparison, reference back-gated transistors have also been integrated on a Smart-Cut SOI commercial substrate and electrically characterized using the same measurement protocol. Although very similar, VLS and SOI reference devices differ in their buried (back-gate) oxide thickness, 95 and 145 nm, respectively. At this point, it is important to note that PtSi S/D contacts selectively give access to the hole mobility without suffering from a prohibitive access resistance. Alternatively, the extraction of electron mobility would require a contact material featuring a low Schottky barrier to electrons, such as $ErSi_{2-x}$ or $YbSi_{2-x}$ [64, 65]. Based on the Y function extraction methodology $Y(V_{gs}) = I_d(V_{gs})/\sqrt{g_m(V_{gs})}$ [66], the hole low field mobility has been extracted for both substrate flavours. Figure 13 provides a complete comparison of electrical characteristics including $I_{ds} - V_{ds}$ and $I_{ds} - V_{gs}$ curves. A hole mobility of 91 and 53 $cm^2 s^{-1} V^{-1}$ has been obtained for the SOI reference and for the VLS approach, respectively. This difference can be explained by several factors affecting the silicon crystal quality. First, mobility is known to depend on the crystal orientation because transport is modified due to asymmetry of the carrier effective masses in the silicon crystal lattice [67].

Secondly, although the Σ3 twin boundary is generally considered to be electrically inactive, the presence of higher order boundaries (e.g. Σ9) is clearly identified as a mobility degradation factor. Finally, contamination by catalyst gold clusters constitutes a source of carriers recombination and can contribute to screen out the electric field originating from the back-gate. By way of conclusion, the mobility performance of VLS thin-film NRs is compared in Fig. 14 with its counterpart obtained by solid-phase (SPC) [3–9] and metal-induced laser (MILC) [10–18] crystallization. It appears that a hole mobility of 53 $cm^2 s^{-1} V^{-1}$ constitutes an excellent compromise for a processing temperature less or equal to 500°C [68].

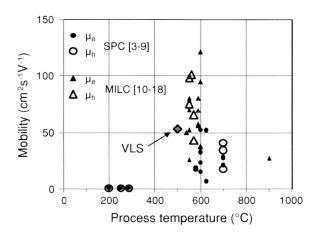

**Fig. 14** State-of-the-art mobility performance obtained in silicon thin-film obtained by solid phase crystallisation (SPC) [3–9], metal induced lateral crystallization (MILC) [10–18] and confined/guided vapor–liquid–solid (VLS) growth (this work)

## 4 Conclusion

In this chapter, guided and confined VLS growth has been considered for the synthesis of horizontal SOI-like layers at low thermal budget. In other words, the innovation effort focused on the adaptation of a technique widely used to grow vertical nanowires to the elaboration of thin crystalline silicon films. Based on this approach, a major prerequisite for implementing a guided growth technique was to propose an accurate cavity fabrication technique. For that sake, a new cavity process scheme has been developed in which the inner volume of cavities is defined by using a sacrificial germanium layer and a highly selective etching step. A preliminary study has been performed on the localized VLS growth of Si NWs on an amorphous substrate. These experiments showed that Au catalytic ingots generate crystalline silicon. However, their lateral dimensions have to be less or equal to 200 nm to control the NW diameter. Similar observations have been done for the guided VLS growth. A careful analysis of the confined growth mechanism has revealed that a single NR fits the whole volume of the cavity and forms a regular domain with controlled width and thickness when the cavity width is reduced to 200 nm. For the first time, the crystalline nature of NRs has been revealed by EBSD and TEM characterizations. Structural analysis corroborates the formation of $\Sigma 3$ twinned domains and random crystal orientation. VLS grown NRs have also been integrated in a back-gated MOS structure to demonstrate device functionality. Considering that VLS growth features a low temperature process without starting from an epitaxially organized substrate, a low field mobility as large as 53 $cm^2 \ s^{-1} \ V^{-1}$ places the proposed synthesis technique in favourable position to develop a high performance TFT technology.

## References

1. Sakuma, K., Andry, P.S., Tsang, C.K., Wright, S.L., Dang, B., Patel, C.S., Webb, B.C., Maria, J., Sprogis, E.J., Kang, S.K., Polastre, R.J., Horton, R.R., Knickerbocker, J.U.: 3D Chip-stacking technology with through-silicon vias and low-volume lead-free interconnections. IBM J. Res. Dev. **52**, 611–622 (2008)
2. Lim, H., Jung, S.M., Rah, Y., Ha, T., Park, H., Chang, C., Cho, W., Park, J., Son, B., Jeong, J., Cho, H., Choi, B., Kim, K.: 65 nm high performance SRAM technology with 25F2 0.16 µm2 S3 (stacked single-crystal Si) SRAM cell. Solid-State Device Research Conference, ESSDERC, Proc IEEE, pp. 549–552 (2005)
3. Mimura, A., Konishi, N., Ono, K., Ohwada, J.-I., Hosokawa, Y., Ono, Y.A., Suzuki, T., Miyata, K., Kawakami, H.: High-performance low temperature poly-Si n-channel TFT's for LCD's. IEEE Trans. Electron. Devices **36**, 351–359 (1989)
4. Subramanian, V., Dankoski, P., Degertekin, L., Khuri-Yakub, B.T., Saraswat, K.C.: Controlled two-step solid-phase crystallization for high performance polysilicon TFT's. IEEE Trans. Electron. Devices **18**, 378–381 (1997)
5. Hatalis, M.K., Greve, D.W.: High-performance thin-film transistors in low-temperature crystallized LPCVD amorphous silicon films. IEEE Electron. Device Lett. **8**, 361–364 (1987)
6. Sheu, J.T., Huang, P.C., Sheu, T.S., Chen, C.C., Chen, L.A.: Characteristics of GAA twin poly-Si NW TFT. IEEE Electron. Device Lett. **30**, 139–141 (2009)

7. Hekmatshoar, B., Cherenack, K.H., Kattamis, A.Z., Long, K., Wagner, S., Stur, J.C.: Highly stable amorphous-silicon thin-film transistors on clear plastic. APL **93**, 032103 (2008)
8. Yamauchi, N., Hajjar, J.J., Reif, R.: Drastically improved performance in poly-Si TFT with channel dimensions comparable to grain size. Technical Digest of International Electron Devices Meeting 1989, IEDM'89, pp. 353–356 (1989)
9. Pan, T.M., Chan, C.L., Wu, T.W.: High-performance poly-silicon TFTs using a high-k $PrTiO_3$ gate dielectric. IEEE Electron. Device Lett. **30**, 39–41 (2009)
10. Oh, J.H., Kang, D.H., Park, W.H., Jang, J., Chang, Y.J., Choi, J.B., Kim, C.W.: A center-offset polycrystalline-Si TFT with n+ amorphous Si contacts. IEEE Electron. Device Lett. **30**, 36–38 (2009)
11. Zhang, D., Kwok, H.S.: A reduced mask-count technology for complementary polycrystalline silicon TFT with self-aligned metal electrodes. IEEE Electron. Device Lett. **30**, 33–35 (2009)
12. Chang, C.P., Wu, Y.C.S.: Improved electrical performance and uniformity of MILC poly-Si TFTs manufactured using drive-in nickel-induced lateral crystallization. IEEE Electron. Device Lett. **30**, 1176–1178 (2009)
13. Lee, S.-W., Joo, S.-K.: Low temperature poly-Si thin film transistor fabrication by metal induced lateral crystallization. IEEE Electron. Device Lett. **17**, 160–162 (1996)
14. Meng, Z., Wang, M., Wong, M.: High performance low temperature metal-induced unilaterally crystallized polycrystalline silicon thin film transistors for system-on-panel applications. IEEE Trans. Electron. Devices **47**, 404–409 (2000)
15. Kim, J.C., Choi, J.H., Kim, S.S., Jang, J.: Stable polycrystalline silicon TFT with MICC. IEEE Electron. Device Lett. **25**, 182–184 (2004)
16. Hu, C.M., Wu, Y.S., Lin, C.C.: Improving the electrical properties of NILC poly-Si films using a gettering substrate. IEEE Electron. Device Lett. **28**, 1000–1003 (2007)
17. Chang, C.P., Wu, Y.S.: Improved electrical characteristics and reliability of MILC poly-Si TFTs using fluorine-ion implantation. IEEE Electron. Device Lett. **28**, 990–992 (2007)
18. Song, N.K., Kim, Y.S., Kim, M.S., Han, S.H., Joo, S.K.: A fabrication method for reduction of silicide contamination in polycrystalline silicon thin-film transistors. Electrochem. Solid-State Lett. **10**, H142–H144 (2007)
19. Sameshima, T., Usui, S., Sekiya, M.: XeCl excimer laser annealing used in the fabrication of poly-Si TFT's. IEEE Electron. Device Lett. **7**, 276–278 (1986)
20. Brunets, I., Holleman, J., Kovalgin, A.Y., Boogaard, A., Schmitz, J.: Low-temperature fabricated TFTs on polysilicon stripes. IEEE Trans. Electron. Devices **56**, 1637–1644 (2009)
21. Hara, A., Takeuchi, F., Takei, M., Suga, K., Yoshino, K., Chida, M., Sano, Y., Sasaki, N.: High-performance polycrystalline silicon thin film transistors on non alkali glass produced using continuous wave laser crystallization. Jpn. J. Appl. Phys. **41**, L311–L313 (2002)
22. Ishihara, R., Matsumura, M.: Excimer-laser-produced single-crystal silicon thin film transistors. Jpn. J. Appl. Phys. **36**, 6167–6170 (1997)
23. Hara, A., Takeuchi, F., Sasaki, N.: Mobility enhancement limit of excimer-laser-crystallized. J. Appl. Phys. **91**, 708–714 (2002)
24. Uchikoga, S., Ibaraki, N.: Low temperature poly Si TFT-LCD by excimer laser anneal. Thin Solid Films **383**, 19–24 (2001)
25. Im, J.S., Kim, H.J., Thompson, M.O.: Phase transformation mechanisms involved in excimer laser crystallization of amorphous silicon films. Appl. Phys. Lett. **63**, 1969–1971 (1993)
26. Im, J.S., Kim, H.J.: On the super lateral growth phenomenon observed in excimer laser-induced crystallization of thin Si films. Appl. Phys. Lett. **64**, 2303–2305 (1994)
27. Crowder, M.A., Carey, P.G., Smith, P.M., Sposili, R.S., Cho, H.S., Im, J.S.: Low-temperature single-crystal Si TFT's fabricated on Si films processed via sequential lateral solidification. IEEE Electron. Device Lett. **19**, 3006–3008 (1998)
28. Crowder, M.A., Voutsas, A.T., Droes, S.R., Moriguchi, M., Mitani, Y.: Sequential lateral solidification processing for polycrystalline Si TFTs. IEEE Electron. Device Lett. **51**, 558–560 (2004)

29. Yin, H., Xianyu, W., Cho, H., Zhang, X., Jung, J., Kim, D., Lim, H., Park, K., Kim, J., Kwon, J., Noguchi, T.: Advanced poly-Si TFT with fin-like channels by ELA. IEEE Electron. Device Lett. **27**, 357–359 (2006)
30. Yin, H., Xianyu, W., Tikhonovsky, A., Park, Y.S.: Scalable 3-D finlike poly-Si TFT and its nonvolatile memory application. IEEE Trans. Electron. Devices **55**, 578–584 (2008)
31. Van der Wilt, P.C., van Dijk, B.D., Bertens, G.J., Ishihara, R., Beenakker, C.I.M.: Formation of location-controlled crystalline islands using substrate-embedded-seeds in excimer-laser crystallization of silicon film. Appl. Phys. Lett. **79**, 1819–1822 (2001)
32. Baiano, A., Danesh, M., Saputra, N., Ishihara, R., Long, J., Metselaar, W., Beenakker, C.I.M., Karaki, N., Hiroshima, Y., Inoue, S.: Single-grain Si thin-film transistors SPICE model, analog and RF circuit applications. Solid-State Electron. **52**, 1345–1352 (2008)
33. Rana, V., Ishihara, R., Hiroshima, Y., Abe, D., Inoue, S., Shimoda, T., Metselaar, W., Beenakker, K.: Dependence of single-crystalline Si TFT characteristics on the channel position inside a localisation-controlled grain. IEEE Trans. Electron. Devices **52**, 2622–2628 (2005)
34. Ishihara, R., van der Wilt, P.C., van Dijk, B.D., Burtsev, A., Voogt, F.C., Bertens, G.J., Metselaar, J.W., Beenakker, C.I.M., Edward, T.V., Kelley, F.: Advanced excimer laser crystallization techniques of Si thin-film for location-control of large grain on glass. Flat Panel Disp. Technol. Disp. Metrol. II **4295**, 14–23 (2001)
35. Baiano, A., Ishihara, R., van der Cingel, J., Beenakker, K.: Strained single-grain silicon n- and p-channel TFT by excimer laser. IEEE Electron. Device Lett. **31**, 308–310 (2010)
36. Sato, T., Yamamoto, K., Kambara, J., Kitahara, K., Hara, A.: Fabrication of large lateral polycrystalline silicon film by laser dehydrogenation and lateral crystallization of hydrogenated nanocrystalline silicon films. Jpn. J. Appl. Phys. **48**, 121201–121206 (2009)
37. Han, S.M., Lee, M.C., Shin, M.Y., Park, J.H., Han, M.K.: Poly-Si TFT fabricated at 150°C using ICP-CVD and excimer laser annealing. Proc. IEEE **93**, 1297–1305 (2005)
38. Ishihara, R., He, T.M., Rana, V., Hiroshima, Y., Inoue, S., Shimoda, T., Metselaar, J.W., Beenakker, C.I.M.: Electrical property of coincidence site lattice grain boundary in location-controlled Si island by excimer-laser crystallisation. Thin Solid Films **487**, 97–101 (2005)
39. Wagner, R.S., Ellis, W.C.: Vapor–liquid–solid mechanism of crystal growth and its application to silicon. Appl. Phys. Lett. **4**, 89–90 (1964)
40. Ke, Y., Weng, X., Redwing, J.M., Eichfeld, C.M., Swisher, T.R., Mohney, S.E., Habib, Y.M.: Fabrication and electrical properties of Si nanowires synthesized by Al catalyzed VLS growth. Nano Lett. **9**, 4494–4499 (2009)
41. Lu, W., Lieber, C.M.: Topical review: semiconductor nanowires. J. Phys. D Appl. Phys. **39**, R387 (2006)
42. Schmidt, V., Wittemann, J.V., Senz, S., Gösele, U.: Silicon nanowires: a review on aspects of their growth and their electrical properties. Adv. Mater. **21**, 2681–2702 (2009)
43. Quitoriano, N.J., Kamins, T.I.: Integrable nanowire transistors. Nano Lett. **8**, 4410–4414 (2008)
44. Fan, H.J., Werner, P., Zacharias, M.: Semiconductor nanowires: from self-organization to patterned growth. Small **2**, 700–717 (2006)
45. Schmidt, V., Senz, S., Gösele, U.: Diameter-dependent growth direction of epitaxial silicon nanowires. Nano Lett. **5**, 931–935 (2005)
46. Persson, A.I., Larsson, M.L., Stenström, S., Ohlsson, B.J., Samuelson, L., Wallenberg, L.R.: Solid-phase diffusion mechanism for GaAs nanowire growth. Nat. Mater. **3**, 677–681 (2004)
47. Lugstein, A., Steinmair, M., Hyun, Y.J., Hauer, G., Pongratz, P., Bertagnolli, E.: Pressure-induced orientation control of the growth of epitaxial silicon nanowires. Nano Lett. **8**, 2310–2314 (2008)
48. Shimizu, T., Zhang, Z., Shingubara, S., Senz, S., Gösele, U.: Vertical epitaxial wire-on-wire growth of Ge/Si on Si(100) substrate. Nano Lett. **9**, 1523–1526 (2009)
49. Lew, K.K., Redwing, J.M.: Growth characteristics of silicon nanowires synthesized by vapor–liquid–solid growth in nanoporous alumina templates. J. Cryst. Growth **254**, 14–22 (2003)

50. Shan, Y., Ashok, S., Fonash, S.J.: Unipolar accumulation-type transistor configuration implemented using Si nanowires. Appl. Phys. Lett. **91**, 093518–093520 (2007)
51. Shan, Y., Kalkan, A.K., Peng, C.Y., Fonash, S.J.: From Si source gas directly to positioned, electrically contacted Si nanowires: the self-assembling "Grow-in-place" approach. Nano Lett. **4**, 2085–2089 (2004)
52. Shan, Y., Fonash, S.J.: Self-assembling silicon nanowires for device applications using the nanochannel-guided "Grow-in-place" approach. ACS Nano **2**, 429–434 (2008)
53. Lecestre, A., Dubois, E., Villaret, A., Coronel, P., Skotnicki, T., Delille, D., Maurice, C., Troadec, D.: Confined and guided catalytic growth of crystalline silicon films on a dielectric substrate. IOP Conf. Ser. Mater. Sci. Eng. **6**, 012022 (2009)
54. Hannon, J.B., Kodambaka, S., Ross, F.M., Tromp, R.M.: The influence of the surface migration of gold on the growth of silicon nanowires. Nature **440**, 69–71 (2006)
55. Allen, J.E., Hemesath, E.R., Perea, D.E., Lensch-Falk, J.L., Li, Z.Y., Yin, F., Gass, M.H., Wang, P., Bleloch, A.L., Palmer, R.E., Lauhon, L.J.: High-resolution detection of Au catalyst atoms in Si nanowires. Nat. Nanotechnol. **3**, 168–173 (2008)
56. den Hertog, M.I., Rouviere, J.L., Dhalluin, F., Desre, P.J., Gentile, P.P., Ferret, P., Oehler, F., Baron, T.: Control of gold surface diffusion on Si nanowires. Nano Lett. **8**, 1544–1550 (2008)
57. Lecestre, A., Dubois, E., Villaret, A., Skotnicki, T., Coronel, P., Patriarche, G., Maurice, C.: Confined VLS growth and structural characterization of silicon nanoribbons. Microelectron. Eng. **87**, 1522–1526 (2010)
58. Cristoloveanu, S., Munteanu, D., Liu, M.S.T.: A review of the pseudo-MOS transistor in SOI wafers: operation, parameter extraction, and applications. IEEE Trans. Electron. Devices **47**, 1018–1027 (2000)
59. Sato, S., Komiya, K., Bresson, N., Omura, Y., Cristoloveanu, S.: Possible influence of the Schottky contacts on the characteristics of ultrathin SOI pseudo-MOS transistors. IEEE Trans. Electron. Devices **52**, 1807–1814 (2005)
60. Larrieu, G., Dubois, E., Wallart, X., Baie, X., Katcki, J.: Formation of Pt-based silicide contacts: kinetics, stoichiometry and current drive capabilities. J. Appl. Phys. **94**, 7801–7810 (2003)
61. Dubois, E., Larrieu, G.: Measurement of low Schottky barrier heights applied to metallic source/drain MOSFETs. J. Appl. Phys. **96**, 729–737 (2004)
62. Breil, N., Dubois, E., Halimaoui, A., Pouydebasque, A., Larrieu, G., Òaszcz, A., Ratajcak, J., Skotnicki, T.: Integration of PtSi in p-type MOSFETs using a sacrificial low-temperature germanidation process. IEEE Electron. Device Lett. **29**, 152–154 (2008)
63. Breil, N., Halimaoui, A., Dubois, E., Larrieu, G., Òaszczc, A., Ratajczakc, J., Rolland, G., Pouydebasquee, A., Skotnicki, T.: Selective etching of Pt with respect to PtSi using a sacrificial low temperature germanidation process. Appl. Phys. Lett. **91**, 232112 (2007)
64. Reckinger, N., Tang, X., Bayot, V., Yarekha, D., Dubois, E., Godey, S., Wallart, X., Larrieu, G., Laszcz, A., Ratajczak, J., Jacques, P., Raskin, J.P.: Schottky barrier lowering with the formation of crystalline Er silicide on n-Si upon thermal annealing. Appl. Phys. Lett. **94**, 191913 (2009)
65. Yarekha, D., Larrieu, G., Breil, N., Dubois, E., Godey, S., Wallart, X., Soyer, C., Remiens, D., Reckinger, N., Tang, X., Laszcz, A., Ratajczak, J., Halimaoui, A.: UHV fabrication of the ytterbium silicide as potential low Schottky barrier S/D contact material for n-type MOSFET. ECS Trans. **19**, 339–344 (2009)
66. Ghibaudo, G.: New method for the extraction of MOSFET parameters. Electron. Lett. **24**, 543–545 (1988)
67. Chang, L., Ieong, M., Yang, M.: CMOS circuit performance enhancement by surface orientation optimization. IEEE Trans. Electron. Devices **51**, 1621–1627 (2004)
68. Lecestre, A., Dubois, E., Villaret, A., Coronel, P., Skotnicki, T., Delille, D., Maurice, C., Troadec, D.: Synthesis and characterization of crystalline silicon ribbons on insulator using catalytic vapor–liquid–solid growth inside a cavity. Proc of the sixth workshop of the Thermatic Network on Silicon–On–Insulator Tecnology, Devices and Circuits, EUROSOI'10 (2010)

# SOI CMOS: A Mature and Still Improving Technology for RF Applications

Jean-Pierre Raskin

**Abstract** This last decade Silicon-on-Insulator (SOI) MOSFET technology has demonstrated its potentialities for high frequency (reaching cut-off frequencies close to 500 GHz for nMOSFETs) and for harsh environments (high temperature, radiation) commercial applications. For RF and system-on-chip applications, SOI also presents the major advantage of providing high resistivity substrate capabilities, leading to substantially reduced substrate losses. Substrate resistivity values higher than 1 k$\Omega$ cm can easily be achieved and high resistivity silicon is commonly foreseen as a promising substrate for radio frequency integrated circuits and mixed signal applications. In this chapter, based on several experimental and simulation results, the interest, limitations but also possible future improvements of the SOI MOS technology are presented.

## 1 Introduction

The semiconductor technology has been progressing enormously these last decades, such evolution has been driven by the continuous look for the increase of the operation speed and the integration density of complex digital circuits [1]. In the early 70's a scaling-down procedure of the transistor dimensions established by Dennard et al. [2] was proposed to pave the way to reaching both objectives. From those days to now, the keystone of the semiconductor industry has been the optimization of this scaling-down procedure.

---

J.-P. Raskin (✉)
Institute of Information and Communication Technologies, Electronics and Applied Mathematics (ICTEAM), Université catholique de Louvain (UCL), Place du Levant, 3, B-1348, Louvain-la-Neuve, Belgium
e-mail: jean-pierre.raskin@uclouvain.be

The communication industry has always been a very challenging and profitable market for the semiconductor companies. The new communication systems are today very demanding; high frequency, high degree of integration, multi-standards, low power consumption, and they have to present good performance even under harsh environment such as high temperature, radiation, etc. The integration and power consumption reduction of the digital part will further improve with the continued downscaling of technologies. The bottleneck for further advancement is the analogue front-end. Present-day transceivers often consist of three or four chip-set solutions combined with several external components. A reduction of the external components is essential to obtain lower cost, power consumption and weight, but it will lead to a fundamental change in the design of analogue front-end architectures. The analogue front-end requires a high performance technology, like GaAs or silicon bipolar, with devices that can easily achieve operating frequencies in the GHz range. For the digital signal processor a small device feature size is essential for the implementation of complex algorithms. Therefore, it appears that only the best submicron CMOS technologies could provide a feasible and cost-effective integration of the communication systems.

This last decade MOS transistors have reached amazingly high operation speed and the semiconductor community has started to notice the radio frequency possibilities of such mainstream devices. Silicon-on-Insulator (SOI) MOSFET technology has demonstrated its potentialities for high frequency (reaching cut-off frequencies close to 500 GHz for nMOSFETs [3]) and for harsh environments (high temperature, radiations) commercial applications. From its early development phase till recent years, SOI has grown from a mere scientific curiosity into a mature technology. Partially depleted (PD) SOI is now massively serving the 45-nm digital market where it is seen as a low cost—low power alternative to bulk silicon. Fully depleted (FD) devices are also widely spread as they outperform existing semiconductor technologies for extremely low power analogue applications [4]. For RF and system-on-chip applications, SOI also presents the major advantage of providing high resistivity substrate capabilities, leading to substantially reduced substrate losses. Substrate resistivity values higher than 1 k$\Omega$ cm can easily be achieved and high resistivity silicon (HRS) is commonly foreseen as a promising substrate for radio frequency integrated circuits (RFIC) and mixed signal applications [5].

In this chapter, based on several experimental and simulation results the interest, limitations but also possible future improvements of the SOI MOSFET technology for microwave and millimetre-wave applications are presented.

## 2 State-of-the-Art RF Performance

Since the invention of the bipolar transistor in 1947, the operating frequencies of integrated transistors have been improved every year. In 1958, a cut-off frequency above 1 Giga-Hertz is reached with a germanium bipolar transistor [6]. Since that

# SOI CMOS: A Mature and Still Improving Technology for RF Applications

date, several integrated technologies have been investigated and improved to further increase the operating frequency of transistors. In 1965, a GaAs metal semiconductor field effect transistor (MESFET) appears in the literature [7]. In 1973, a maximum oscillation frequency ($f_{max}$) of 100 GHz is measured for a FET [8]. In 1980, a new architecture of field effect transistor with high electron mobility (HEMT) is proposed and fabricated [9]. In 1995, a cut-off frequency $f_{max}$ higher than 500 GHz is extrapolated for a HEMT [10]. In 2000, the limit of 1 Tera-Hertz is reached with III-V heterostructure bipolar transistor (HBT) [11] and even overpassed by HEMT in 2007 [12]. It is only in 1996, thanks to the successful downscaling of the silicon MOSFET gate, that cut-off frequencies higher than 200 GHz are presented [13]. Since that date, the interest in MOSFETs for low voltage, low power, high integration mixed-mode ICs (digital and analogue parts on the same chip) in the field of microwave and millimetre-wave applications has been constantly growing. MOSFET is a well-known, well-controlled and mature technology, as well as cost effective, which makes it the key technology for mass production. Nowadays, thanks to the introduction of mobility booster such as strained silicon channel, cut-off frequencies close to 500 and 350 GHz are achieved, respectively, for n- and p-MOSFETs [3] with the channel length of 30 nm.

Figure 1 presents the state-of-the-art current gain cut-off frequency ($f_T$) for n-type MOSFETs as a function of gate length. In that graph, the continuous line represents the prediction from the International Technology Roadmap for Semiconductors (ITRS) published in 2006 [14]. Despite the poor carrier mobility of electrons in silicon compared to III-V materials, silicon MOSFET can be considered as a competitive technology for high frequency applications. It is worth to notice that strained channel silicon MOSFETs even overcome the ITRS roadmap values which gives quite good prospects for silicon technology still for certainly more than 15 years from now on.

**Fig. 1** State-of-the-art current gain cut-off frequency as a function of gate length for unstrained and strained Si and SOI nMOSFETs

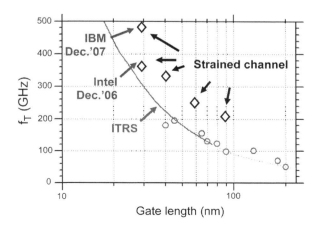

## 3 Main Limiting Factors

Historically, device scaling remains the primary method by which the semiconductor industry has improved productivity and performance. From the 100-nm technology node, CMOS technologies have been facing many grand technological challenges. In this context, the most critical issue consists in the so-called short-channel effects (SCE). These parasitic effects tend to degrade the subthreshold characteristic, increase the leakage current and lead to a dependence of threshold voltage with respect to the channel length. Those static SCE have been reported theoretically and experimentally in the literature and solutions have been proposed. However, only a few publications have analyzed the limitation or degradation of high frequency characteristics versus the downscaling of the channel length. Considering a classical small-signal equivalent circuit for MOSFET as presented in Fig. 2, we can define the cut-off frequencies $f_c$, $f_T$ and $f_{max}$ representing the intrinsic (related to the useful MOSFET effect), the current gain and the available power gain cut-off frequencies, by expressions (1)–(3), respectively.

$$f_c = \frac{g_m}{2\pi C_{gs}} \tag{1}$$

$$f_T \approx \frac{f_c}{\left(1 + \frac{C_{gd}}{C_{gs}}\right) + (R_s + R_d)\left(\frac{C_{gd}}{C_{gs}}(g_m + g_d) + g_d\right)} \tag{2}$$

$$f_{max} \approx \frac{f_c}{2 \cdot \left(1 + \frac{C_{gd}}{C_{gs}}\right)\sqrt{g_d(R_g + R_s) + \frac{1}{2}\frac{C_{gd}}{C_{gs}}\left(R_s g_m + \frac{C_{gd}}{C_{gs}}\right)}} \tag{3}$$

with $g_m$, the gate transconductance, $g_d$, the ouput conductance, $C_{gs}$, $C_{gd}$ and $C_{ds}$, the gate-to-source, gate-to-drain, and drain-to-source capacitances, $R_g$, $R_d$ and $R_s$, the gate, drain and source access resistances, respectively.

Figure 3 represents a schematic cross-section of a Si MOSFET where the different components of source and drain resistances and capacitances are illustrated.

**Fig. 2** Small-signal lumped equivalent circuit of MOSFET

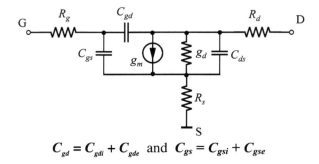

$C_{gd} = C_{gdi} + C_{gde}$ and $C_{gs} = C_{gsi} + C_{gse}$

**Fig. 3** Schematic cross-section of a Si MOSFET illustrating the different access resistances to the channel and the surrounding overlap and fringing capacitances

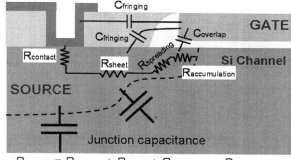

$$R_{source} = R_{contact} + R_{sheet} + R_{spreading} + R_{accumulation}$$

The intrinsic cut-off frequency, $f_c$, measures the intrinsic ability of a field effect transistor (FET) to amplify high frequency signals. As reported in [15], the $f_c$ values are a factor of 1.5 to 2 higher for HEMTs than for silicon MOSFETs with comparable gate length, and this is mainly explained by the respective dynamic properties of the two types of semiconductors (difference of $g_m$ which is directly proportional to the carrier mobility). In order to enhance the carrier mobility in silicon channel and then to improve the current drive and high frequency characteristics [3] of MOSFETs, strained n- and p-MOSFETs have been investigated these last years. Beside the carrier mobility difference between Si and III-V materials, it has been demonstrated that the $f_{max}/f_T$ ratio is lower in the case of Si devices. As explained in [15], besides the well-know degradation of high frequency characteristics due to access resistances ($R_g$, $R_d$ and $R_s$), the decrease of the ratios $g_m/g_d$ and $C_{gs}/C_{gd}$ in CMOS technology strongly contributes to the limiting improvement of $f_T$ and $f_{max}$ with the transistor channel length skrinkage. The increase of the output conductance, $g_d$, with the reduction of gate channel length is one of the well-known short channel effects of FET devices. The degradation of the ratio $C_{gs}/C_{gd}$ means a loss of channel charge control by the gate and an increase of the direct coupling capacitance between gate (input) and drain (output) terminals. The self-alignment of the source and drain regions, one of the main advantages of MOSFET structure, is also a reason for the increase of parasitic capacitances between source and gate and more importantly drain and gate. As demonstrated in [15] from extraction results, the $C_{gs}/C_{gd}$ ratio is equal to 7.8 for the HEMT and only 1.5–1.6 in the case of a 90-nm MOSFET. It is therefore obvious that the optimization of these internal parameters will be crucial in order to further improve cut-off frequencies of ultra deep submicron MOSFETs.

The impact of lightly doped drain (LDD) dose and energy implant as well as annealing temperature and time on $C_{gs}/C_{gd}$ ratio, $g_m$ and $g_d$ and then on $f_{max}$ has been investigated in [16]. The results demonstrate that LDD implant can indeed be considered as an optimization parameter for improving $f_{max}$ and especially the ratio $G_{ass}/NF_{min}$ ($G_{ass}$ and $NF_{min}$ being, respectively, the associated power gain and the minimum RF noise figure), which is the most important figure of merit for low noise microwave applications. However, the optimization window is quite

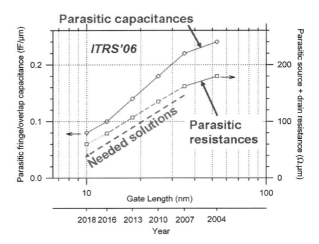

Fig. 4 Parasitic capacitances and source and drain resistances as a function of the gate length published in ITRS'06 [12]

narrow and it seems difficult for a given technological node to get higher $C_{gs}/C_{gd}$ and $g_m/g_d$ ratios than 2 and 6, respectively, for a classical sub-100-nm gate length MOSFET structure. It is the main reason why $f_{max}$ is almost equal to $f_T$ in the case of MOSFETs and not 1.5–2 times higher as in the case of HEMTs with similar gate length and characterized by $C_{gs}/C_{gd}$ and $g_m/g_d$ ratios of 8 and 20, respectively. In order to further improve the microwave performance of deep sub-micrometer MOSFETs, it seems crucial to keep the parasitic resistances and capacitances as low as possible, as predicted by ITRS and shown in Fig. 4 and to consider alternative MOS structures for which the $C_{gs}/C_{gd}$ and $g_m/g_d$ ratios (analogue SCE) are improved.

Several technological options have been presented in the literature those last years to push further the digital and analogue performance limits of single gate Si MOSFETs such as:

- Move from bulk Si MOSFETs to partially depleted (PD) [17] or fully depleted (FD) [18] SOI MOSFETs to enhance the gate electrostatic control on the channel carriers and thus minimize the SCE. Nowadays, ultra-thin body (UTB) MOSFET in SOI technology with a silicon body thickness less that 10 nm has been proposed [19, 20]. Thanks to the buried oxide layer (BOX) underneath the SOI transistors, their junction capacitances (noted Junction capacitance in Fig. 3) to the Si substrate are drastically reduced;
- Strained MOSFETs have been recently largely investigated to improve the carrier mobility. The mechanical stress in the channel originates from specific process steps [21] added into the classical CMOS process flow. Nowadays, strained SOI wafers are produced as well for which the top silicon layer is under a certain level of stress [22, 23];
- Low Schottky barrier contacts [24–28] are foreseen as a very interesting candidate to lower the source/drain contact resistances, to form abrupt junctions (no overlap), and drastically reduce the thermal budget of CMOS process;

- Metal gate allows to get rid of loss of electrostatic gate control related to the polysilicon gate depletion [29, 30], as well as to reduce the gate sheet resistance;
- Low-k and air gap [31, 32] should be introduced to reduce fringing capacitances between gate-to-source and gate-to-drain electrodes;
- SOI wafers with thin BOX have been proposed these last years to reduce SCE (for instance, DIBL) but also to lower self-heating issues [19, 20], [33, 34];
- High resistivity silicon substrate has demonstrated superior characteristics for the integration of high quality passive elements such as transmission lines [35], inductors [36], etc., as well as for reduction of the crosstalk between circuit blocks integrated on the same silicon chip [5].

This last point will be developed in details in Sect. 6. Sections 4 and 5 present the static and high frequency characteristics of two advanced MOS devices often cited as solutions to reduce SCE, respectively, low Schottky Barrier (SB) ultra-thin SOI MOSFET and FinFET.

## 4 Schottky Barrier MOSFETs

Since Schottky barrier (SB) MOSFETs with metal source/drain (S/D) combine low extrinsic resistances with superior scalability, they are regarded as a potential replacement for sub-32-nm MOSFETs with doped S/D [37]. However, the SB formed at the metal/channel interfaces results in an electrical performance inferior to conventional MOSFETs. Thus, SB-MOSFETs suffer from low ON current, an ambipolar switching behavior, and a poor subthreshold swing. Recently, simulations have demonstrated that SB-MOSFETs having a SB lower than 0.1 eV are capable of outperforming conventional MOSFETs [38, 39]. In addition to the use of silicides with low SB heights either for holes or electrons [40, 41], dopant segregation (DS) is a promising solution to lower the effective SB (eSB) [42, 43]. A thin highly doped layer, formed at the silicide/silicon interface during silicidation, causes strong band bending at the interface. As a consequence, the tunnelling probability of carriers through the effectively lowered SB is significantly increased. This technique facilitates a reduction of the eSB at the NiSi/Si interface from 0.65 to sub-0.1 eV for electrons, depending on the implantation dose [44].

RF characteristics of p- and n-SB MOSFETs have been recently presented in the literature. A $f_T$ of 280 GHz is observed for 22-nm devices using PtSi [45] and 180 GHz for 30-nm MOSFETs when combining PtSi with DS [27, 46]. The gate-length dependence of the direct current and RF characteristics of p- and n-type SB-MOSFETs with DS is studied in [47]. Hereafter, the impact of the SB height on the electrical performance of n-type NiSi SB-MOSFETs with DS using different As implantation doses presented in [48] is summarized. The key device parameters are extracted from scattering-parameter (S-parameter) measurements using a small-signal equivalent circuit and compared with the SB-MOSFETs having different SB heights.

**Fig. 5** Schematic cross-section view of a ultra-thin SB SOI MOSFET lying on a high-resistivity Si substrate

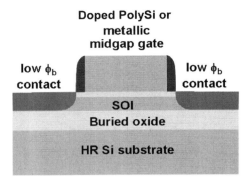

SB-MOSFETs with channel lengths $L_G$ from 80 to 380 nm are fabricated on 20-nm-thick intrinsic SOI substrate. A 3.5-nm-thick $SiO_2$ gate oxide is grown, followed by deposition of 160 nm n + poly-Si. Electron-beam lithography and dry etching are used to define the gate stack. For the S/D implantation, doses ranging from $5 \times 10^{13}$ to $3 \times 10^{15}$ As/cm² are used at an energy of 5 keV. Subsequently, $SiO_2$ gate spacers (<20 nm) are formed using an improved etching process compared to [47] with nearly no loss of highly doped Si at S/D, and Ni is deposited. During the following silicidation step at 450°C for 30 s, the initial Si layer is fully silicided, and the NiSi encroaches under the spacer to the edge of the gate stack. The unreacted Ni is selectively removed using SPM ($H_2SO_4:H_2O_2$ = 4:1). A schematic of the SOI SB-MOSFET is shown in Fig. 5. The SB-MOSFETs consisting of two parallel gate fingers with a total gate width of $W_G = 2 \times 40$ μm are embedded in coplanar-waveguide (CPW) transmission lines for on-wafer microwave measurements.

Figure 6 shows the transfer characteristics of SB-MOSFETs ($L_G$ = 180 nm) with DS using different As doses. The devices exhibit ambipolar switching, typical of SB-MOSFETs [49]. For negative $V_{gs}$, holes tunnel through the drain-side (as defined for an nFET) barrier into the channel, resulting in the p-branch, whereas for the n-branch, electrons are injected through the source-side barrier for positive $V_{gs}$. In case of the SB-MOSFET with the lowest dose of $5 \times 10^{13}$ As/cm², we observe almost equal ON currents for the p- and n-branches, indicating that the eSB for electrons is still close to the original value, whereas a significantly reduced p-branch is apparent for devices with higher dose. Moreover, the ON current is drastically improved by more than two orders of magnitude between the lowest and the highest dose, and the inverse subthreshold slope gets steeper with higher As concentration. This improved gate control can be explained by an enhanced tunnelling probability through the effectively lowered SB for higher implantation doses. Whereas the current flow of SB-MOSFETs with a high SB is determined by the carrier injection through the source SB which is modulated by $V_{gs}$, it is mainly controlled by the potential barrier in the channel if the SB is lowered, and the tunnelling probability approaches unity [49, 50]. Using appropriate on-wafer open test-structures to de-embed the CPW pads and feed lines, Cold-FET S-parameter

**Fig. 6** Transfer characteristics of SB-MOSFETs with DS using different As doses from $5 \times 10^{13}$ to $3 \times 10^{15}$ As/cm² ($L_G = 180$ nm)

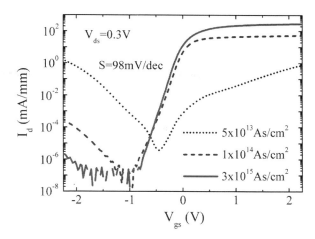

measurements [51] show a strong decrease of the extrinsic resistance $R_{S/D}$ from 5.2 kΩ μm down to 600 Ω μm with increasing As dose from $1 \times 10^{14}$ to $3 \times 10^{15}$ As/cm², i.e., decreasing eSB (inset Fig. 7). Figure 7 shows $f_T$ which is extracted from the current gain versus frequency plots and the saturation currents $I_{on}$ ($V_{gs} - V_t = 1.5$ V, $V_{ds} = 1.5$ V) of the 180-nm channel devices. Note that although $I_{on}$ varies by a factor of approximately two between the lowest and the highest dose, a change of only 17% is evident for $f_T$. The reason is the improved carrier injection through the SB at the source side which simultaneously impacts the extrinsic transconductance $G_M$ and the amount of charge in the channel. Figure 8 shows $G_M$ and the total gate-to-source $C_{GS}$, gate-to-drain $C_{GD}$, and channel capacitance $C_{GG} = C_{GS} + C_{GD}$ which are extracted for the same 180-nm-channel devices at peak $f_T$ biases, like in Fig. 7 [28, 52]. $G_M$ increases when the implantation dose is increased from $1 \times 10^{14}$ to $5 \times 10^{14}$ As/cm² as well as $C_{GS}$ and $C_{GG}$, whereas $C_{GD}$ remains constant at a value of 0.25 fF/μm. The latter might be related to the fringing capacitance between the drain region and the poly-Si gate stack and drain-induced control over carriers in the channel region due to short-channel effects [15]. The increase of $G_M$ with the As dose can be explained by drastically reduced S/D resistances due to the decreasing eSB. At the same time, the tunnelling probability of carriers through the lower eSB is significantly enhanced, resulting in an increased amount of charge injected into the channel. Therefore, we observe a higher value of $C_{GG}$. Since the cut-off frequency is given by $f_T \approx G_M/(2\pi C_{GG})$, it varies only slightly for different As doses due to the similar dependence of $G_M$ and $C_{GG}$ on the SB. As a result, although the SB strongly deteriorates the dc performance, it has only little impact on the RF performance of the SB-MOSFETs.

A cut-off frequency of 140 GHz is extracted at $V_{gs} - V_t = 0.34$ V and $V_{ds} = 1.2$ V, which is the highest value reported for n-type SB-MOSFETs so far (Fig. 9). The inset shows a perfect $1/L_G$ dependence of $f_T$ for decreasing $L_G$ from 380 to 80 nm, suggesting an impressive RF performance increase when the devices are further scaled down.

**Fig. 7** Unity-gain cut-off frequency $f_T$ and saturation current $I_{on}$ ($V_{gs} - V_t = 1.5$ V, $V_{ds} = 1.5$ V) of the 180-nm-channel devices versus As implantation dose

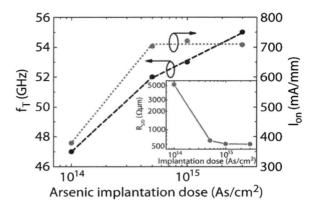

**Fig. 8** Maximum extrinsic transconductance $G_M$ as well as $C_{GS}$ and $C_{GG}$ extracted at peak $f_T$ biases

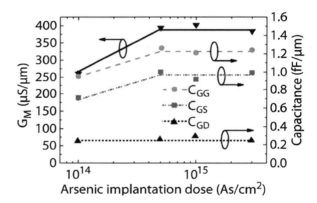

## 5 RF Performance of a Multigate MOSFET: FinFET

To reduce the SCE in nanometre scale MOSFETs, multiple-gate architectures emerge as one of the most promising novel device structures, thanks to the simultaneous control of the channel by more than one gate. The idea of the double-gate (DG) MOSFET was first introduced by Colinge [53]. Starting by the FinFET [54], other multiple-gate SOI MOSFETs have been introduced since [55] such as triple-gate (TG), FinFET, pi-gate (PG), quadruple-gate (QG), omega-gate ($\Omega$-G), etc. Many works have investigated and demonstrated the great potential of multiple-gate devices to comply with the $I_{on}/I_{off}$ requirements of the ITRS for logic operation [55, 56]. Indeed, FinFETs are known to be promising devices for high density digital applications in the sub-65 nm nodes due to their high immunity to SCE and their excellent compatibility with planar CMOS process. Most of the investigations performed on FinFETs have focused on their technological aspects and perspectives for digital applications [57, 58], while only a few have assessed

**Fig. 9** Current gain versus frequency for n-type SB-MOSFETs with $L_G = 80$ nm ($3 \times 10^{15}$ As/cm$^2$) showing a cut-off frequency of 140 GHz at $V_{gs} - V_t = 0.34$ V and $V_{ds} = 1.2$ V. The inset shows the linear dependence of peak $f_T$ versus $1{,}000/L_G$

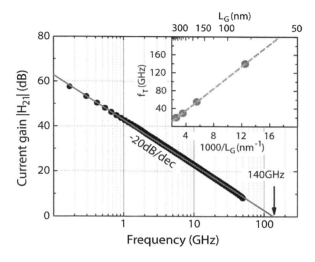

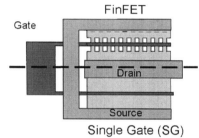

**Fig. 10** Schematic top view of a FinFET composed of 10 fins (*upper*) and planar (SG) MOSFET (*lower*) occupying the same active silicon foot print

their analogue figures of merit [59, 60]. In this section, the RF performance of FinFETs with various geometries is presented.

FinFETs are fabricated on a SOI wafer with 60-nm-thick Si film on 145-nm of buried oxide, with (100) and (110) Si planes for top and lateral channels, respectively. The silicon active area is patterned using 193-nm lithography with aggressive resist and oxide hard mask trimming to define narrow silicon fins. A hydrogen anneal and a sidewall oxidation are used for surface smoothening and corner rounding. The fin patterning resulted in a fin height ($H_{fin}$) of 60 nm, fin width ($W_{fin}$) of 22, 32 and 42 nm, and fin spacing ($S_{fin}$) of 328 nm. The gate stack consisting of a plasma nitrided oxide with equivalent oxide thickness equal to 1.8 nm and 100-nm polysilicon is deposited. Gate lengths ($L_g$) of 40, 60 and 120 nm are fabricated. High angle As/BF$_2$ extensions are then implanted and a 40 nm-thick selective epitaxial growth (SEG) is performed on the source and drain regions. After the heavily doped drain (HDD) implantations and rapid thermal annealing (RTA), NiSi is used as silicide and only one metal level is deposited.

The DC and RF analyses are performed on RF FinFETs (Fig. 10) composed of 50 gate fingers ($N_{finger}$) controlling 6 fins ($N_{fin}$) each. As shown in Fig. 11a the 60-nm technology investigated here outlines a good control over SCE, with a

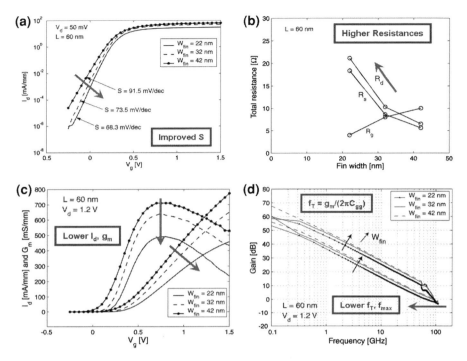

**Fig. 11** DC and RF characteristics of 60 nm-gate length FinFET for various fin widths ($W_{fin}$): **a** transfer characteristic in log scale, **b** extracted access resistances, **c** transfer characteristic in linear scale and gate transconductance and **d** current gain and maximum available power gain vs. frequency

subthreshold slope ($S$) close to 73.5 mV/dec. This value is even closer to ideal for $L_g = 120$ nm ($S = 62.9$ mV/dec). Data in Fig. 11 are normalized by considering the total gate width $W_{tot} = N_{finger} \times N_{fin} (W_{fin} + 2H_{fin})$. No threshold voltage ($V_t$) roll off was observed with respect to $L_g$ ($V_t \sim 260$ mV) and only small $V_t$ variations (within 30 mV) are recorded as a function of $W_{fin}$. As expected, the devices also exhibit reduced SCE as the fin width is reduced. This is shown in Fig. 11a, which indicates lower $S$ values for narrower fins. However, reducing $W_{fin}$ is also expected to increase the source ($R_s$) and drain ($R_d$) resistance [61], as shown in Fig. 11b, which leads to a reduction of the normalized drain current as well as the effective gate transconductance (Fig. 11c).

The S-parameters of the devices are measured with a 110 GHz vectorial network analyzer (VNA) from Agilent. An open-short de-embedding step is performed to remove the parasitics associated with the access pads. The current gain ($|H_{21}|$) as a function of frequency which yields the device transition frequency ($f_T$) is presented in Fig. 11d for FinFETs with different fin widths. Unfortunately, we can observe a reduction of the cut-off frequency with the shrinkage of $W_{fin}$. This degradation is mainly related to the increase of the source and drain resistances with the thinning down of the fin width (Fig. 11b).

**Fig. 12** Extracted intrinsic ($f_{Ti}$) and extrinsic ($f_{Te}$) current gain cutoff frequencies for a SG MOSFET and FinFET as a function of the channel length

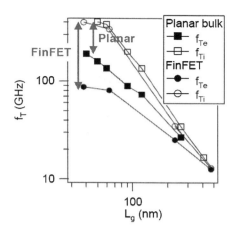

The DC and RF performances of planar MOSFETs with similar dimensions (Fig. 10) have been measured for comparison purposes. Figure 12 presents the extracted RF cut-off frequencies of planar and FinFET devices as a function of channel length. The so-called intrinsic ($f_{Ti}$) and extrinsic ($f_{Te}$) cut-off frequencies stand, respectively, for the current gain cut-off frequency related to only the intrinsic lumped parameter elements ($g_m$, $g_d$, $C_{gsi}$ and $C_{gdi}$) and the complete small-signal equivalent circuit presented in Fig. 2 (including the parasitic capacitances, $C_{gse}$ and $C_{gde}$, as well as the access resistances $R_s$, $R_d$, and $R_g$). It is quite interesting to see that both devices present similar intrinsic cut-off frequencies (around 400 GHz for a channel length of 60 nm) but the extrinsic cut-off frequency, $f_{Te}$, of FinFET (90 GHz) is nearly twice lower than that of the planar MOSFET (180 GHz). A possible explanation for the latter effect might be the more relevant impact of extrinsic capacitances and resistances in the case of short gate length FinFETs.

Based on a wideband analysis, the lumped small-signal equivalent circuit parameters (Fig. 2) are extracted from the measured S-parameters according to the methods described in [51, 52]. Figure 13 shows the relative impact of each parasitic parameter on the current gain ($f_T$, Fig. 13a) and maximum available power gain ($f_{max}$, Fig. 13b) cut-off frequencies of a 60 nm-long FinFET. As expected from the expressions (1–3) and the published results for SG MOSFETs [62] the gate resistance has an important impact on $f_{max}$ whereas $f_T$ is unchanged. The sum of fringing capacitances $C_{inner}$ directly linked to the FinFET three-dimensional (3-D) architecture has a huge impact on both cut-off frequencies. In fact, $f_T$ and $f_{max}$ drop down, respectively, by a factor of 3 and 2. Finally, the source and drain resistances as well as the parasitic capacitances related to the feed connections outside the active area of the transistor slightly decrease both cut-off frequencies. Based on that analysis, it is quite clear that the fringing capacitances inside the active area of the FinFET are the most important limiting factor for this type of non-planar multiple gate transistors.

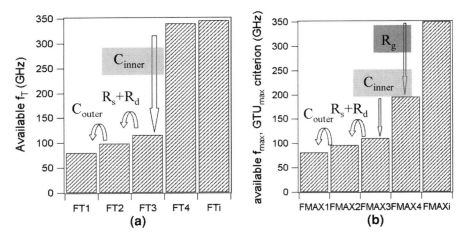

**Fig. 13** Relative impact of each lumped extrinsic parameters on **a** the current gain cut-off frequency ($f_T$) and on **b** the maximum available gain cut-off frequency ($f_{max}$) for a 60 nm-long FinFET

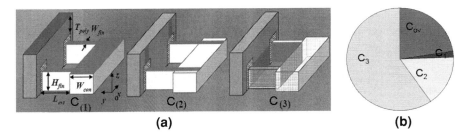

**Fig. 14 a** Three-dimensional schematic presentation of the various contributions for the fringing capacitances of a FinFET ($C_1$, $C_2$, $C_3$), **b** relative importance of each fringing capacitance ($C_1$, $C_2$, $C_3$) and overlap capacitance ($C_{ov}$)

In [63], Wu and Chan analyze the geometry-dependent parasitic components in multifin FinFETs. Parasitic fringing capacitance and overlap capacitance are physically modelled as functions of gate geometry parameters using the conformal mapping method. The relative contribution from each part of the 3-D geometry of the FinFET is calculated. They subdivide the fringing capacitances in 3 distinct components noted $C_1$, $C_2$ and $C_3$ in Fig. 14a. They demonstrate the importance of the fringing capacitance $C_3$ (Fig. 14b) which originates from the capacitive coupling between the source and drain regions of the fins (side walls) and the gate electrode located between fins assuring the electrical connection between the gates wrapping the different fins connected in parallel through the source and drain contacts. In [64, 65], the authors have demonstrated based on finite element numerical simulations the possibility to reduce $C_{inner}$ and thus its impact on the

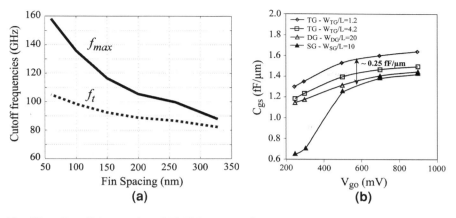

**Fig. 15 a** Cut-off frequencies of FinFETs versus fin spacing, **b** effect of W/L ratio on the normalized $C_{gs}$ extracted at $V_d = 1$ V at various $V_{go}$ and $L = 100$ nm

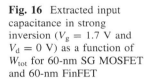

**Fig. 16** Extracted input capacitance in strong inversion ($V_g = 1.7$ V and $V_d = 0$ V) as a function of $W_{tot}$ for 60-nm SG MOSFET and 60-nm FinFET

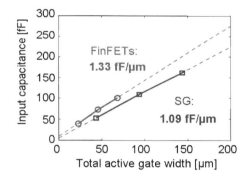

FinFET cut-off frequencies by reducing the fin spacing (Fig. 15a) or by increasing the aspect ratio of the fin (higher $H_{fin}/W_{fin}$, Fig. 15b), respectively.

Figure 16 shows the extracted input capacitance ($C_{gg} = C_{gs} + C_{gd}$) in strong inversion ($V_g = 1.7$ V and $V_d = 0$ V) as a function of the active gate width ($W_{tot}$) for a FinFET and a SG MOSFET with 60 nm gate length. Both devices are built simultanuously on the same SOI wafer. A first order extrapolation of the measured data yields $C_{gg}$ values of 1.33 fF/μm for the FinFET devices and only 1.09 fF per μm of active gate width for the SG, indicating a 20% increase of input capacitance in the case of FinFETs. Assuming that the normalized oxide capacitance is equal in both SG and FinFET devices, this increase is solely due to additional fringing in FinFETs. Using additional capacitance data measured in deep depletion, the extrinsic gate capacitance is actually found to be 40% higher for FinFETs. As explained above, this higher normalized input capacitance for FinFET can be explained by the fact that the gate fingers must run over non active area between each pair of parallel fins, a situation that is not encountered in SG MOSFETs.

To summarize, the simulation and experimental results indicate that FinFET is a multiple gate structure of interest to reduce digital short channel effects and then assure a lower threshold voltage roll-off, a better subthreshold slope and then higher $I_{on}/I_{off}$ ratio, but the high frequency performance such as the cut-off frequencies as well as RF noise figure as presented in [66] are degraded compared to its planar MOSFET counterpart because of the increased fringing capacitance linked to its complex 3-D non-planar architecture. Consequently, a trade-off exists regarding $W_{fin}$ between high $f_T$ and $f_{max}$ (large $W_{fin}$) and good control of SCE (small $W_{fin}$).

## 6 High Resistivity SOI Substrate

### 6.1 CPW Transmission Lines

The use of high resistivity silicon substrate is mandatory to reduce as much as possible the high frequency losses associated with the substrate conductivity. High resistivity silicon substrate cannot be introduced in the case of bulk Si MOSFETs due to the problem related to latch-up between devices. In SOI technology, thanks to the buried oxide (BOX), the thin top silicon layer, in which the transistors are implemented, is electrically isolated from the Si substrate which can have high resistivity without impacting the good behaviour of the MOS integrated circuits (ICs). Recently, high quality coplanar waveguides presenting insertion loss of less than 2 dB/mm at 200 GHz as well as low- and high-pass filters at millimetre-waves have been successfully built in an industrial SOI CMOS process environment [67].

The insertion loss of a CPW line lying on a lossy silicon substrate depends on the conductor loss ($\alpha_{cond}$) and the substrate loss ($\alpha_{sub}$) which is inversely proportional to the effective resistivity of the substrate. The effective resistivity represents the value of the substrate resistivity that is actually seen by the coplanar devices. This parameter accounts for the wafer inhomogeneities (i.e., oxide covering and space charge effects) and corresponds to the resistivity that a uniform (without oxide nor space charge effects) silicon wafer should have in order to sustain identical RF substrate losses. The effective resistivity is extracted from the measured S-parameters of the CPW line with a method depicted in [68]. Simulation results displayed in Fig. 17 outline how this parameter affects substrate and total losses for a 50-$\Omega$ CPW with 1 μm-thick Al line, the central conductor width of 40 μm and spacing between conductors of 24 μm. These data are obtained with analytical formulas presented in [69] and assuming metal conductivity of $3 \times 10^7$ S/m. It is seen that substrate losses ($\alpha_{sub}$) are small ($\sim 0.1$ dB/cm) when $\rho_{eff}$ is close to 3 k$\Omega$ cm and become clearly meaningless compared to conductor losses ($\alpha_{cond}$) when $\rho_{eff}$ reaches 10 k$\Omega$ cm. Keeping substrate losses at low levels is a priority target when designing high performance integrated silicon systems. In

**Fig. 17** Total ($\alpha_{tot}$) and conductor ($\alpha_{cond}$) losses as a function of $\rho_{eff}$ at 20 GHz for a CPW line geometry according to [69]

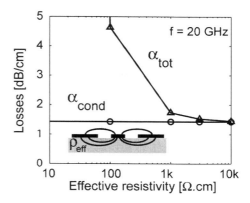

this field, high resistivity (>3 k$\Omega$ cm) silicon wafers are foreseen as promising candidates for radio frequency integrated circuits (RFIC) [70] and mixed signal applications [71]. However, oxide passivated high resistivity (HR) wafers are known to suffer from parasitic surface conduction due to fixed charges ($Q_{ox}$) in the oxide [72]. Indeed, charges within the oxide attract free carriers near the substrate surface, reducing the effective resistivity ($\rho_{eff}$) seen by coplanar devices and increasing substrate losses. It has been recently shown in [73] that values as low as $Q_{ox} = 10^{10}$ cm$^{-2}$ could lower the value of resistivity by more than one order of magnitude in the case of 50-$\Omega$ CPW transmission line. The parasitic surface conduction can also be formed underneath metallic lines with the application of a DC bias ($V_a$) [74]. The extracted line loss and effective substrate resistivity as a function of the DC bias applied to the central conductor of a CPW line are, respectively, presented in Fig. 18a and b for different substrates, oxide layers and metallic lines as summarized in Table 1. Techno A and B are wafers coming from the industry while the three other wafers named C, D and E are home processes with one metal layer. In all cases, the metallic structures are patterned on either oxidized p-type HR Unibond SOI (techno A, B, C) or oxidized p-type HR bulk Si (techno D and E) substrates.

The total RF losses ($\alpha_{tot}$) of the CPW lines are extracted from the measured S-parameters with a Thru-Line-Reflect method [75]. They are reported at 10 GHz in Fig. 18a as a function of $V_a$, where it is seen that $\alpha_{tot}$ may be significantly affected by $V_a$ when the oxide thickness ($t_{ox}$) is in the several hundreds of nm (techno C). Indeed, in that case highly positive or negative biases have a large impact on the free carrier concentration below the oxide, thereby strongly affecting substrate losses. This effect is attenuated for thicker oxides (techno A, B and D). The $V_a$ value for which losses are minimum ($V_{a,min}$) corresponds to the state of deep depletion underneath the oxide. As shown in Fig. 18, $V_a$ depends on the flatband voltage of the structure and is therefore dependent on $t_{ox}$ as well as the oxide charge density ($Q_{ox}$).

The parasitic surface conduction can be reduced or even suppressed if the silicon substrate is passivated before oxidation with a trap-rich, highly resistive

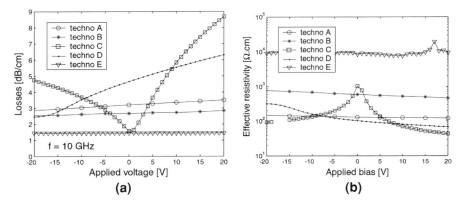

**Fig. 18** a CPW losses and **b** effective substrate resistivity measured for different technologies described in Table 1 as a function of DC bias applied to the CPW central conductor

**Table 1** Additional information on the different technologies investigated in Fig. 18

| Techno | Starting wafer | Metal layers | Oxide thickness (μm) | Si passivation | Oxide type |
|---|---|---|---|---|---|
| A | HR SOI | M3 | 3 | No | BOX + oxidized SOI + interlayer dielectrics |
| B | HR SOI | M5–M6 | 4.1 | No | BOX + oxidized SOI + interlayer dielectrics |
| C | HR SOI | M1 | 0.3 | No | BOX + oxidized SOI |
| D | HR Si bulk | M1 | 1 | No | PECVD |
| E | HR Si bulk | M1 | 1 | Polysilicon | PECVD |

The data in columns 3 and 4, respectively indicate the metal levels that were used and the total equivalent oxide thickness for CPW lines

layer. Figure 19 illustrates the impact of trap density ($D_{it}$) at the HR Si substrate/$SiO_2$ interface on the value of $\rho_{eff}$ at 0 V for several $Q_{ox}$ densities. It is seen with no surprise that the minimum $D_{it}$ level ($D_{it10k}$) that is required to obtain lossless substrates (i.e., $\rho_{eff}$ = 10 kΩ cm) is an increasing function of the fixed charge density in the oxide. This is because for higher positive densities, a higher concentration of electrons is attracted near the substrate surface and a higher density of traps is required to absorb those charges.

The introduction of a high density of traps at the Si/$SiO_2$ interface has been successfully achieved using low-pressure chemical vapour-deposited (LPCVD) polysilicon (PolySi) and amorphous silicon (α-Si) in [76] and [77], respectively. In the context of SOI technology, substrate passivation could also be an efficient technique to reduce substrate losses. To be compatible with a HR SOI wafer fabrication process, the passivation layer should be included within the SOI structure by bonding an oxidized silicon wafer with a passivated HR substrate. In [78], the proposed method consists in the LPCVD-deposition of amorphous silicon

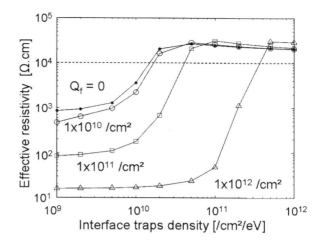

**Fig. 19** Simulated effective resistivity values $\rho_{\text{eff}}$ as a function of the trap density $D_{\text{it}}$ for several fixed charges densities $Q_{\text{ox}}$ at an applied bias value of 0 V

followed by Si-crystallization at 900°C with RTA. This method was compared with previously published techniques (passivation with amorphous silicon in [77] or LPCVD-polysilicon in [76]) and was demonstrated to perform better in terms of substrate loss reduction: effective resistivity values higher than 10 kΩ cm were reported, compared to 3 and 6 kΩ cm in the case of amorphous Si and LPCVD PolySi passivation, respectively. The new passivation method was also shown to present better rms surface roughness ($\sigma = 0.37$ nm) and to remain effective after long thermal anneals (4 h at 900°C). A successful bonding of this layer with an oxidized substrate was achieved, showing that this new passivation technique could be introduced at reduced cost inside a Smartcut or BESOI process in order to fabricate SOI wafers with enhanced resistivity, i.e., higher than 10 kΩ cm.

Figure 18a indicates that substrate passivation with polysilicon (techno E) significantly reduces RF losses while getting rid of the $V_a$ influence. This is because traps present inside the PolySi layer can absorb free carriers and pin the surface potential to a value independent on $V_a$ [76]. Figure 18b presents the effective resistivity ($\rho_{\text{eff}}$) extracted according to a method depicted in [68]. Not surprisingly, the highest $\rho_{\text{eff}}$ value is observed for the passivated substrate, while at 0 V, the lowest value is obtained for the low quality ($Q_{\text{ox}}$-rich) PECVD oxide. It should also be noticed that due to the inverted layer underneath the BOX in techno A and B, the extracted values of $\rho_{\text{eff}}$ do not exceed 130 and 580 Ω cm, respectively. These values are both more than one order of magnitude lower than the nominal substrate resistivity.

## 6.2 Crosstalk

In recent years, rapid progress of integrated circuit technology has enabled the co-integration of analogue front-end and digital baseband processing circuits of communication systems onto the same chip. Such mixed-signal systems-on-chip (SoC) allow more functionality, higher performance, lower power and higher reliability than non-integrated solutions, where at least two chips are needed, one for digital and one for the analogue applications. Moreover, thanks to CMOS technology scaling and its associated increasing integration level, SoCs have become the way to achieve cost effectiveness for demanding applications such as home entertainment and graphics, mobile consumer devices, networking and storage equipment. Such a rising integration level of mixed-signal ICs raises new issues for circuit designers. One of these issues is the substrate noise (see Fig. 22a) generated by switching digital circuits, called digital substrate noise (DSN), which may degrade the behaviour of adjacent analogue circuits [79]. DSN issues become more and more important with IC evolution as: (1) digital parts get more noisy due to increasing complexity and clock frequencies (2) digital and analogue parts get closer and (3) analogue parts get more sensitive because of $V_{dd}$ scaling for power concern issues.

In general, substrate noise can be decomposed in three different mechanisms: noise generation, injection/propagation into the substrate and reception by the analogue part [80]. Improvement in the reduction of any of these three mechanisms, or in all of them, will lead to a reduction of the DSN and in a relaxation of the design requirements. Typically, guard rings and overdesigned structures are adopted to limit the effect of substrate noise, thereby reducing the advantages of the introduction of new technologies. It is thus a major issue for the semiconductor industry to find area-efficient design/technology solutions to reduce the impact of substrate noise in mixed-signal ICs.

This last decade several publications have demonstrated theoretically and experimentally the interest of high resistivity SOI substrate to greatly reduce the crosstalk level between integrated circuits [5]. Figure 20 shows how the crosstalk between two 50 µm-spaced metallic pads is affected by $\rho_{eff}$ and indicates that $\rho_{eff}$ must be at least in the several kΩ cm range to get rid of conductive coupling inside the substrate for frequencies around 100 MHz and lower. The result of the substrate crosstalk measurements using a classical double-pad structure in which both pads are connected to separate RF probing pads [5] is shown in Fig. 21 in the form of $|S_{21}|$ versus frequency curves. The measurements are performed by using the low-frequency VNA up to 4 GHz and by applying various bias conditions on the coupling pads. The figure shows significantly higher ($\sim 13$ dB at 0 V) crosstalk level below 1 GHz for the standard HR SOI wafer, due to conductive effects in the substrate associated with parasitic surface conduction [81]. It also highlights a significant dependence with respect to the applied bias. The crosstalk level is strongly reduced for negative bias and when deep depletion is formed below the BOX, whereas it is enhanced and exhibits higher cut-off frequencies for positive bias and increased inversion below the oxide. On the other hand, the passivated wafer exhibits: (1) no effect of the applied bias due to the presence of the trap-rich polysilicon layer below the BOX [68] and (2) a perfect 20-dB/dec slope which shows that purely capacitive coupling occurs in the measurable frequency range (i.e., above the VNA noise floor). A reduction of crosstalk below 1 GHz is of

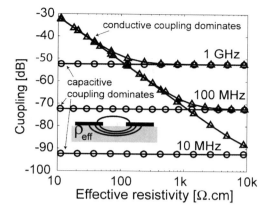

**Fig. 20** Simulated crosstalk level at 10, 100 MHz and 1 GHz as a function of $\rho_{eff}$ according to model presented in [5]

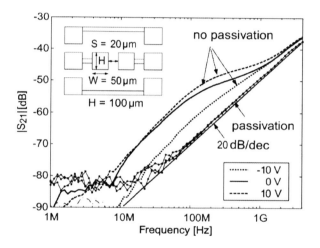

**Fig. 21** Crosstalk measured as a function of frequency and under distinct bias conditions on the unpassivated and passivated HR Si wafers

particular interest for mixed-signal applications, since it is known from previous studies that the frequency spectrum of the noise generated by digital logic typically expands to several hundreds of megahertz, corresponding to multiples of the clock signal [82, 83] or circuit internal resonance frequencies [84]. The generation of noise in that frequency range has also been shown to strongly increase the jitter in phase-locked loops (PLLs), which seem to be particularly sensitive to substrate noise injected at the PLL reference frequency, i.e., in the few hundreds of megahertz range [85]. It is further believed that in terms of crosstalk, the benefits gained by substrate passivation will even increase in the future. Indeed, a reduction of the BOX thickness for the next generations of active SOI devices will be required to reduce short channel effects and self-heating [86].

In [87], experimental DSN characterizations of CMOS circuits lying on SOI and bulk Si substrates are compared. Current injected into the substrate creates substrate voltage fluctuations (substrate noise). It is mainly created by two

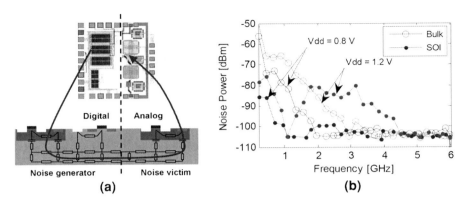

**Fig. 22** a Schematic representation of the substrate crosstalk between digital and analogue parts of a SoC, b comparison of the frequency envelope of the measured DSN (clock frequency = 225 MHz, 8 inverter trees)

mechanisms [79]: coupling from the noisy digital power supply circuit and from switching drains.

The DSN for 8 switching inverter trees biased at either 0.8 or 1.2 V and for an input clock frequency of 225 MHz has been measured in the case SOI and Si bulk substrates. DSN for SOI circuit presents a quite different frequency response (Fig. 22b). At low frequency, SOI and Si bulk present the same kind of response, with the SOI DSN level decreasing faster with increasing frequency. At higher frequency, the SOI DSN presents a kind of "pass-band filter" shape, which is not visible in the case of the bulk circuit. We have shown in our previous work that this second part of the frequency response is due to ringing on supply rail, due to parasitic capacitances and inductances [88]. For the 1.2 V supply voltage, the SOI technology allows an important reduction of DSN up to 1 GHz. At higher frequency, the noise due to ringing on supply rail becomes dominant, and the bulk circuit shows a lower DSN level. This conclusion is in agreement with the results of studies on the supply noise showing that special attention should be paid to supply rail for SOI technology, due to lower intrinsic decoupling capacitances [89]. At lower power supply (0.8 V), as for the bulk, high frequency noise generation decreases. The ringing supply noise tends thus to be negligible. The SOI technology presents then better DSN results than bulk for frequency up to 2 GHz, and similar DSN level for upper frequency.

## 6.3 Non-Linearities Along CPW Lines

High-resistivity silicon substrates are promising for RF applications due to their reduced substrate loss and coupling, as presented in the two previous paragraphs, which helps to enable RF cellular transmit switches on SOI using HRS handle wafers [90, 91]. RF switches have high linearity requirements: for instance, a

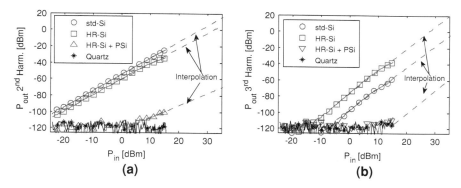

**Fig. 23** **a** Second and **b** third harmonic power of a 3,385 μm-long CPW line lying on different substrates

recent III–V RF switch product specifies less than −45 and −40 dBm for 2nd and 3rd harmonic power (H2 and H3), respectively, at +35 dBm input power [92]. As requirements become even more stringent for advanced multimode phones and 3G standards, it is important to investigate even small contributions to harmonic distortion (HD).

As explained above, when the CPW line is biased the distribution of potential and free carriers inside the Si substrate changes like in the case of a classical MOS capacitor. The variation of carrier distribution in the Si substrate with the applied bias or large RF signal will thus lead to the existence of non-linear capacitance ($C$) and conductance ($G$) associated with the Si substrate. Those variables $C$ and $G$ are at the origin of the harmonics formation inside the Si substrate.

Figure 23 shows the harmonic distortion (HD) of CPW Al metal lines on Si p-type substrates of different resistivities with 50 nm of top $SiO_2$. HR-Si substrate presents lower HD than the 20 Ω cm (std-Si) substrate over most of the power sweep. Measurements are made up to +15 dBm input power, and linearly interpolated up to +35 dBm. It can be observed that non-passivated Si substrates have HD levels higher than 45 dBc at +15 dBm, already 25 dB higher than the switch specifications at +35 dBm. The introduction of a 300-nm Polysilicon (PSi) layer reduces HD levels by more than 60 dB, compared to non-passivated HR-Si. As explained above, thanks to the high density of traps in the polycrystalline silicon or as-deposited amorphous silicon layer located at the Si-$SiO_2$ interface, the surface potential at this interface is nearly fixed, and the external DC bias or large amplitude RF signal applied to the line does not impact the distribution of carriers inside the Si substrate.

# 7 Conclusions

The performance of SOI MOSFET technology in microwaves and millimetre-waves has been presented. Nowadays, strained SOI n-MOSFET which exhibits a cut-off frequency close to 500 GHz is really competing with the III-V technologies. Thanks to the introduction of high resistivity SOI substrate, the integration of high quality passives is a reality and the reduction of the substrate crosstalk is a real advantage compared to Si bulk for the development of high integration low voltage mixed-mode applications. Major semiconductor companies such as ST-Microelectronics, IBM, RFMD, Honeywell, OKI, etc., have already produced several products for the telecommunication market based on SOI RF technologies.

As demonstrated in this chapter, by the introduction of a trap-rich layer underneath the BOX, HR SOI substrate can still be improved. Having a polysilicon-based layer with the thickness of approximately 300 nm sandwiched between the BOX and the HR Si substrate, CPW insertion loss, crosstalk, DSN, as well as harmonic distortion are greatly reduced.

To summarize, present and future HR SOI MOSFET technologies are very good candidates for mixed-mode low voltage low power RF and even millimetre-waves applications.

**Acknowledgments** I would like to thank all the Ph.D. students, senior researchers, and professors who have actively participated to the simulation and experimental results presented in this chapter: Mr. M. Emam, Mr. C. Roda Neve, Mr. R. Ambroise, Dr. M. El Kaamouchi, Dr. M. Si Moussa, Dr. R. Valentin, Dr. D. Bol, Dr. V. Kilchytska, Dr. A. Kranti, Dr. D. Lederer, Dr. C. Urban, Dr. Qing-Tai Zhao, Dr. O. Moldovan, Prof. B. Iniguez, Prof. D. Flandre, Prof. F. Danneville, Prof. E. Dubois and Prof. S. Mantl. I would like to also thank Mr. P. Simon (Welcome) for performing most of the RF measurements, the UCL clean rooms team (Winfab), as well as Dr. Jurczak Malgorzata's group and Dr. S. Decoutere's group (especially Dr. B. Parvais, Dr. M. Dehan, Dr. A. Mercha, Dr. Subramanian Vaidy), IMEC, Leuven, Belgium, for providing FinFETs. This research has been financially supported by the European Networks of Excellence SINANO, NANOSIL and EuroSOI+.

# References

1. Moore, G.E.: Cramming more components onto integrated circuits. Electronics **38**, 114–117 (1965)
2. Dennard, R.H., Gaensslen, F.H., Hwa-Nien, Yu., Rideout, V.L., Bassous, E., Leblanc, A.R.: Design of ion-implanted MOSFET's with very small physical dimensions. IEEE J. Solid State Circuits **9**, 256–268 (1974)
3. Lee, S., Jagannathan, B., Narasimha, S., Chou, A., Zamdmerm N., Johnson, J., Williams, R., Wagner, L., Kim, J., Plouchart, J.-O., Pekarik, J., Springer, S., Freeman, G.: Record RF performance of 45-nm SOI CMOS technology. IEEE Int. Electron Devices Meeting IEDM, pp. 255–258 (2007)
4. Sakurai, T., Matsuzawa, A., Douseki, T.: Fully-depleted SOI CMOS circuits and technology for ultralow-power applications. Springer XV ISBN:978-0-387-29217-5 (2006)
5. Raskin, J.-P., Viviani, A., Flandre, D., Colinge, J.-P.: Substrate crosstalk reduction using SOI technology. IEEE Trans. Electron Devices **44**, 2252–2261 (1997)
6. Cooke, H.F.: Microwave transistors: theory and design. Proc. IEEE **59**, 1163–1181 (1971)
7. Mead, C.A.: Schottky barrier gate field effect transistor. Proc. IEEE **59**, 307–308 (1966)
8. Baechtold, W., Daetwyler, K., Forster, T., Mohr, T.O., Walter, W., Wolf, P.: Si and GaAs 0.5 μm gate Schottky-barrier field-effect transistors. Electron Lett. **9**, 232–234 (1973)

9. Mimura, T., Hiyamizu, S., Fujii, T., Nanbu, K.: A new field-effect transistor with selectively doped GaAs/n-AlxGa1-xAs heterojunctions. Jpn. J. Appl. Phys. **19**, L225–L227 (1980)
10. Smith, P.M., Liu, S.-M.J., Kao, M.-Y., Ho, P., Wang, S.C., Duh, K.H.G., Fu, S.T., Chao, P.C.: W-band high efficiency InP-based power HEMT with 600 GHz $f_{max}$. IEEE Microw. Guid. Wave Lett. **5**, 230–232 (1995)
11. Rodwell, M.J.W., Urteaga, M., Mathew, T., Scott, D., Mensa, D., Lee, Q., Guthrie, J., Betser, Y., Martin, S.C., Smith, R.P., Jaganathan, S., Krishnan, S., Long, S.I., Pullela, R., Agarwal, B., Bhattacharya, U., Samoska, L., Dahlstrom, M.: Submicron scaling of HBTs. IEEE Trans. Electron Dev. **48**, 2606–2624 (2001)
12. Lai, R., Mei, X.B., Deal, W.R., Yoshida, W., Kim, Y.M., Liu, P.H., Lee, J., Uyeda, J., Radisic, V., Lange, M., Gaier, T., Samoska, L., Fung, A.: Sub 50 nm InP HEMT device with Fmax greater than 1 THz. IEEE Int. Electron Devices Meet., pp. 609–611 (2007)
13. Momose, H.S., Morifuji, E., Yoshitomi, T., Ohguro, T., Saito, I., Morimoto, T., Katsumata, Y., Iwai, H.: High-frequency AC characteristics of 1.5 nm gate oxide MOSFETs. IEEE Int. Electron Devices Meet., pp. 105–108 (1996)
14. International Technology Roadmap for Semiconductors. http://www.itrs.net/Common/2006ITRS/Home2006.html
15. Dambrine, G., Raynaud, C., Lederer, D., Dehan, M., Rozeaux, O., Vanmackelberg, M., Danneville, F., Lepilliet, S., Raskin, J.P.: What are the limiting parameters of deep-submicron MOSFETs for high frequency applications? IEEE Electron Device Lett. **24**, 189–191 (2003)
16. Pailloncy, G., Raynaud, C., Vanmackelberg, M., Danneville, F., Lepilliet, S., Raskin, J.-P., Dambrine, G.: Impact of down scaling on high frequency noise performance of bulk and SOI MOSFETs. IEEE Trans. Electron Devices **51**, 1605–1612 (2004)
17. Kilchytska, V., Nève, A., Vancaillie, L., Levacq, D., Adriaensen, S., van Meer, H., De Meyer, K., Raynaud, C., Dehan, M., Raskin, J.-P., Flandre, D.: Influence of device engineering on the analog and RF performances of SOI MOSFETs. IEEE Trans. Electron Devices **50**, 577–588 (2003)
18. Vanmackelberg, M., Raynaud, C., Faynot, O., Pelloie, J.-L., Tabone, C., Grouillet, A., Martin, F., Dambrine, G., Picheta, L., Mackowiak, E., Llinares, P., Sevenhans, J., Compagne, E., Fletcher, G., Flandre, D., Dessard, V., Vanhoenacker, D., Raskin, J.-P.: 0.25 µm fully-depleted SOI MOSFET's for RF mixed analog-digital circuits, including a comparison with partially-depleted devices for high frequency noise parameters. Solid State Electron. **46**, 379–386 (2002)
19. Burignat, S., Flandre, D., Kilchytska, V., Andrieux, F., Faynot, O., Raskin, J.-P.: Substrate impact on sub-32 nm ultra thin SOI MOSFETs with thin buried oxide. Fifth workshop of the thematic network on silicon on insulator technology, devices and circuits, EUROSOI'09, Göteborg, Sweden, pp. 27–28 (2009)
20. Rudenko, T., Kilchytska, V., Burignat, S., Raskin, J.-P., Andrieu, F., Faynot, O., Nazarov, A., Flandre, D.: Transconductance and mobility behaviors in UTB SOI MOSFETs with standard and thin BOX. Fifth workshop of the thematic network on silicon on insulator technology, devices and circuits, EUROSOI'09, Göteborg, Sweden, pp. 111–112 (2009)
21. Kah-Wee, Ang., Jianqiang, Lin., Chih-Hang, Tung., Balasubramanian, N., Samudra, G.S., Yee-Chia, Yeo.: Strained n-MOSFET with embedded source/drain stressors and strain-transfer structure (STS) for enhanced transistor performance. IEEE Trans. Electron Devices **55**, 850–857 (2008)
22. Néau, G., Martinez, F., Valenza, M., Vildeuil, J.C., Vincent, E., Boeuf, F., Payet, F., Rochereau, K.: Impact of strained-channel n-MOSFETs with a SiGe virtual substrate on dielectric interface quality evaluated by low frequency noise measurements. Microelectron. Reliab. **47**, 567–572 (2007)
23. Olsen, S.H., Escobedo-Cousin, E., Varzgar, J.B., Agaiby, R., Seger, J., Dobrosz, P., Chattopadhyay, S., Bull, S.J., O'Neill, A.G., Hellstrom, P.-E., Edholm, J., Ostling, M., Lyutovich, K.L., Oehme, M., Kasper, E.: Control of self-heating in thin virtual substrate strained Si MOSFETs. IEEE Trans. Electron Devices **53**, 2296–2305 (2006)

24. Larson, J.M., Snyder, J.: Overview and status of metal S/D Schottkybarrier MOSFET technology. IEEE Trans. Electron Devices **53**, 1048–1058 (2006)
25. Pearman, D.J., Pailloncy, G., Raskin, J.P., Larson, J.M., Whall, T.E.: Static and high-frequency behavior and performance of Schottky Barrier p-MOSFET devices. IEEE Trans. Electron Devices **54**, 2796–2802 (2007)
26. Raskin, J.P., Pearman, D.J., Pailloncy, G., Larson, J.M., Snyder, J., Leadley, D.L., Whall, T.E.: High-frequency performance of Schottky Barrier p-MOSFET devices. IEEE Electron Device Lett. **29**, 396–398 (2008)
27. Larrieu, G., Dubois, E., Valentin, R., Breil, N., Danneville, F., Dambrine, G., Raskin, J.-P., Pesant, J.-C.: Low temperature implementation of dopant-segregated band-edge metallic S/D junctions in thin-body SOI p-MOSFETs. IEEE Int. Electron Devices Meet., Washington, DC, USA, pp. 147–150 (2007)
28. Valentin, R., Dubois, E., Raskin, J.-P., Larrieu, G., Dambrine, G., Lim Tao, C., Breil, N., Danneville, F.: RF small signal analysis of Schottky-Barrier p-MOSFET. IEEE Trans. Electron Devices **55**, 1192–1202 (2008)
29. Ricco, B., Versari, R., Esseni, D.: Characterization of polysilicon-gate depletion in MOS structures. IEEE Electron Device Lett. **17**, 103–105 (1996)
30. Vandooren, A., Thean, A.V.Y., Du, Y., To, I., Hughes, J., Stephens, T., Huang, M., Egley, S., Zavala, M., Sphabmixay, K., Barr, A., White, T., Samavedam, S., Mathew, L., Schaeffer, J., Triyoso, D., Rossow, M., Roan, D., Pham, D., Rai, R., Nguyen, B.-Y., White, B., Orlowski, M., Duvallet, A., Dao, T., Mogab, J.: Mixed-signal performance of sub-100 nm fully-depleted SOI devices with metal gate, high K ($HfO_2$) dielectric and elevated source/drain extensions. IEEE Int. Electron Devices Meet., Washington, DC, USA, pp. 11.5.1–11.5.3 (2003)
31. Ko, C.H., Kuan, T.M., Zhang, K., Tsai, G., Seutter, S.M., Wu, C.H., Wang, T.J., Ye, C.N., Chen, H.W., Ge, C.H., Wu, K.H., Lee, W.C.: A novel CVD-SiBCN low-K spacer technology for high-speed applications. Symposium on VLSI Technology, Honolulu, Hawaii, USA, pp. 108–109 (2008)
32. Bao, T.I., Chen, H.C., Lee, C.J., Lu, H.H., Shue, S.L., Yu, C.H.: Low capacitance approaches for 22 nm generation Cu interconnect. International symposium on VLSI technology, systems, and applications, Hsinchu, Taiwan, pp. 51–56 (2009)
33. Ernst, T., Tinella, C., Raynaud, C., Cristoloveanu, S.: Fringing fields in sub-0.1 μm fully depleted SOI MOSFET's: optimization of the device architecture. Solid State Electron. **46**, 373–378 (2002)
34. Fujiwara, M., Fujiwara, M., Morooka, T., Yasutake, N., Ohuchi, K., Aoki, N., et al.: Impact of BOX scaling on 30 nm gate length FD SOI MOSFET. IEEE international SOI conference, Honolulu, Hawaii, USA, pp. 180–182 (2005)
35. Gianesello, F., Gloria, D., Raynaud, C., Montusclat, S., Boret, S., Clement, C., Benech, P.H., Fournier, J.M., Dambrine, G.: State of the art 200 GHz passive components and circuits integrated in advanced thin SOI CMOS technology on high resistivity substrate. IEEE international SOI conference, Niagara Falls, New York, USA, pp. 121–122 (2006)
36. Gianesello, F., Gloria, D., Raynaud, C., Montusclat, S., Boret, S., Touret, P.: On the design of high performance RF integrated inductors on high resistively thin film 65 nm SOI CMOS technology. IEEE topical meeting on silicon monolithic integrated circuits in RF systems, pp. 98–101 (2008)
37. Kinoshita, T., Hasumi, R., Hamaguchi, M., Miyashita, K., Komoda, T., Kinoshita, A., Koga, J., Adachi, K., Toyoshima, Y., Nakayama, T., Yamada, S., Matsuoka, F.: Ultra low voltage operations in bulk CMOS logic circuits with dopant segregated Schottky source/drain transistors. IEEE Int. Electron Devices Meet., pp. 1–4 (2006)
38. Connelly, D., Faulkner, C., Grupp, D.E.: Optimizing Schottky S/D Offset for 25-nm dual-gate CMOS performance. IEEE Electron Device Lett. **24**, 411–413 (2003)
39. Xiong, S., King, T.-J., Bokor, J.: A comparison study of symmetric ultrathin-body double-gate devices with metal source/drain and doped source/drain. IEEE Trans. Electron Devices **52**, 1859–1867 (2005)

40. Jang, M., Kim, Y., Shin, J., Lee, S., Park, K.: A 50-nm-gate-length erbium-silicided n-type Schottky barrier metal-oxide-semiconductor field-effect transistor. Appl. Phys. Lett. **84**, 741–743 (2004)
41. Zhu, S., Chen, J., Li, M.-F., Lee, S.J., Singh, J., Zhu, C.X., Du, A., Tung, C.H., Chin, A., Kwong, D.L.: N-type Schottky barrier source/drain MOSFET using Ytterbium silicide. IEEE Electron Device Lett. **25**, 565–567 (2004)
42. Kedzierski, J., Xuan, P., Anderson, E.H., Bokor, J., King, T.-J., Hu, C.: Complementary silicide source/drain thin-body MOSFETs for the 20 nm gate length regime. IEEE Int. Electron Devices Meet., pp. 57–60 (2000)
43. Luo, J., Qiu, Z.-J., Zhang, D.W., Hellström, P.-E., Östling, M., Zhang, S.-L.: Effects of carbon on Schottky Barrier heights of NiSi modified by dopant segregation. IEEE Electron Device Lett. **30**, 608–610 (2009)
44. Urban, C., Zhao, Q.T., Sandow, C., Müller, M., Breuer, U., Mantl, S.: Schottky barrier height modulation by arsenic dopant segregation. Proceedings of 9th international conference ULIS, pp. 151–154 (2008)
45. Fritze, M., Chen, C.L., Calawa, S., Yost, D., Wheeler, B., Wyatt, P., Keast, C.L., Snyder, J., Larson, J.: High-speed Schottky-barrier pMOSFET with $f_T = 280$ GHz. IEEE Electron Device Lett. **25**, 220–222 (2004)
46. Valentin, R., Dubois, E., Larrieu, G., Raskin, J.-P., Dambrine, G., Breil, N., Danneville, F.: Optimization of RF performance of metallic source/drain SOI MOSFETs using dopant segregation at the Schottky interface. IEEE Electron Device Lett. **30**, 1197–1199 (2009)
47. Urban, C., Emam, M., Sandow, C., Zhao, Q.T., Fox, A., Raskin, J.-P., Mantl, S.: High-frequency performance of dopant-segregated NiSi S/D SOI SB-MOSFETs. Proceedings of 38th ESSDERC, pp. 149–152 (2009)
48. Urban, C., Emam, M., Sandow, C., Knoch, J., Zhao, Q.T., Raskin, J.-P., Mantl, S.: Radio frequency study of dopant-segregated n-type SB-MOSFETs on thin-body SOI. IEEE Electron Device Lett. **31**, 537–539 (2010)
49. Knoch, J., Zhang, M., Zhao, Q.T., St. Lenk, J., Mantl, S., Appenzeller, J.: Effective Schottky barrier lowering in silicon-oninsulator Schottky-barrier metal–oxide–semiconductor field-effect transistors using dopant segregation. Appl. Phys. Lett. doi:10.1063/1.2150581 (2005)
50. Zhang, M., Knoch, J., Zhao, Q.T., Breuer, U., Mantl, S.: Impact of dopant segregation on fully depleted Schottky-barrier SOI-MOSFETs. Solid State Electron **50**(4), 594–600 (2006)
51. Bracale, A., Ferlet-Cavrois, V., Fel, N., Pasquet, D., Gautier, J.L., Pelloie, J.L., Du Port De Pon-charra, J.: A new approach for SOI devices small-signal parameters extraction. Analog Integr. Circuits Signal Process **25**(2), 157–169 (2000)
52. Raskin, J.-P., Gillon, R., Chen, J., Vanhoenacker-Janvier, D., Colinge, J.-P.: Accurate SOI MOSFET characterization at microwave frequencies for device performance optimization and analog modelling. IEEE Trans. Electron Devices **45**(5), 1017–1025 (1998)
53. Colinge, J.-P., Gao, M.-H., Romano, A., Maes, H., Claeys, C.: Silicon-on-insulator 'gate-all-around' MOS device. Proceedings of IEEE SOS/SOI technical conference, Key West, Florida, USA, pp. 137–138 (1990)
54. Hisamoto, D., Lee, W.-C., Kedzierski, J., Takeuchi, H., Asano, K., Kuo, C., Anderson, E., King, T.-J., Bokor, J., Hu, C.: FinFET—a self-aligned double-gate MOSFET scalable to 20 nm. IEEE Trans. Electron Devices **47**(12), 2320–2325 (2000)
55. Cristoloveanu, S.: Silicon on insulator technologies and devices: from present to future. Solid State Electron **45**(8), 1403–1411 (2001)
56. Park, J.-T., Colinge, J.-P.: Multiple-gate SOI MOSFETs: device design guidelines. IEEE Trans. Electron Devices **49**(12), 2222–2229 (2002)
57. Kedzierski, J., Fried, D.M., Nowak, E.J., et al.: High performance symmetric-gate and CMOS-compatible $V_t$ asymmetric-gate FinFET devices. IEEE International Electron Devices Meeting—IEDM, Washington, DC, USA, pp. 437–440 (2001)
58. Woo, D.-S., Lee, J.-H., Choi, W.Y., Choi, B.Y., Choi, Y.J., Lee, J.D., Park, B.-G.: Electrical characteristics of FinFET with vertically nonuniform source/drain profile. IEEE Trans. Nanotech. **1**(4), 233–237 (2002)

59. Kilchytska, V., Collaert, N., Rooyackers, R., Lederer, D., Raskin, J.-P., Flandre, D.: Perspective of FinFETs for analog applications. 34th European solid-state device research conference—ESSDERC 2004, Leuven, Belgium, pp. 65–68 (2004)
60. Lederer, D., Kilchytska, V., Rudenko, T., Collaert, N., Flandre, D., Dixit, A., De Meyer, K., Raskin, J.-P.: FinFET analog characterization from DC to 110 GHz. Solid State Electron. **49**, 1488–1496 (2005)
61. Dixit, A., Kottantharayil, A., Collaert, N., Goodwin, M., Jurczak, M., De Meyer, K.: Analysis of the parasitic source/drain resistance in multiple gate field effect transistors. IEEE Trans. Electron Dev. **52**(6), 1131–1140 (2005)
62. Razavi, B., Ran-Hong, Y., Lee Kwing, F.: Impact of distributed gate resistance on the performance of MOS devices. IEEE Trans. Circuits Syst. I Fundam. Theory Appl. **41**(11), 750–754 (1994)
63. Wen, W., Mansun, C.: Analysis of geometry-dependent parasitics in Multifin double-gate FinFETs. IEEE Trans. Electron Devices **54**(4), 692–698 (2007)
64. Moldovan, O., Lederer, D., Iniguez, B., Raskin, J.-P.: Finite element simulations of parasitic capacitances related to multiple-gate field-effect transistors architectures. The 8th topical meeting on silicon monolithic integrated circuits in RF systems—SiRF 2008, Orlando, FL, USA, pp. 183–186 (2008)
65. Raskin, J.-P., Chung, T.M., Kilchytska, V., Lederer, D., Flandre, D.: Analog/RF performance of multiple-gate SOI devices: wideband simulations and characterization. IEEE Trans. Electron Devices **53**, 1088–1094 (2006)
66. Raskin, J.-P., Pailloncy, G., Lederer, D., Danneville, F., Dambrine, G., Decoutere, S., Mercha, A., Parvais, B.: High frequency noise performance of 60 nm gate length FinFETs. IEEE Trans. Electron Devices **55**, 2718–2727 (2008)
67. Gianesello, F., Montusclat, S., Martineau, B., Gloria, D., Raynaud, C., Boret, S., Dambrine, G., Lepilliet, S., Pilard, R.: 1.8 dB insertion loss 200 GHz CPW band pass filter integrated in HR SOI CMOS technology. IEEE radio frequency integrated circuits (RFIC) symposium, Hawaï, USA, pp. 555–558 (2007)
68. Lederer, D., Raskin, J.-P.: Effective resistivity of fully-processed high resistivity wafers. Solid State Electron. **49**, 491–496 (2005)
69. Heinrich, W.: Quasi-TEM description of MMIC coplanar lines including conductor-loss effects. IEEE Trans. Microw. Theory Tech. **41**, 45–52 (1993)
70. Reyes, A.C., El-Ghazaly, S.M., Dom, S.J., Dydyk, M., Schroeder, D.K., Patterson, H.: Coplanar waveguides and microwave inductors on silicon substrates. IEEE Trans. Microw. Theory Tech. **43**, 2016–2021 (1995)
71. Benaissa, K., Yuan, J.-T., Crenshaw, D., Williams, B., Sridhar, S., Ai, J., Boselli, G., Zhao, S., Tang, S., Ashbun, S., MAdhani, P., Blythe, T., Mahalingam, N., Schichijo, H.: RF CMOS high-resistivity substrates for systems-on-chip applications. IEEE Trans. Electron Devices **50**, 567–576 (2003)
72. Wu, Y., Gamble, H.S., Armstrong, B.M., Fusco, V.F., Stewart, J.A.C.: SiO2 interface layer effects on microwave loss of high-resistivity CPW line. IEEE Microw. Guid. Wave Lett. **9**, 10–12 (1999)
73. Lederer, D., Desrumeaux, C., Brunier, F., Raskin, J.-P.: High resistivity SOI substrates: how high should we go? In: Proceedings of IEEE international SOI conference, Newport Beach, CA, USA, pp. 50–51 (2003)
74. Schollhorn, C., Zhao, W., Morschbach, M., Kasper, E.: Attenuation mechanisms of aluminum millimeter-wave coplanar waveguides on silicon. IEEE Trans. Electron Devices **50**, 740–746 (2003)
75. Lu, H.-C., Chu, T.-H.: The thru-line-symmetry (TLS) calibration method for on-wafer scattering matrix measurement of four-port networks. IEEE MTT-S international microwave symposium digest, Fort Worth, TX, USA, pp. 1801–1804 (2004)
76. Gamble, H., Armstrong, B.M., Mitchell, S.J.N., Wu, Y., Fusco, V.F., Stewart, J.A.C.: Low-loss CPW lines on surface stabilized high resistivity silicon. IEEE Microw. Guid. Wave Lett. **9**, 395–397 (1999)

77. Wong, B., Burghartz, J.N., Natives, L.K., Rejaei, B., van der Zwan, M.: Surface-passivated high resistivity silicon substrates for RFICs. IEEE Electron Device Lett. **25**, 176–178 (2004)
78. Lederer, D., Raskin, J.-P.: New substrate passivation method dedicated to high resistivity SOI wafer fabrication with increase substrate resistivity. IEEE Electron Device Lett. **26**, 805–807 (2005)
79. Calmon, F., Andrei, C., Valorge, O., Perez, J.-C., Verdier, J., Nunez, J., Gontrand, C.H.: Impact of low-frequency substrate disturbances on a 4.5 GHz VCO. Microelectron. J. **37**, 1119–1127 (2006)
80. Van Heijningen, M., Badaroglu, M., Donnay, S., Engels, M., Bolsen, I.: High-level simulation of substrate noise generation including power supply noise coupling. 37th conference on design automation—DAC 2000, Los Angeles, CA, USA, pp. 46–451 (2000)
81. Lederer, D., Raskin, J.-P.: Bias effects on RF passive structures in HR Si substrates. In: Proceedings of 6th topical meeting silicon microwave. integrative circuits RF systems, pp. 8–11 (2006)
82. Van Heijningen, M., Compiet, J., Wambacq, P., Donnay, S., Engels, M.G.E., Bolsens, I.: Analysis and experimental verification of digital substrate noise generation for epi-type substrates. IEEE J. Solid State Circuits **35**, 1002–1008 (2000)
83. Van Heijningen, M., Badaroglu, M., Donnay, S., Gielen, G.G.E., De Man, H.J.: Substrate noise generation in complex digital systems: efficient modeling and simulation methodology and experimental verification. IEEE J. Solid State Circuits **37**, 1065–1072 (2002)
84. Badaroglu, M., Donnay, S., De Man, H.J., Zinzius, Y.A., Gielen, G.G.E., Sansen, W., Fonden, T., Signell, S.: Modeling and experimental verification of substrate noise generation in a 220-kgates wlan system-on-chip with multiple supplies. IEEE J. Solid State Circuits **38**, 1250–1260 (2003)
85. Jenkins, K.A., Rhee, W., Liobe, J., Ainspan, H.: Experimental analysis of the effect of substrate noise on PLL. In: Proceedings of 6th topical meeting silicon monolithic integrative circuits RF systems, San Diego, CA, pp. 54–57 (2006)
86. ITRS Roadmap: Front end processes. http://www.itrs.net/Common/2005ITRS/FEP2005.pdf (2005)
87. Neve, C., Roda, B.D., Ambroise, R., Flandre, D., Raskin, J.-P.: Comparison of digital substrate noise in SOI and Bulk Si CMOS technologies. 7th workshop on low-voltage low power design, Louvain-la-Neuve, Belgium, pp. 23–28 (2008)
88. Bol, D., Ambroise, R., Neve, C., Roda, C., Raskin, J.-P., Flandre, D.: Wide-band simulation and characterization of digital substrate noise in SOI technology. IEEE international SOI conference, Indian Wells, CA, USA, pp. 133–134 (2007)
89. Chen, H.H., Ling, D.D.: Power supply noise analysis methodology For deep-submicron VLSI chip design proceedings of the 34th design automation, Anaheim, CA, USA, pp. 638–643 (1997)
90. Tinella, C., Richard, O., Cathelin, A., Reaute, F., Majcherczak, S., Blanchet, F., Belot, D.: 0.13 µm CMOS SOI SP6T antenna switch for multi-standard handsets. Topical meeting on silicon monolithic integrated circuits in RF systems, pp. 58 (2006)
91. McKay, T.G., Carroll, M.S., Costa, J., Iversen, C., Kerr, D.C., Remoundos, Y.: Linear cellular antenna switch for highly integrated SOI front-end. IEEE Intl. SOI Conf. (2007)
92. Single-pole four-throw high-power switch RF1450 Data sheet. http://www.rfmd.com/pdfs/1450DS.pdf

# Part II
# Physics of Modern SemOI Devices

# Silicon-based Devices and Materials for Nanoscale FETs

Francis Balestra

**Abstract** Silicon on insulator (SOI)-based devices seem to be the best candidates for the ultimate integration of integrated circuits on silicon down to nm structures. An overview of the performance of nanoscale FETs, based on innovative concepts, technologies and device architectures, is addressed. The impact of alternative channel materials, source-drain contacts and multi-gates/channels on the performance and physical mechanisms in ultimate MOSFETs is highlighted. The interest of multi gate emerging and beyond-CMOS nanodevices for long term applications, based on nanowires or small slope switch structures for ultra low power applications is also presented. Finally, the flexibility of multi-gate and nanowire SOI structures for boosting the scalability and performance of DRAM, SRAM and flash memories is outlined.

## 1 Introduction

The minimum critical feature size of the elementary CMOS devices (physical gate length of the transistors) will drop from 45 nm in 2010 to 9 nm in 2024. In the sub-10 nm range, Beyond-CMOS devices will certainly play an important role and could be integrated on CMOS platforms in order to pursue integration down to nm structures. Si will remain the main semiconductor material in a foreseeable future, but the needed performance improvements for the end of the ITRS Roadmap [1, 2] will lead to a substantial enlargement of the number of materials, technologies and device architectures.

F. Balestra (✉)
IMEP-LAHC/Sinano Institute, CNRS-Grenoble INP Minatec, 3 Parvis Louis Neel, 38016, Grenoble, France
e-mail: balestra@minatec.grenoble-inp.fr

Silicon on insulator (SOI)-based devices seem to be the best candidates for the ultimate integration of integrated circuits on silicon [3, 4]. An overview of the main advantages of SOI for the Nanoelectronics of the next two decades is presented in this paper. Nanoscale CMOS, emerging and beyond-CMOS nanodevices, based on innovative concepts, technologies and device architectures, are addressed. The flexibility of the SOI structure and the possibility to realize new architectures allow to obtain optimum electrical properties for both low power and high performance circuits. These transistors are also very interesting for memory applications. The performance and physical mechanisms are investigated in single- and multi-gate/multi-channel thin film MOSFETs. A comparison between the performance of Si, Ge and III–V semiconductors MOSFETs and nanowires on insulator is given. The impact of strains in the channel and metallic Schottky source-drain architectures is discussed. The interest of advanced emerging and beyond-CMOS nanodevices on SOI for long term applications, based on nanowires or small slope switch structures are also presented. Finally, the advantage of using multi-gate and nanowire architectures for scaling down DRAM, SRAM and non-volatile memories is outlined.

## 2 Nanoscale CMOS

### 2.1 New Channel Materials for Ultimate CMOS

As simple scaling of silicon CMOS becomes increasingly complex and expensive there is considerable interest in increasing performance by using strained channels, which can improve carrier mobility and drive current in a device. The combination of the advantages of SOI structures for improving the electrostatic control and of strained semiconductors is very promising. An original method for strained SOI (sSOI) is based on thin SiGe buffer layers relaxed by $He^+$ ions implantation and post-implantation annealing. If the Si cap layer thickness lies below a critical thickness, an induced strain is transferred during the relaxation of the SiGe layer. The wafer is then cleaved with hydrogen implantation according to the Smart-Cut® process and the stack layer is bonded to an oxidized handle wafer. Split-CV mobility measurements have shown enhanced electron mobility values as high as 1,200 $cm^2$/Vs [5].

Using this biaxial tensile stress as starting material, uniaxial tensile strain can be obtained by lateral strain relaxation of patterned structures (Fig. 1a, b). Figure 1c shows the transfer characteristics of a strained and an unstrained nanowire-nFET. The NWs have a square cross-section of 40 × 40 $nm^2$ and a length of 3 μm. The subthreshold slope of both devices is between 70 and 80 mV/dec. Both devices have a very low off-current and a high $I_{on}/I_{off}$ ratio. The on-current of the uniaxially strained device is enhanced by a factor of 2.5 due to uniaxial tensile strain. The inset of Fig. 1c shows a plot of $I_d/gm^{1/2}$ against $V_g$ for a strained and unstrained device. The slope of the linear part of the curves, related to the carrier mobility, shows a mobility enhancement due to uniaxial tensile strain of a factor 2.3. For maximum performance enhancement due to lateral strain relaxation sSOI devices must have a large length to width ratio [6].

Strain engineering is also useful for mobility enhancement for Si film thickness in the sub-10 nm range [7], which will be needed for ultimate MOSFETs. A similar enhancement of electron mobility in 3.5 nm SOI devices under biaxial and uniaxial tensile strain has been obtained. The electron mobility is also enhanced in 2.3 nm Si layer under uniaxial tensile strain (Fig. 2), and the hole mobility increases in 2.5 nm film under uniaxial compressive strain.

The impact of the strain for ultra-short and ultra-thin FDSOI is shown in Fig. 3. A 40% gain of the driving current at given $I_{off}$ and 18 nm $L_g$ is obtained for various $t_{Si}$. The mobility increases down to ultra-thin film and up to high $N_{inv}$ and a similar ballisticity is shown for SOI and sSOI. The improvement of the injection velocity (about 35%) in sSOI nMOS, which is demonstrated down to 2.5 nm, explains the substantial $I_{on}$ increase exemplified in this figure [8].

## 2.2 Metallic Schottky Source/Drain MOSFETs

As CMOS technology is entering the decananometre era, the contact resistance associated with the silicide/silicon interface is identified as one of the biggest challenges to solve in order to preserve the driving current. Therefore, source/drain engineering takes an increasing importance in the development of leading edge CMOS generations. In order to further pursue down-scaling of MOSFETs in the sub-32 nm gate length range, one alternative is to implement metallic S/D

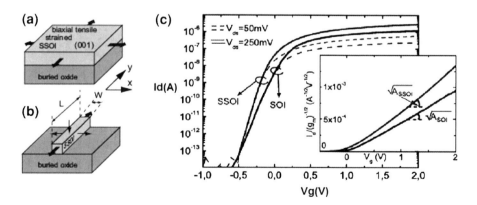

**Fig. 1** a 50 nm sSOI with 1 GPa biaxial tensile stress is used as starting material. b Uniaxial tensile strain is obtained by lateral strain relaxation of patterned structures. c Transfer characteristics of two NW-FETs, one fabricated on SOI and one on uniaxial sSOI. The channel length of both devices is L = 3 μm and the gate oxide thickness $t_{ox}$ = 5 nm. The *inset* shows the $I_d/g_m^{1/2}$-plot for the devices. The slope of the linear region is related to carrier mobility [6]

**Fig. 2** Electron mobility in 2.3-nm ultra-thin-body MOSFET under ⟨110⟩ uniaxial strain [7]

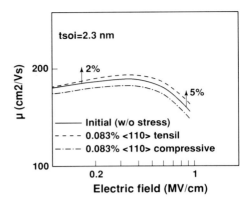

**Fig. 3** $I_{on}$–$I_{off}$ plot showing the impact of tensile strain obtained by smart-cut process from strained Si/relaxed $Si_{0.8}Ge_{0.2}$ [8]

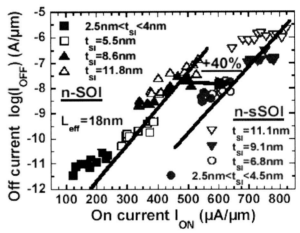

combined to a dopant segregation technique at reduced thermal budget. The expected benefit is to considerably reduce the specific contact resistance of the metal/semiconductor junction while keeping activated dopants sharply localized at the interface. The efficiency of dopant segregated Schottky contacts has been demonstrated for n- and p-MOS [9–11]. It has been shown that a sub-100 meV barrier can be obtained consistently with the boron or arsenic pile-up observed at the PtSi/Si or YbSi interfaces, respectively. Figure 4 shows the substantial increase of $I_d$ in n-channel MOSFETs realized with YbSi contacts and As segregation at the interface with the Si channel [11].

## 2.3 Multi-gate and Multi-channel Devices

Multi-gate MOSFETs realized on thin films are the most promising devices for the ultimate integration of MOS structures due to the volume inversion or volume accumulation in the thin layer (for enhancement- and depletion-type devices,

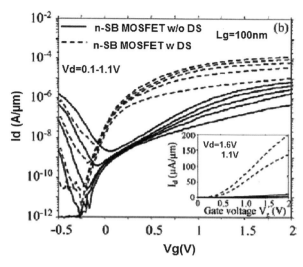

**Fig. 4** Comparison of $I_d(V_g)$ in n-channel Schottky (YbSi) MOSFET with and without As segregation [11]

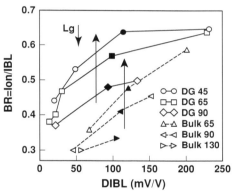

**Fig. 5** Ballisticity ratio at $V_g = V_d = V_{dd}$ versus DIBL. Filled symbols represent transistors with the nominal gate length for the high-performance MOSFET of each technology node [14]

respectively) [12], leading to an increase of the number and the mobility of electrons and holes as well as driving current (additional gain in performance in a loaded environment), optimum subthreshold swing and the best control of short channel effects and off-state current, which is the main challenge for future nanodevices due to the power consumption crisis and the need to develop green/sustainable ICs.

A comparison of short channel effect for bulk, single gate SOI, single gate SON with thin buried oxide and double gate MOSFETs has been done. For sub-30 nm gate length, the advantages of SOI structures, thin film and BOX as well as multi-gate architectures has been demonstrated for reducing the threshold voltage variations versus $L_g$ [13].

The on-current $I_{on}$ of the MOSFET is limited to a maximum value $I_{BL}$ that is reached in the ballistic transport regime. Figure 5 reports the self-consistent Monte Carlo simulation of the ballistic ratio BR = $I_{on}/I_{BL}$ versus Drain-Induced-Barrier-

Lowering, which is one of the main short channel effects, showing that one can increase the BR by scaling the gate length, thus increasing the longitudinal field at the source, but this comes at the cost of a larger DIBL. For a given DIBL, an increased ballisticity is obtained for low doping double gate SOI devices [14].

The transfer characteristics of several multiple-gate (1–4 gates) MOSFETs, calculated using the 3D Schrödinger–Poisson equation and the Non-Equilibrium Green's Function formalism for the ballistic transport or MC simulations, have shown similar trends. The best performance (drain current, subthreshold swing) is outlined for the four-gates [Quadruple Gate or Gate-All-Around (GAA)] structure [15, 16].

In decananometer MOSFETs, gate underlap is also a promising solution in order to reduce the DIBL effect. Figure 6 presents the variations of the driving current $I_{on}$, the subthreshold current $I_{sub}$ and the gate direct tunneling current $I_{gdt}$ versus gate underlap [17]. The on-current is only slightly affected by the gate underlap whereas the leakage currents are substantially reduced due to a decrease in DIBL and drain to gate tunneling current. A reduction of the effective gate capacitance $C_g$ for larger underlap values at similar $I_{on}$ has also been shown. This reduction of $C_g$ leads to a decrease in the propagation delay and power.

Another important issue deals with the possible interest in alternative channel materials. Figure 7 is a plot of the driving current as a function of gate length for Si, strain Si, Ge and GaAs n-channel double-gate MOSFETs, for a given $I_{off}$ according to ITRS for the next generations of HP ICs. The four materials are chosen in their optimized orientation and short channel effects and access resistance are included. When neglecting source to drain tunneling, Ge and GaAs devices lead to the best $I_d$ for sub-10 nm gate length. However, when source-drain tunneling is included, only 2G strained Si MOSFETs satisfy the need for the sub-decananometre range [18].

In order to reach very high performance at the end of the roadmap, multi-bridge-channel MOSFETs or multi-channel MOSFETs (MCFETs) present very high driving current larger than those of GAA devices and exceeding the ITRS requirements (Fig. 8) [19].

**Fig. 6** $I_{on}$, subthreshold ($I_{sub}$) and gate direct tunneling ($I_{gdt}$) currents as a function of gate underlap [17]

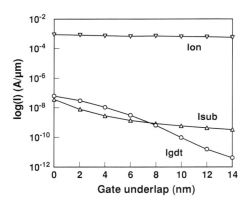

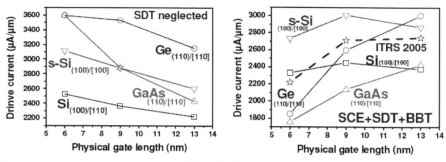

**Fig. 7** Driving current of n-channels 2G MOSFETs obtained by simulations as a function of channel materials (Si, strain Si, Ge, GaAs) and source-drain tunneling, taking into account short channel effects and band-to-band tunneling [18]

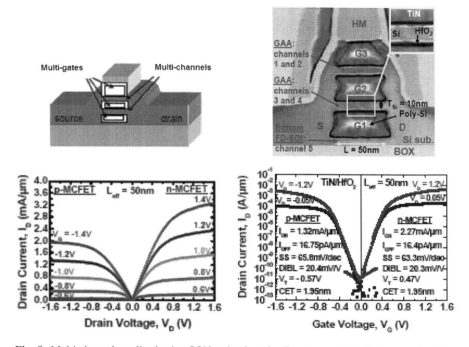

**Fig. 8** Multi-channels realized using SON technology leading to very high drain currents with a very good control of leakage currents [19]

## 2.4 Variability and Low Frequency Noise in Bulk, Single and Multi-gate Structures

Silicon on insulator devices can also be interesting for reducing the variability in decananometre MOSFETs, which also represents a major challenge at the end of the roadmap. Sources responsible for local and inter-die threshold voltage ($V_t$)

variability in undoped ultra-thin FDSOI MOSFETs with a high-k/metal gate stack have been experimentally discriminated. Charges in the gate dielectric and/or TiN gate workfunction fluctuations are determined as major contributors to the local $V_t$ variability and it is found that SOI thickness variations have a negligible impact down to $t_{Si} = 7$ nm. Moreover, $t_{Si}$ scaling is shown to limit both local and inter-die $V_t$ variability induced by gate length fluctuations. The highest matching performance ever reported for 25 nm gate length MOSFETs has been achieved (Pelgrom coefficient Avt = 0.95 mV μm), demonstrating the effectiveness of the undoped ultra-thin FDSOI architecture in terms of $V_t$ variability control [20].

High immunity in $V_t$ variability due to the possible use of undoped or lightly doped ultra-thin SOI has also been demonstrated in multi-gate SOI devices (Fig. 9), which are also more tolerant to line edge roughness induced variability [21].

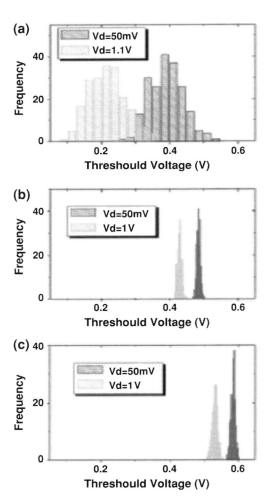

**Fig. 9** Impact of intrinsic parameter fluctuations due to random discrete dopant **a** bulk 45 nm, **b** 32 nm UTB SOI ($10^{15}$ cm$^{-3}$ to 7 nm Si), **c** 22 nm DG ($10^{15}$ cm$^{-3}$ to 10 nm Si) [21]

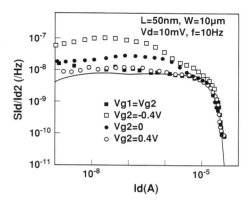

**Fig. 10** Normalized drain current noise spectral density of a double gate N-MOSFET for different back gate voltages. *Solid line*: $SV_g \times (G_m/I_d)^2$ for double gate mode [22]

Low frequency noise is an important issue especially for analog and RF applications. In a double gate MOSFET, the application of a back gate voltage can lead to the volume inversion and to a screening, reducing dynamic carrier trapping in the gate oxides (reduction of carrier number fluctuations) and the impact of these trapped carriers on mobility fluctuations (reduction of Coulomb scattering fluctuations). This phenomenom induces a reduction of the low frequency noise in multi-gate devices exemplified in Fig. 10 [22].

## 3 Nanoscale Beyond-CMOS Devices

The objectives in this domain are to explore the horizon beyond conventional CMOS, or beyond Moore, in order to overcome possible downscaling and performance limits of CMOS, or increase circuits functionalities in the next two decades. In this field, SOI structures can also be considered as a platform for pursuing the integration in the nanoworld.

### 3.1 Nanowires

It has been shown previously that multi-gate architectures based on the concept of volume inversion are very promising in order to overcome the number of challenges for CMOS integration down to at least decananometre gate length devices. GAA semiconductor nanowires can be seen as the ultimate integration of these innovative nanodevices and present very interesting properties down to the sub-10 nm range.

The combination of strain effects with NWs can lead to very high performance ICs for the end of the roadmap. Top-down bended GAA nanowires have been fabricated in order to improve carrier mobility and driving current [23]. A bending induced by thermal oxidation in suspended nanowires has been demonstrated. A maximum tensile strain is obtained in the middle of the wire. Tensile stresses from 200 Mpa to 2 Gpa can build-up in these suspended nanowires. The substantial enhancement of electron mobility in these structures has been shown.

Three-dimensional integration of nanowires is a promising solution to boost the driving current and keep a very low off-state current. Using SON process, three stacked GAA sub-15 nm nanowires (Fig. 11), with 100 nm length and 1.8 nm EOT (high k/metal gate stack) has shown extremely high driving current and very low leakage currents ($I_{on}/I_{off}$: 6.5 mA/μm to 27 nA/μm for nMOS—$I_{on}/I_{off}$: 3.3 mA/μm-0.5 nA/μm for pMOS). A new optional independent gate nanowire with a FinFET-like structure named ΦFET has also been reported, leading to new design flexibility (Fig. 11). These 3D structures can be extended to a combined vertical and lateral integration for logic, memories and NEMS applications (Fig. 11). The 3D-NWFET and ΦFET, compared to a co-processed FinFET, relaxes the channel width requirement for a targeted DIBL and improves transport properties (Fig. 12). ΦFET also exhibits significant performance boosts compared to Independent-Gate FinFET (IG-FinFET): a two-decade smaller $I_{off}$ current and a lower subthreshold slope (82 mV/dec. instead of 95 mV/dec.). This highlights the better scalability of 3D-NWFET and ΦFET compared to FinFET and IG-FinFET [24, 25].

## 3.2 Small Slope Switches

Even though the aggressive scaling will continue to play an important role in the future nanoelectronics, new technology drivers, such as ultra-low power and new functionality will open alternative ways for future high performance systems.

One interesting class of Beyond- or non-CMOS devices are the small slope switches. A small slope electronic switch is defined as a solid-state semiconductor device showing a value of the subthreshold slope smaller than the 60 mV/decade

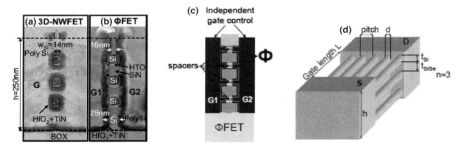

Fig. 11 a, b and c Cross-sectional pictures of 3D-NWFET and ΦFET (spacers are introduced to obtain stacked nanowires with independent gate operation), d lateral and vertical integration of 3D nanowires [24, 25]

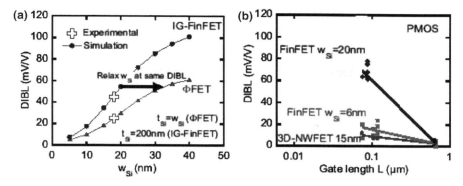

**Fig. 12** a Simulated and measured DIBL versus Si width $w_{Si}$ for ΦFET and IG-FinFET in the 3TFET configuration; L = 50 nm; $t_{Si}$ is the Si thickness. b DIBL versus gate length L for 3D-NWFET compared with FinFET. These *plots* show the strong reduction of SCEs in 3D NWs and ΦFET [24]

limit for a conventional MOSFET, set by the Boltzmann distribution at room temperature. The smaller the value of S, the more abrupt the transition from the off- to the on-state and closer the switch to the ideal case. Benefits of small slope switches are the ultra-low standby power due to a very low $I_{off}$ but also the high-speed potential and dynamic power savings, since less power is drawn per transition when the subthreshold slope is more abrupt. Development of small slope switches requires exploration of new physical principles for very abrupt off–on transition, such as impact ionization, band-to-band tunneling, electro-mechanical effect, or ferroelectric gate. It is worth noting that all these small slope device architectures can be implemented as extensions of advanced silicon CMOS or by hybridization of silicon CMOS with other compatible technologies (SiGe, nanowire, CNT, etc.).

A very promising device is the Tunnel FET [26–28], which is a gated p–i–n diode with a gate over the intrinsic region; it exploits the gate controlled electron tunneling from the valence band of the p-region to the conduction band of the i-region for reversed biases, resulting in very abrupt off-on transition. They have been reported on Si, III–V and CNT structures.

Multi-gates TFETs have recently shown to present very interesting performance. Figure 13 shows the transfer characteristics of various device architectures: single gate SOI, stress at the source junction, high k dielectric, double gate, oxide aligned to i-region, very short channel. The short channel 2G device lead to a strong enhancement of the driving current [29].

Figure 14 demonstrates the interest of using ultra thin semiconductor film for reducing the subthreshold swing of TFETs. The atomistic full-band Schrödinger–Poisson simulation shows that S decreases when the InAs film thickness is reduced down to 2 nm. Better performances are obtained for double-gate MOSFETs and GAA nanowires (S down to 20 mV/decade) compared with single gate MOSFETs. On the other hand, InAs film thicknesses lager than 5 nm lead to the best

ON-current, which shows that DG MOSFET or GAA nanowire are the best candidates for high TFET performance [30].

## 4 Nanoscale Memories

It is becoming difficult for memories to be scaled down. Indeed, traditional embedded DRAM requires a complicated stack capacitor or a deep trench capacitor in order to obtain a sufficient storage capacitance in smaller cells. This leads to more process steps and thus less process compatibility with logic devices. Capacitor-less 1T-DRAM or Floating body cells have shown promising results. The operation principle is based on excess holes which can be generated either by impact ionization or by Gate-Induced Leakage Current in partially-depleted SOI MOSFETs. The GIDL current is due to band-to-band tunneling and occurs in accumulation leading to a low drain current writing and reduced power consumption together with a high speed operation. However, conventional PD SOI MOSFETs require high channel doping to suppress short-channel effects, which induces a degradation in retention characteristics. In order to overcome this problem, DG-FinDRAMs have been proposed showing superior memory characteristics [31].

Conventional floating-gate flash memory has also scaling difficulties due to nonscaling of gate-insulator stack and inefficient hot carrier injection processes at sub-50 nm gate dimensions. Back-floating gate flash memory overcomes these limitations by decoupling read and write operations and independent positioning and/or sizing of the storage element (back-floating gate) under the Si channel. The charge in the back gate affects the field and the potential at the bottom interface and thus changes the threshold voltage of the device. The back-floating gate is charged by applying $-10$ V to the source, the drain and the front gate simultaneously, and the charges are removed from the back floating gate (erasing) with the same method but with a bias of $+10$ V [32].

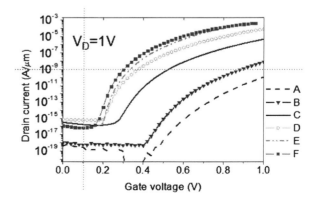

**Fig. 13** (A) single gate SOI ($L_g = 100$ nm, 3 nm $SiO_2$), (B) stress (4 GPa) at the source junction, (C) high-k dielectric, (D) double gate, (E) oxide aligned to i-region, (F) 2G with very short channel (30 nm)

FinFlash, SiN-based FinFlash and Nano-crystals Finflash on SOI (Fig. 15) present also many advantages: increased drive current and improved access time, reduced SCEs leading to integration capability down to at least 20 nm, higher

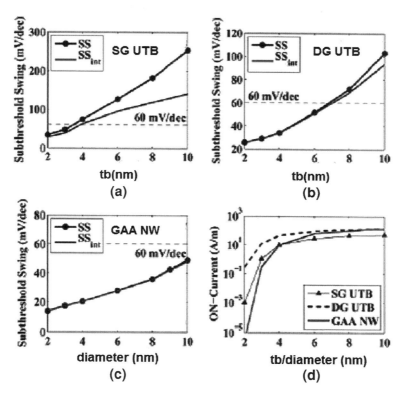

**Fig. 14** Atomistic full-band Schrödinger–Poisson solver for SG, DG, GAA NW InAs TFETs (*SS* total swing, $SS_{int}$ intrinsic swing without the contribution of the capacitive ratio) [30]

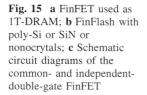
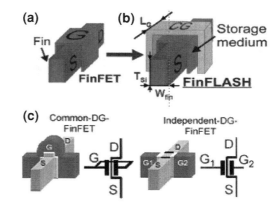

**Fig. 15 a** FinFET used as 1T-DRAM; **b** FinFlash with poly-Si or SiN or nonocrytals; **c** Schematic circuit diagrams of the common- and independent-double-gate FinFET

**Fig. 16** 1T nonvolatile SRAM on Si NW SONOS: NVM property with O/N/O gate stacks and SRAM functionality using latch phenomena of floating body in Si NW [35]

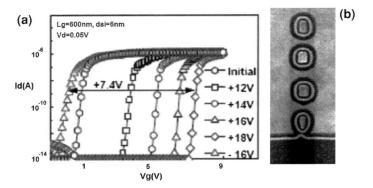

**Fig. 17 a** $I_d(V_g)$ characteristics showing large programming windows and fast programming times in three channels 3D SONOS nanowires, **b** 3D nanowire integration [36]

number of stored electrons and thus less fluctuation problems, compatibility with future multi-gate CMOS (SoC, embedded) [33].

Concerning SRAM cells, a reduction of leakage current and an enhancement of read and write noise margins have been successfully demonstrated by introducing the independent double-gate FinFETs (Fig. 15) [34].

A 1-transistor non volatile SRAM realized on Si nanowire with a SONOS gate stack has been realized (Fig. 16). This nanostructure shows non-volatile memory properties with the O/N/O stack and SRAM functionalities using the latch phenomena of the floating body in the Si nanowire [35].

A 3D integration of SONOS nanowires has also been recently demonstrated, showing large programming windows and fast programming times (Fig. 17) [36].

Multi-gate SOI MOSFET and nanowire structures seem therefore to be very promising for extending the scalability and the performance of DRAM, SRAM and flash memories.

# 5 Conclusion

An overview of the performance of nanoscale FETs, based on innovative concepts, technologies and device architectures, has been addressed. The impact of alternative channel materials, source-drain contacts and multi-gates/channels on the performance and physical mechanisms in ultimate MOSFETs has been highlighted. The interest of multi-gate emerging and beyond-CMOS nanodevices for long term applications, based on nanowires or small slope switch structures, has also been presented. Finally, the flexibility of multi-gate and nanowire SOI structures for boosting the scalability and performance of DRAM, SRAM and flash memories has been outlined.

**Acknowledgments** This work was partially supported by the European Network of Excellence NANOSIL (FP7) devoted to Silicon-based Nanodevices.

# References

1. ITRS Roadmap. http://www.itrs.net/
2. ENIAC Strategic Research Agenda. http://www.eniac.eu/web/SRA/local_index.php
3. Balestra, F.: SOI Devices. Wiley Encyclopedia of Electrical and Electronics Engineering. Wiley, New York (1999)
4. Cristoloveanu, S., Balestra, F.: Silicon on insulator: technology and devices. In: Morkoc, H. (ed.) Advanced Semiconductor and Organic Nano-techniques. Academic Press, New York (2003)
5. Driussi, F., Esseni, D., Selmi, L., Schmidt, M., Lemme, M.C., Kurz, H., Buca, D., Mantl, S., Luysberg, M., Loo, R., Nguyen, D., Reiche, M.: Fabrication, characterization and modeling of strained SOI MOSFETs with very large effective mobility. In: Proceeding of the European Solid State Device Research Conference (ESSDERC), Munich, p. 315 (2007)
6. Feste, S.F., Knoch, J., Habicht, S., Buca, D., Zhao, Q.T., Mantl, S.: Performance enhancement of uniaxially-tensile strained Si NW-nFETs fabricated by lateral strain relaxation of SOI. In: Proc. ULIS, Juliech, p. 109 (2009)
7. Uchida, K., Zednik, R., Lu, C.H.: Experimental study of uniaxial and biaxial strain effects on carrier mobility in bulk and ultra-thin-body SOI MOSFETs. In: Proc. IEDM, San Francisco, p. 229 (2004)
8. Barral, V., Poiroux, T., Andrieu, F., Buj-Dufournet, C., Faynot, O., Ernst, T., Brevard, L., Fenouillet-Beranger, C., Lafond, D., Hartmann, J.M., Vidal, V., Allain, F., Daval, N., Cayrefourcq, I., Tosti, L., Munteanu, D., Autran, J.L., Deleonibus, S.: Strained FDSOI CMOS technology scalability down to 2.5 nm film thickness and 18 nm gate length with a TiN/HfO$_2$ gate stack. In: Proc. IEDM, Washington, p. 61 (2007)
9. Larrieu, G., Dubois, E., Valentin, R., Breil, N., Danneville, F., Dambrine, G., Raskin, J.P., Pesant, J.C.: Low temperature implementation of dopant-segregated band-edge metallic S/D junctions in thin-body SOI p-MOSFETs. In: Proc. IEDM, Washington, p. 147 (2007)
10. Zhang, Z., Qiu, A., Liu, R., Ostling, M., Zhang, S.L.: Schottky-barrier height tuning by means of ion implantation into preformed silicide films followed by drive-in anneal. Electron Device Lett. **28**, 565 (2007)
11. Larrieu, G., Yarekha, D.A., Dubois, E., Breil, N., Faynot, O.: Arsenic-segregated rare-earth silicide junctions: reduction of Schottky barrier and integration in metallic n-MOSFETs on SOI. IEEE Electron Device Lett. **30**, 1266–1268 (2009)
12. Balestra, F., Cristoloveanu, S., Benachir, M., Brini, J.: Double-gate silicon-on-insulator transistor with volume inversion: a new device with greatly enhanced performance. IEEE Electron Device Lett. **8**, 410 (1987)
13. Skotnicki, T.: Silicon-on-nothing devices. International Summer School MIGAS'2008, Grenoble (2008)

14. Eminente, S., Esseni, D., Palestri, P., Fiegna, C., Selmi, L., Sangiorgi, E.: Enhanced ballisticity in nano-MOSFETs along the ITRS roadmap: a Monte Carlo study. In: Proc. IEDM, San Francisco, p. 609 (2004)
15. Bescond, M., Néhari, K., Autran, J.L., Cavassilas, N., Munteanu, D., Lannoo, M.: 3D quantum modeling and simulation of multi-gate nanowire MOSFETs. In: Proc. IEDM, San Francisco, p. 617 (2004)
16. Saint Martin, J., Bournel, A., Dollfus, P.: Comparison of multiple-gate MOSFET architectures using Monte-Carlo simulation. In: Proc. ULIS, Bologna, p. 61 (2005)
17. Bansal, A., Paul, B.C., Roy, K.: Impact of gate underlap on gate capacitance and gate tunneling current in 16 nm DGMOS devices. In: Proceedings of the IEEE International SOI Conference, Charleston, p. 94 (2004)
18. Raphay, Q., Clerc, R., Ghibaudo, G., Pananakakis, G.: Impact of source-to-drain tunnelling on the scalability of arbitrarily oriented alternative channel material nMOSFETs. Solid-State Electron. **52**, 1474–1481 (2008)
19. Bernard, E., Ernst, T., Guillaumot, B., Vulliet, N., Barral, V., Maffini-Alvaro, V., Andrieu, F., Vizioz, C., Campidelli, Y., Gautier, P., Hartmann, J., Kies, R., Delaye, V., Aussenac, F., Poiroux, T., Coronel, P., Souifi, A., Skotnicki, T., Deleonibus, S.: Novel integration process and performances analysis of low standby power (LSTP) 3D multi-channel CMOSFET (MCFET) on SOI with metal/high-K gate stack. In: Proceedings of Symposium on VLSI Technology, Honolulu, p. 16 (2008)
20. Weber, O., Faynot, O., Andrieu, F., Buj-Dufournet, C., Allain, F., Scheiblin, P., Foucher, J., Daval, N., Lafond, D., Tosti, L., Brevard, L., Rozeau, O., Fenouillet-Beranger, C., Marin, M., Boeuf, F., Delprat, D., Bourdelle, K., Nguyen, B.-Y., Deleonibus, S.: High immunity to threshold voltage variability in undoped ultra-thin FDSOI MOSFETs and its physical understanding. In: Proc. IEDM, San Francisco, p. 245 (2008)
21. Cheng, B., Roy, S., Brown, A.R., Millar, C., Asenov, A.: Evaluation of intrinsic parameter fluctuations on 45, 32 and 22 nm technology node LP N-MOSFETs. In: Proc. ESSDERC, Edinburgh, p. 47 (2008)
22. Balestra, F., Jomaah, J.: Performance and new effects in advanced SOI devices and materials. Microelectron. Eng. **80**, 230–240 (2005)
23. Moselund, K.E., Dobrosz, P., Olsen, S., Pott, V., De Michielis, L., Tsamados, D., Bouvet, D., O'Neill, A., Ionescu, A.M.: Bended Gate-All-Around Nanowire MOSFET: a device with enhanced carrier mobility due to oxidation-induced tensile stress. In: Proc. IEDM, Washington, p. 191 (2007)
24. Dupre, C., Hubert, A., Becu, S., Jublot, M., Maffini-Alvaro, V., Vizioz, C., Aussenac, F., Arvet, C., Barnola, S., Hartmann, J.-M., Garnier, G., Allain, F., Colonna, J.-P., Rivoire, M., Baud, L., Pauliac, S., Loup, V., Chevolleau, T., Rivallin, P., Guillaumot, B., Ghibaudo, G., Faynot, O., Ernst, T., Deleonibus, S.: 15 nm-diameter 3D stacked nanowires with independent gates operation: FET. In: Proc. IEDM, San Francisco, p. 549 (2008)
25. Ernst, T., Duraffourg, L., Dupre, C., Bernard, E., Andreucci, P., Becu, S., Ollier, E., Hubert, A., Halte, C., Buckley, J., Thomas, O., Delapierre, G., Deleonibus, S., de Salvo, B., Robert, P., Faynot, O.: Novel Si-based nanowire devices: will they serve ultimate MOSFETs scaling or ultimate hybrid integration. In: Proc. IEDM, San Francisco, p. 745 (2008)
26. Baumgärtner, H.: Performance enhancement of vertical tunnel FET with SiGe in the p+ layer. Jpn. J. Appl. Phys. **43**, 4073–4078 (2004)
27. Wang, P.-F., Hilsenbeck, K., Nirschl, T., Oswald, M., Stepper, C., Weis, M.: Complementary tunneling transistor for low power application. Solid-State Electron. **48**, 2281–2286 (2004)
28. Boucart, K., Ionescu, A.M.: Double-gate tunnel FET with high-k gate dielectric. IEEE Trans. Electron. Devices (2007). doi:10.1109/ESSDER.2006.307718
29. Boucart, K., Riess, W., Ionescu, A.M.: Asymmetrically strained all-silicon tunnel Fets featuring 1 V operation. In: Proc. ESSDERC'2009, Athens (2009)
30. Luisier, M., Klimeck, G.: Atomistic full-band design study of InAs band-to-band tunneling field-effect transitor. IEEE Electron Device Lett. **30**, 602–604 (2009)

31. Tanaka, T.: Scalability study on a capacitorless 1T-DRAM: from single-gate PD-SOI to double-gate FinDRAM. In: Proc. IEDM, San Francisco (2004). doi:10.1109/IEDM. 2004.1419332
32. Song, K.-W., Jeong, H., Lee, J.-W.: 55 nm capacitor-less 1T DRAM cell transistor with non-overlap structure. In: Proc. IEDM, San Francisco (2008). doi:10.1109/IEDM.2008.4796818
33. Nowak, E., Boquet, M., Perniola, L., Ghibaudo, G., Molas, G., Jahan, C., Kies, R., Reimbold, G., De Salvo, B., Boulanger, F.: New physical model for ultra-scaled 3D nitride-trapping non volatile memories. In: Proc. IEDM, San Francisco (2008). doi:10.1109/IEDM.2008.4796750
34. Endo, K., O'uchi, S.-I., Ishikawa, Y.: Enhancing SRAM cell performance by using independent double-gate FinFET. In: Proc. IEDM, San Francisco (2008). doi:10.1109/IEDM. 2008.4796833
35. Ryu, S.-W., Han, J.-W., Moon, D.-I., Choi, Y.-K.: One-transistor nonvolatile SRAM (ONSRAM) on silicon nanowire SONOS. In: Proc. IEDM, Washington, p. 27.5.1–27.5.4 (2009)
36. Hubert, A., Nowak, E., Tachi, K., Maffini-Alvaro, V., Vizioz, C., Arvet, C., Colonna, J.-P., Hartmann, J.-M., Loup, V., Baud, L., Pauliac, S., Delaye, V., Carabasse, C., Molas, G., Ghibaudo, G., De Salvo, B., Faynot, O., Ernst, T.: A stacked SONOS technology, up to 4 levels and 6 nm crystalline nanowires, with gate-all-around or independent gates (Φ-Flash), suitable for full 3D integration. In: Proc. IEDM, Washington, p. 637 (2009)

# FinFETs and Their Futures

N. Horiguchi, B. Parvais, T. Chiarella, N. Collaert, A. Veloso,
R. Rooyackers, P. Verheyen, L. Witters, A. Redolfi, A. De Keersgieter,
S. Brus, G. Zschaetzsch, M. Ercken, E. Altamirano, S. Locorotondo,
M. Demand, M. Jurczak, W. Vandervorst, T. Hoffmann and S. Biesemans

**Abstract** FinFET is a promising device structure for scaled CMOS logic/memory applications in 22 nm technology and beyond, thanks to its good short channel effect (SCE) controllability and its small variability. Scaled SRAM and analog circuit are promising candidates for finFET applications and some demonstrations for them are already reported. On the other hand, for finFETs production, quite a lot of process challenges are required due to difficult fin/gate patterning in the 3D structure, conformal doping to fin and high access resistance in extremely thin body, etc. The fin/gate patterning can be improved by optimization of patterning stack, patterning scheme and etch chemistry. Alternative doping techniques show good conformal doping in 3D structure in finFETs. High access resistance is reduced by junction optimization and strain booster technique.

## 1 Introduction

VLSI performance is improved by planar device scaling according to Moore's law for past decades. However, for 22 nm technology and beyond, it is very difficult to meet ITRS targets, especially short channel effect (SCE) control and suppression

---

N. Horiguchi (✉), B. Parvais, T. Chiarella, N. Collaert, A. Veloso,
R. Rooyackers, P. Verheyen, L. Witters, A. Redolfi, A. De Keersgieter, S. Brus,
M. Ercken, E. Altamirano, S. Locorotondo, M. Demand, M. Jurczak, T. Hoffmann
and S. Biesemans
IMEC VZW, Leuven, Belgium
e-mail: horiguc@imec.be

G. Zschaetzsch and W. Vandervorst
Instituut voor Kern-en Stralingsfysica, K. U. Leuven, Leuven, Belgium

of device performance variability, by planar device scaling. Therefore, there is a strong interest in new device architectures. FinFET is one of the most promising device structures for scaled CMOS/memory applications thanks to its good SCE controllability and its small variability. However, the future of finFET is not so obvious due to difficult patterning (3D structure), difficult doping on fin structure and high access resistance in extremely thin body, etc. This paper describes the finFET device characteristics, promising finFET applications and process challenges for the finFET future.

## 2 FinFET Device Characteristics

FinFET has a so-called double gate structure (Fig. 1). Thanks to the double gate structure, finFET has a good channel potential control by gate electrode [1].

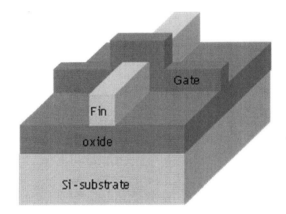

**Fig. 1** Device structure of finFET. FinFET has a double gate structure, which enables better SCE control

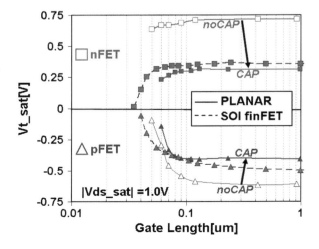

**Fig. 2** $V_t$ roll-off comparison between finFETs and planar FETs. FinFETs show better SCE control and comparable $V_t$ at long gate length without hik capping such as LaO or AlO

**Fig. 3** Matching characteristics of planar FETs, SOI finFETs (FF) and bulk finFETs (BFF). SOI FF shows better matching thanks to low channel dopant concentration

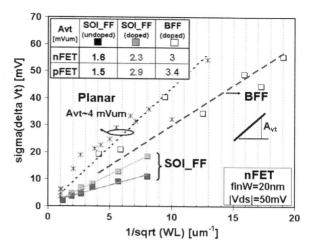

**Fig. 4** Ion-Ioff in finFETs and planar FETs. Ion difference between nMOS and pMOS is smaller due to fin sidewall crystal orientation

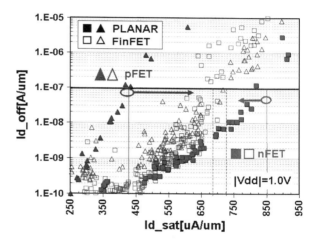

This enables better SCE control in finFET than that in single gate planar FET. This is a big advantage in finFET for the device scaling.

Figure 2 shows the Vt roll-off comparison between finFET and planar FET [2, 3]. FinFETs show better SCE control as expected. Another advantage in finFET is the threshold voltage (Vt) control in high-k/metal gate stack. No high-k capping layer ($La_2O_3$ for nMOS or $Al_2O_3$ for pMOS) is necessary for finFETs to obtain comparable Vt at long gate length with planar FETs that use dual capping layers. This enables simpler process integration in finFETs.

Thanks to the good channel potential control by the double gate structure, high channel dopant concentration is not necessary for finFET to suppress the SCE. The low channel dopant concentration makes the finFET device variability smaller than planar FET in scaled technology. This can be seen in the good matching

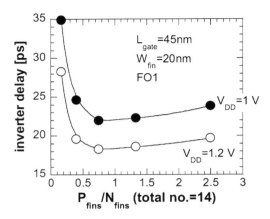

**Fig. 5** Ring oscillator (RO) delay versus fin number ratio between nfinFET and pfinFET. Minimum RO delay is achieved around the fin number ratio of 1

characteristics (Fig. 3). Once the channel dopant concentration is increase, the matching degradation can be seen clearly even in finFET device structures.

In finFETs fabricated on conventional (100)/<110> substrate, the Ion difference between nMOS and pMOS becomes smaller (Fig. 4) since the (110)/<110> fin sidewall orientation increases pMOS mobility significantly and decreases nMOS mobility slightly [4]. The smaller Ion difference in finFET impacts directly on the optimum NMOS/PMOS ratio in a CMOS layout (Fig. 5) [5].

## 3 Promising Applications for finFETs

### 3.1 SRAM

Device variability is a big concern for further scaling of planar bulk 6T-SRAM. FinFET has a chance to break through the barrier by its good SCE controllability and its good matching. Superior Vt-matching, Vdd-scalability and read current improvement in finFET SRAMs are reported [6]. Sub 0.1 $\mu m^2$ finFET 6T-SRAMs are demonstrated already (Figs. 6, 7) [7–9].

### 3.2 Analog

The good SCE control in finFETs is beneficial not only for digital but also for analog applications. By optimizing the process, finFETs show higher voltage gain and higher transconductance than planar FETs (Fig. 8) [2]. Several analog circuits, such as operational amplifiers, comparators and VCOs [10], have been demonstrated.

**Fig. 6** SEM picture of 0.089 μm² finFET 6T-SRAM

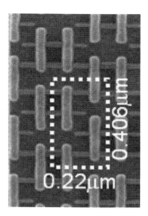

**Fig. 7** Butterfly curves in 0.089 μm² SRAM cell using finFETs

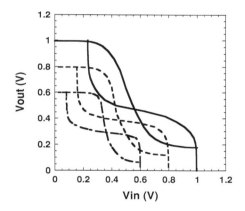

## 4 Process Challenges for finFET Future

### 4.1 Fin Patterning

Although finFETs have an intrinsically small variability due to the good channel potential control by the double gate structure and the low channel dopant concentration, their variability is dramatically influenced by fin/gate patterning. And the patterning becomes more difficult in the dens finFET structures such as scaled SRAMs. Therefore, the precise fin/patterning control is a key for finFET future.

Figure 9 shows fin patterning improvement in finFET 6 transistor (6T)-SRAM by using thinner BARC/photo resist/hard mask stack and a new etch chemistry/sequence [11]. The thinner resist/hard mask reduces the aspect ratio (Total thickness of fin, hard mask and BARC/photo resist/fin width or fin space) for the fin patterning. This improves the fin cd/profile control.

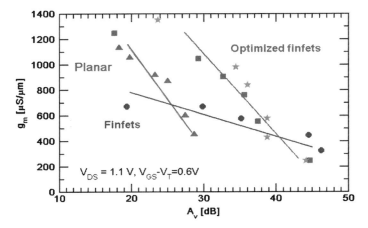

**Fig. 8** Voltage gain ($A_v$) versus transconductance ($g_m$) in planar FETs and finFETs. Optimized finFETs show higher $A_v$ and higher $g_m$

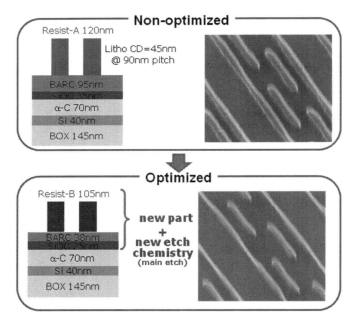

**Fig. 9** Fin patterning optimization for dense SRAM. Thinner photo resist/hard mask and new etch chemistry improve fin cd/line edge roughness (LER) control

In addition to that, the etch by-product control is important for fin cd/profile control. Generally speaking, if too much non-volatile by-product is created during fin etch and deposited fin side wall, the narrow fin cd control becomes very difficult. This has to be taken into account when the etch chemistry is chosen.

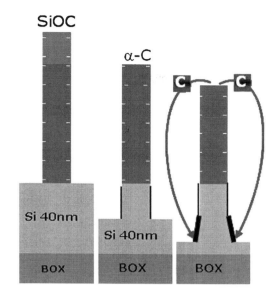

**Fig. 10** Possible mechanism for fin profile degradation by carbon containing etch by-product. If α-C is exposed to the etch chemistry, a carbon containing etch by-product is created. The by-product attached to fin foot. This causes the fin footing. To avoid the fin footing, precise hard mask thickness control during fin etch is necessary

To get a straight fin profile in the narrow fin, the remaining hard mask thickness control during fin etch is important. Figure 10 shows the fin profile degradation by carbon containing etch by-product. If α-C hard mask is exposed to the etch chemistry (after full consumption of SiOC hard mask), the carbon containing by-product is created and deposited at fin foot. This causes the fin footing. On the other hand, if the remaining SiOC hard mask is too thick, fin cd loss during SiOC hard mask removal becomes significant. This makes the fin cd control difficult, especially in dense fin patterns. To avoid these problems, the remaining SiOC hard mask thickness has to be controlled precisely during fin etch. Figure 11 shows the comparison between optimized and non-optimized fin etch from the remaining SiOC hard mask thickness control view point. The optimized fin etch gives better fin profile.

The line edge roughness (LER)/line width roughness (LWR) improvements in the fin by using a spacer defined patterning are reported [12, 13].

## 4.2 Gate Patterning

In the gate etch in finFETs, both the gate cd control and the gate end of line control are important, especially dense finFET patterns such as 6T-SRAM. The standard single mask/single etch approach for the gate patterning can make the gate cd on target but it suffers from the end of line shortening (Fig. 12). The end of line control is improved by introducing a double mask/double exposure (gate line and gate cut) (Fig. 12) [11]. Further end of line control becomes possible to introduce double mask/double exposure and double etch scheme (so-called double patterning

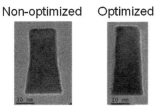

**Fig. 11** Fin profile comparison between optimized and non-optimized carbon containing etch by-product control during fin etch. In optimized fin etch, SiOC hard mask is fully consumed at the end of fin etch. Therefore, $\alpha$-C is not exposed to the etch chemical and the carbon containing by-product is not created

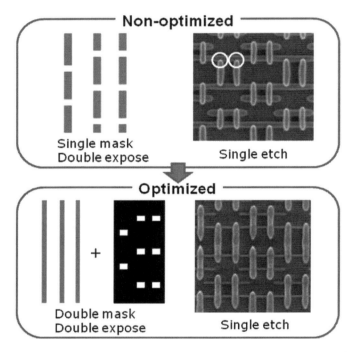

**Fig. 12** Gate patterning optimization. Gate/fin overlap control (*circles* in non-optimized pictures) is improved by double mask/double exposure

scheme). For the double patterning, SiN hard mask is introduced and BARC/resist thickness is optimized (Fig. 13). In the double gate patterning scheme, gate line patterns are printed and transferred to SiN hard mask by gate line etch. Then the gate cut patterns are printed and etched (Fig. 14). Figure 15 shows the gate double patterning results in a sub-0.1 $\mu m^2$ finFET 6T-SRAM. The end of line gap and shape control becomes more robust.

The spacer defined patterning shows better LER/LWR also in the gate [12, 13].

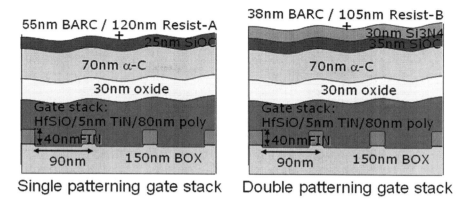

**Fig. 13** Gate stacks for the single gate patterning and the double gate patterning. The SiN hard mask is introduced for the double patterning and the BARC/photo resist thickness is optimized

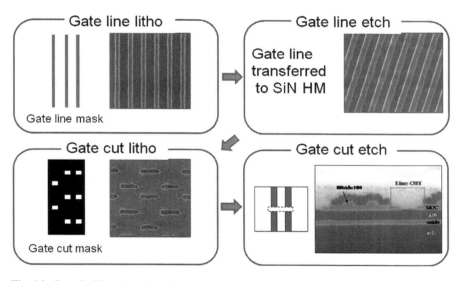

**Fig. 14** Gate double patterning scheme

## 4.3 Conformal Doping

For planar devices, the junction depth and the sheet resistance are the figure of merit for junction. For finFETs, due to their 3D structure, "conformality" is added as a new figure of merit for junction. It is very difficult to make the conformal doping profile with the conventional ion implantation, since only a small angle implant ($\sim 10°$) is acceptable for dense finFET structures and the small angle implant causes low ion-incorporation efficiency on fin sidewalls due to reflection, self-sputtering and/or the geometry effect, etc. [14]. Several alternative doping techniques are investigated to improve the conformality (plasma doping and vapor phase doping). By using BF3 plasma doping, a good conformal p-doping is reported (Fig. 16) [15]. The advantage of conformal doping in pMOS finFET extension is demonstrated (Fig. 17) [15]. On the other hand, n-doping for finFET suffers from less conformality or low dopant activation (high sheet resistance). The improvement of n conformal doping is a key for finFET future.

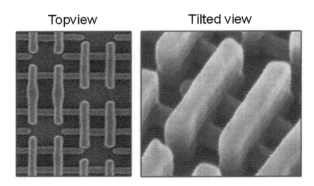

**Fig. 15** *Top view* and *tilted view* SEM pictures of gate double patterning in sub 0.1 μm² finFET 6T-SRAM

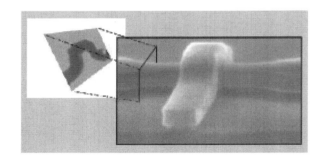

**Fig. 16** Conformal junction in finFET by using BF3 plasma doping

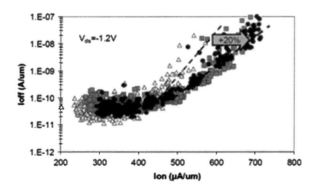

**Fig. 17** pFinFET ion improvement by BF3 plasma doping. (*Open* ion implantation reference, *filled* BF3 plasma doping)

## 4.4 Strain Engineering

The mobility enhancement on finFETs is a key technology to meet the high drive current requirement. Several strain engineering techniques have been reported, such as SSOI and tCESL (Fig. 18) [16], SiGe SD SEG (Figs. 19 and 20) [17–19] and SiC SD SEG. Recently good drive current is reported by combining junction optimization and SiGe SD SEG [19]. Limited performance gain due to small substrate volume is an issue in finFETs.

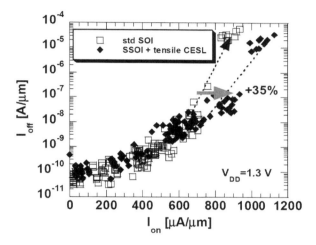

**Fig. 18** nFinFET performance boast by SSOI and tCESL. 35% performance improvement is achieved

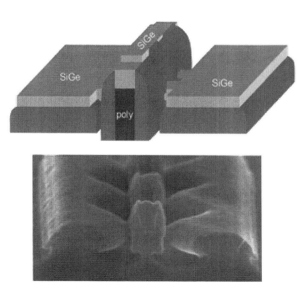

**Fig. 19** *Schematic* and *top-view* SEM picture of SiGe SD epi on pFin FET

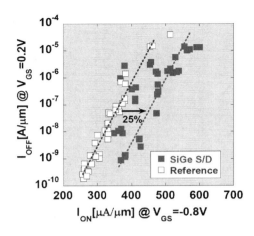

**Fig. 20** FinFET performance improvement by SiGe SD epi. 25% performance improvement is achieved thanks to access resistance reduction and mobility enhancement by SiGe SD epi

## 5 Summary

FinFETs show clear advantages in SCE control and device variability as compared to planar FETs. Thanks to the advantages, scaled SRAM and analog circuit are promising for finFET applications. Several sub-0.1 mm$^2$ 6T SRAMs are demonstrated with reasonable SNM already. Both better voltage gain and better transconductance are reported by optimized finFETs. On the other hand, for finFETs production, quite a lot of process challenges are required due to difficult fin/gate patterning in the 3D structure, conformal doping to fin and high access resistance in extremely thin body, etc. The fin/gate patterning can be improved by optimization of patterning stack, patterning scheme and etch chemistry. Alternative doping techniques show good conformal doping in 3D structure in finFETs. High access resistance is reduced by junction optimization and strain boaster technique.

## References

1. Wong, H.S.P., Frank, D.J., Solomon, P.M., et al.: Device design considerations for double-gate, ground-plane, and single-gated ultra-thin SOI MOSFET's at the 25 nm channel length generation. Int. Electron Dev. Meeting Tech. Digest. 407–410 (1998)
2. Parvais, B., Mercha, A., Collaert, N., et al.: The device architecture dilemma for CMOS technologies: opportunities & challenges of Finfet over planar mosfet. In: International Symposium on VLSI Technology, Systems and Applications, pp. 80–81 (2009)
3. Chiarella, T., Witters, L., Mercha, A., et al.: Migrating from planar to FinFET for further cmos scaling: SOI or bulk? In: ESSDERC Conference Proceedings, pp. 85–88 (2009)
4. Rudenko, T., Kilchytska, V., Collaert, N., et al.: Carrier mobility in undoped triple-gate FinFET structures and limitations of its description in terms of top and sidewall channel mobilities. IEEE Trans. Electron Dev. **55**, 3532–3541 (2008)
5. Collaert, N., von Arnim, K., Rooyackers, R., et al.: Low-voltage 6T FinFET SRAM cell with high SNM using HfSiON/TiN gate stack, fin widths down to 10 nm and 30 nm gate length. In: International Conference on IC Design and Technology, pp. 59–62 (2008)

6. Merelle, T., Curatola, G., Nackaerts, A., et al.: First observation of FinFET specific mismatch behavior and optimization guidelines for SRAM scaling. In: International Electron Devices Meeting Technical Digest, pp. 241–244 (2008)
7. Veloso, A., Demuynck, S., Ercken, M., et al.: Demonstration of scaled 0.099 $\mu m^2$ FinFET 6T-SRAM cell using full-field EUV lithography for (Sub-)22 nm node single-patterning technology. In: International Electron Devices Meeting Technical Digest, pp. 301–304 (2009)
8. Guillorn, M., Chang, J., Pyzyna, A., et al.: Trigate 6T SRAM scaling to 0.06 $\mu m^2$. In: International Electron Devices Meeting Technical Digest, pp. 961–963 (2009)
9. Horiguchi, N., Demuynck, S., Ercken, M., et al.: High yield sub-0.1 $\mu m^2$ 6T-SRAM cells, featuring High-k/Metal-Gate, Finfet devices, double gate patterning, a novel fin etch strategy, full-field EUV lithography and optimized junction design & layout. In: Symposium on VLSI Technology, pp. 23–24 (2010)
10. Wambacq, P., Mercha, A., Scheir, K., et al.: Advanced planar bulk and multigate CMOS technology: analog-circuit benchmarking up to mm-wave frequencies. In: International Solid-State Circuits Conference, pp. 528–529 (2008)
11. Ercken, M., Altamirano-Sanchez, E., Baerts, C., et al.: Challenges in using optical lithography for the building of a 22 nm node 6T-SRAM cell. Microelectronic Eng. **87**, 993–996 (2010)
12. Choi, Y.K., Lindert, N., Xuan, P., et al.: Sub-20 nm CMOS FinFET technologies. In: International Electron Devices Meeting Technical Digest, pp. 421–424 (2001)
13. Rooyackers, R., et al.: (2006) Doubling or quadrupling MuGFET fin integration scheme with higher pattern fidelity, lower CD variation and higher layout efficiency. In: International Electron Devices Meeting Technical Digest, pp. 168–171 (2001)
14. Mody, J., Duffy, R., Eyben, P., et al.: Experimental studies of dose retention and activation in fin field-effect-transistor-based structures. J. Vac. Sci. Technol. B **28**, 1 (2010)
15. Lenoble, D., Anil, K.G., De Keersgieter, A., et al.: Enhanced performance of PMOS MUGFET via integration of conformal plasma-doped source/drain extensions. In: Symposium on VLSI Technology, pp. 168–169 (2006)
16. Collaert, N., Rooyackers, R., Clemente, F., et al.: Performance enhancement of MUGFET devices using super critical strained-SOI (SC-SSOI) and CESL. In: Symposium on VLSI Technology, pp. 176–177 (2006)
17. Verheyen, P., Collaert, N., Rooyackers, R., et al.: 25% Drive current improvement for p-type multiple gate FET (MuGFET) devices by the introduction of recessed $Si_{0.8}Ge_{0.2}$ in the source and drain regions. In: Symposium on VLSI Technology, pp. 194–195 (2005)
18. Kavalieros, J., Doyle, B., Datta, S., et al.: Tri-gate transistor architecture with high-k Gate dielectrics, metal gates and strain engineering. In: Symposium on VLSI Technology, pp. 62–63 (2006)
19. Chang, C.Y., Lee, T.L., Wann, C., et al.: A 25-nm gate-length FinFET transistor module for 32 nm node. In: International Electron Devices Meeting Technical Digest, pp. 293–296 (2009)

# Ultrathin Body Silicon on Insulator Transistors for 22 nm Node and Beyond

T. Poiroux, F. Andrieu, O. Weber, C. Fenouillet-Béranger,
C. Buj-Dufournet, P. Perreau, L. Tosti, L. Brevard and O. Faynot

**Abstract** Ultrathin body silicon on insulator technology has acquired during the last few years a significant maturity. Since it offers breakthroughs in terms of electrostatic control and variability, this technology is today a serious alternative to bulk for the coming technology generations. This technology is indeed likely to be scaled down to the 10 nm range. In addition, several performance booster options can be efficiently implemented to reach very high transistor performances. Furthermore, gate stacks allowing the design of low, medium and high threshold voltage transistors are identified and their integration is demonstrated. Finally, the use of an ultrathin buried oxide together with an implanted back-plane brings additional flexibility in terms of threshold voltage adjustment, and ensures the efficiency of conventional power management techniques based on back-biasing, even in very aggressively scaled devices.

## 1 Introduction

In aggressively scaled technology generations, high-channel doping are required in conventional bulk planar MOSFETs to control short channel effects. Trade-offs have then to be found with mobility degradation and increased power consumption. As mentioned in the short term grand challenges of the International Technology Roadmap for Semiconductors 2009 edition [1], the use of such high doping levels in narrow and short transistors leads to an increase of the threshold voltage

T. Poiroux (✉), F. Andrieu, O. Weber, C. Fenouillet-Béranger,
C. Buj-Dufournet, P. Perreau, L. Tosti, L. Brevard and O. Faynot
CEA-LETI/Minatec, 17 rue des Martyrs, 38054 Grenoble Cedex 9, France
e-mail: thierry.poiroux@cea.fr

variation, which limits strongly the supply voltage downscaling. The main source for this variability is the statistical fluctuation of the number of dopant atoms in the transistor channel, and this relative variation is indeed increased as the device dimensions are scaled down. On the other hand, Fully-Depleted (also called ultrathin body) Silicon On Insulator (FDSOI) technologies, in which the transistor electrostatic integrity is ensured by the thinness of the body, provide a unique solution at the device level, since the transistors can be designed with an undoped channel. Thus, these technologies are among the most promising ways to extend the CMOS scaling below 22 nm. In this chapter, we bring experimental results to answer some key questions about the potential of FDSOI technology: How far can we scale FDSOI devices? Is the film thickness control an issue? How can we implement multi-threshold voltage transistors? How can we boost the performance of FDSOI transistors? In the first part, we review the basic equations that govern the electrostatics of FDSOI transistors, in order to provide a device design guideline, and we detail the fabrication process of such devices. The second part is dedicated to the scalability assessment of this technology, in terms of electrostatic control and variability. Finally, in the third part, we illustrate some possible ways to improve the FDSOI transistors performance.

## 2 Ultrathin Body Silicon on Insulator Technology

The physical behaviour of ultrathin body transistors is different from that of conventional bulk devices. In particular, the body of these devices is generally left undoped, which leads to some specificity in terms of long channel threshold voltage and bi-dimensional electrostatics. In this part, we first derive the basic expressions governing the ultrathin body transistor electrostatics in order to give some simple device design rules. In the second part, we describe the fabrication of such transistors.

### 2.1 Device Design

In ultrathin body transistors, the threshold voltage cannot be adjusted through the doping of the channel. Indeed, this approach would require very high doping levels, in the $10^{19}$ cm$^{-3}$ range. Thus, these transistors can be advantageously designed without intentional channel doping. Then, the threshold voltage of long channel devices has to be adjusted with the gate material workfunction and the bi-dimensional behavior of short channel transistors is mainly controlled by the thinness of the body. Here, we first give simple equations describing the transistor electrostatics. From these expressions, we can derive the gate workfunctions required to address the different technology options. The third paragraph is dedicated to some considerations about source and drain engineering.

### 2.1.1 Electrostatics of Ultrathin Body Transistors

If we consider first a long channel ultra thin body transistor with a thick buried oxide (Fig. 1), the null transverse field condition at the back interface of the body translates into a flat electrostatic potential profile in the subthreshold regime.

Thus, the inversion charge in this regime is given by

$$Q_{inv} = q n_i t_{Si} \exp\left(\frac{\psi}{k_B T / q}\right) \quad (1)$$

where $q$ is the elementary charge, $n_i$ is the carrier concentration in intrinsic silicon, $t_{Si}$ is the silicon film thickness, $\psi$ the electrostatic potential in the body, defined as the difference between the Fermi energy level and the midgap energy level over $q$, $k_B$ is the Boltzmann constant and $T$ is the temperature.

In addition, it can be shown that the inversion charge dependence on the gate voltage changes from an exponential dependence to a linear one for a threshold value of

$$Q_{inv}^{th} = C_{ox} \frac{k_B T}{q} \quad (2)$$

with $C_{ox} = \varepsilon_{ox}/t_{ox}$ the gate dielectric capacitance per unit area. Thus, the threshold value of the electrostatic potential is

$$\psi^{th} = \frac{k_B T}{q} \ln\left[\frac{C_{ox} k_B T / q}{q n_i t_{Si}}\right] \quad (3)$$

It follows that the threshold voltage is given by

$$V_{th} = \Delta\phi_{mi} + \frac{k_B T}{q} \ln\left[\frac{C_{ox} k_B T / q}{q n_i t_{Si}}\right] \quad (4)$$

In (4), $\Delta\phi_{mi}$ is the difference between the gate workfunction and the silicon midgap energy level. Note that in this equation, the quantum confinement effect that occurs in the subthreshold regime is not accounted for. This confinement is due to the potential well formed by the silicon film located between the gate

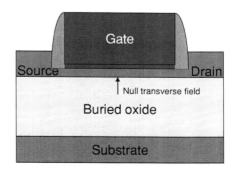

**Fig. 1** Representation of a long channel FDSOI transistor with a thick buried oxide

dielectric and the buried oxide. A first-order approximation of the long channel threshold voltage including quantum confinement is obtained by adding to the previous equation a term corresponding to the energy shift of the first sub-band

$$V_{th} = \Delta\phi_{mi} + \frac{k_B T}{q} \ln\left[\frac{C_{ox} k_B T/q}{q n_i t_{Si}}\right] + \frac{\hbar^2 \pi^2}{2 q m_{conf} t_{Si}^2} \quad (5)$$

where $\hbar$ is the reduced Planck constant and $m_{conf}$ is the carrier confinement effective mass in the considered sub-band. This latter term represents a few millivolts for silicon films thicker than 5 nm and can generally be neglected.

Let us consider now a short channel transistor. The electrostatic potential within the channel results from the capacitive coupling of the channel with the different electrodes. A good electrostatic control of the channel by the gate electrode requires maximizing the following ratio

$$r_{\text{short channel}} = \frac{C_{GC}}{C_{GC} + C_{SC} + C_{DC} + C_{BC}} \quad (6)$$

where $C_{GC}$, $C_{DC}$, $C_{SC}$ and $C_{BC}$ are the capacitive couplings between the channel and the gate, the drain, the source and the substrate, respectively.

In short ultrathin body transistors, the centroid of the inversion charge in the subthreshold regime is located at the interface between the body and the buried oxide (back interface) because of the bi-dimensional electrostatic effects. Thus, the gate to channel capacitance is composed of the gate dielectric capacitance in series with the body capacitance. This capacitance increases consequently when the body is thinned down. Furthermore, the source/drain to channel coupling depends also strongly on the film thickness. As a consequence, the body thickness is the first-order parameter that governs the short channel effects in this technology.

From a deeper analysis of the bi-dimensional electrostatics, an accurate dependence of the short channel effects with the device geometry can be derived. In particular, the Drain Induced Barrier Lowering (DIBL) is given by

$$\left.\frac{dV_{th}}{dV_{DS}}\right|_{V_{DS}=0\,V} = -\frac{1/2}{\cosh(L/2\lambda) - 1} \quad (7)$$

where $\lambda$ is the scaling characteristic length of the technology, given by [2]

$$\lambda = \sqrt{\frac{\varepsilon_{Si}}{\varepsilon_{ox}} t_{Si} \left(t_{ox} + \frac{\varepsilon_{ox}}{\varepsilon_{Si}} \frac{t_{Si}}{2}\right)} \quad (8)$$

From these expressions, we find that a ratio of about 4 between the minimum channel length and the silicon film thickness is required to ensure DIBL values below 100 mV/V.

### 2.1.2 Gate Workfunction

As the channel of FDSOI transistor is undoped, the threshold voltage has to be adjusted by choosing the appropriate metal gate. From the above considerations, we can determine the metal gate workfunction needed to reach the desired off-state current for the nominal gate length transistor (Table 1).

Contrary to bulk technologies, in which the co-integration of band-edge metal gates is a serious challenge, FDSOI technology requires gate workfunctions close to the midgap (±150 meV). In particular, low and high threshold voltages can be achieved for both n- and pMOSFETs with only two metal gates with workfunctions of 4.45 and 4.75 eV.

Alternately, a channel counter-doping can be used to reach low threshold voltage values with a midgap metal gate [3]. Furthermore, the use of a thin buried oxide together with an adequate ground-plane doping located below this thin BOX can be envisaged to obtain multiple-$V_{th}$ with a reduced number of metal gates. This approach will be detailed in the last part of this chapter.

### 2.1.3 Elevated Source and Drain

Since the FDSOI transistors have to be designed with a thin body, elevated source/drain are mandatory in order to achieve acceptable series resistances. Thus, a selective epitaxial growth has to be performed after the formation of an offset spacer and before the extension implantation. That way, series resistance as low as 200 Ω μm (Fig. 2) can be achieved on nMOSFETs with a thin body of 10 nm below the gate. These series resistance can be controlled below 500 Ω μm down to film thicknesses of 3 nm. It should be noted that the presented experimental results are obtained with standard 1,050°C spike anneal. The use of advanced laser or flash anneal can further reduce these parasitic resistance values.

## 2.2 Device Fabrication and Electrical Results

In the first paragraph of this part, we describe the process flow used for the fabrication of FDSOI transistors. Then, we give some electrical results illustrating the maturity level of this technology and we finally describe the possible use of

| Table 1 Required metal gate workfunctions for various threshold voltage options | Device type | Threshold voltage options (eV) | | |
|---|---|---|---|---|
| | | Low-$V_{th}$ | Medium-$V_{th}$ | High-$V_{th}$ |
| | nMOSFET | 4.45 | 4.6 | 4.75 |
| | pMOSFET | 4.75 | 4.6 | 4.45 |

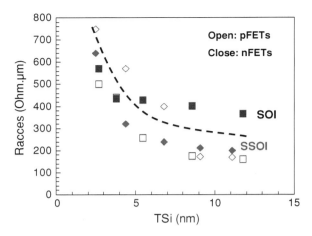

**Fig. 2** Experimental series resistance values extracted on n and pMOSFETs as a function of their channel thickness (TSi)

ultrathin buried oxide in order to improve the device performance and add some design flexibility.

### 2.2.1 Process Description

The fabrication of FDSOI transistors relies on the use of SOI substrates composed of a thin silicon active layer, with a typical initial thickness of 12 nm, that is electrically isolated from the mechanical substrates by a buried oxide with either a standard thickness of 145 nm (thick-BOX option) or a thickness down to 10 nm (Ultrathin Body and BOX, UTB$^2$ option). Thickness dispersions ($6\sigma$) as low as 1 and 0.5 nm are already reached respectively for the silicon film and for the buried oxide on commercially available 300 mm wafers [4].

Then, the silicon film is thinned down to a thickness close to the desired value for the body by sacrificial oxidation. As stated above, a 22 nm node technology requires a body thickness of about 7 nm. This thinning step does not degrade the film thickness dispersion. Next, active areas are patterned, and the isolation between devices can be made by a mesa or with a Shallow Trench Isolation.

The channel being left undoped, the high-k metal gate stack is then deposited and patterned. As explained in the previous part, we can reach well balanced $V_{th}$ of ±0.4 V for nMOSFET and pMOSFET at gate lengths of 30 nm with a single midgap metal gate, such as TiN. This gate stack is completed with a polycrystalline silicon layer and patterned. Afterwards, a thin nitride offset spacer is formed on the silicon source/drain extension regions and a selective epitaxial growth of silicon is performed to control the series resistance. Extensions are then implanted before the formation of a second oxide/nitride spacer. Source/drain high doping and salicidation are carried out, followed by a standard middle and back-end process.

**Fig. 3** Transmission electron microscopy cross-section of a 32 nm gate length FDSOI transistor fabricated with a 20 nm thick buried oxide

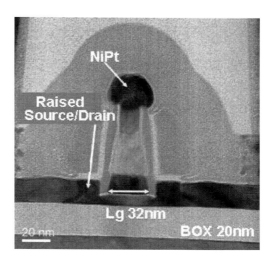

Figure 3 shows the transmission electron microscopy cross section of a 32 nm gate length transistor fabricated on a 20 nm thin buried oxide [5].

When compared with a conventional planar bulk technology, this thin film technology presents a few specificities. First, no channel nor pocket implant is required, which allows simplifications of the process integration in case of standard buried oxide, by suppressing lithography and implantation steps. Other specificities of thin film technologies are the need for a careful control of the silicon consumption during gate and spacer patterning and the requirement of source/drain epitaxial regrowth. The latter process step is no longer specific to thin film technologies, since SEG is used nowadays in bulk technologies in order to incorporate SiGe or Si:C elevated source/drain as performance boosters.

### 2.2.2 Technology Maturity and Benchmark

This technological process leads to very reproducible results, with transistor electrical characteristics well reproduced from one run to the others, and low off-state currents achieved on very aggressive gate length devices.

In terms of dynamic performance, FDSOI ring oscillator trade-offs between propagation delays and dynamic power consumption exhibit significant gains over their bulk counterpart [6] (Fig. 4). These gains can be explained first by a better subthreshold slope in the saturation regime. Thus, at a given off-state current, the gate overdrive is higher for the FDSOI technology, which leads to a larger drive current. This gain is all the more important that the supply voltage ($V_{DD}$) is scaled down, the relative gate overdrive gain over bulk being increased at low $V_{DD}$. In addition, thanks to the presence of the buried oxide, FDSOI transistors exhibit lower junction capacitances.

**Fig. 4** Ring oscillator propagation delays (Tp) as a function of the switching energy ($E_d$). FDSOI technology offers a 20% gain over bulk

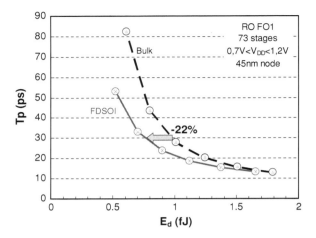

### 2.2.3 Ultrathin Buried Oxide Option

As already mentioned, the buried oxide can be made ultrathin. The advantages of such an option are an improved electrostatic control of the device and the possibility to tune its threshold voltage. This will be detailed afterwards.

In terms of technological integration, the use of a very thin buried oxide requires some specific optimizations. In particular, all cleaning steps have to be optimized in order to limit the buried oxide consumption in the isolation areas, which could lead to short-cuts between the source/drain electrodes and the substrate [7].

In addition, the electrical use of the back bias will be much more efficient if the substrate is locally doped beneath the buried oxide. An implantation step is then required at the beginning of the process to create these doped regions, generally called ground-plane or back-plane.

## 3 Scalability of Ultrathin Body Technology

Two aspects have to be considered regarding the scalability of MOSFETs, namely the electrostatic control of ultrashort channels and the technology potential in terms of threshold voltage mismatch.

### 3.1 Electrostatics

As shown in the first part of this chapter, the most efficient way to improve the electrostatic control of FDSOI transistors is to decrease the body thickness. Then, for a given technology node, a trade-off has to be found with the series resistance. As shown in Fig. 5, DIBL values as low as 80 mV/V can be reached on 20 nm gate length FDSOI transistors with a body thickness of 6 nm [8]. For such a film thickness, the series resistances are about 200 Ω μm.

Furthermore, calibrated simulations show that the planar FDSOI architecture can be scaled down to the 10 nm technology node by using a ground-plane below a buried oxide thinned down to 10 nm, while maintaining a body thickness of 6 nm [9]. In that case, the fringing fields originating from the source and drain electrodes are killed in the ground-plane region, and the capacitive coupling between the channel and the source/drain through the buried oxide is significantly reduced, which leads to an improved electrostatic control by the gate.

## 3.2 Variability

One of the main advantages of the FDSOI technology compared to bulk is the excellent threshold voltage variability that can be obtained (Fig. 6). These outstanding matching factor ($A_{Vt}$) values are reached thanks to the use of an undoped channel, which suppresses the Random Dopant Fluctuation (RDF), strongly detrimental on bulk devices. This yields a record $A_{Vt}$ measured on this architecture by different groups [4, 10, 11] and can be directly translated into the SRAMs stability and their minimum supply voltage ($V_{min}$). Ratios between the Static Noise Margin and its dispersion (SNM/$\sigma_{SNM}$) higher than 6 down to supply voltages as low as 0.7 V have thus been demonstrated on a 32 nm FDSOI technology [7].

In addition, it can be shown that the introduction of thin films does not induce any additional dispersion. In particular, the film or BOX thickness variations or the use of back bias, in the case of ultrathin buried oxide technology, do not degrade significantly the $V_{th}$-variability [4, 7].

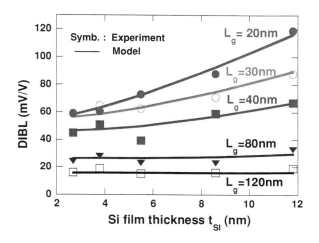

**Fig. 5** Drain Induced Barrier Lowering (DIBL) effect as a function of the channel thickness for various gate lengths

**Fig. 6** Benchmark of the matching factor versus the gate length for bulk or FDSOI devices. See references in [4]

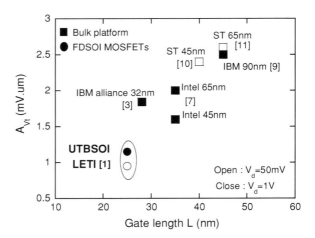

Finally, in these transistors, the matching is mainly governed by the high-k/metal gate stack related variability (especially for long channel devices) and by the Line Edge Roughness (LER) of the gate (especially for short transistors). By the way, it must be noticed that the LER of the transistor width (W) is not so influent in the planar FDSOI architecture as for FinFETs or TriGate transistors because FDSOI devices exhibit a the much smaller $V_{th}(W)$ dependence.

## 4 Performance of Ultrathin Body Transistors

In this last part, we discuss about the possibility to fabricate FDSOI transistors with various threshold voltages, and about the possible implementation of carrier transport boosters.

### 4.1 Threshold Voltage Tuning and Power Management

Addressing several threshold voltages in FDSOI transistors requires the integration of several gate stacks, according to the equations established in part Sect . 15.2. The integration of several gate stacks to fabricate n and pMOSFETs from low-$V_{th}$ to high-$V_{th}$ has already been demonstrated. Interestingly, the gate stack allowing low-$V_{th}$ nMOSFET corresponds to high-$V_{th}$ pMOSFET and vice versa.

In addition, if an ultrathin buried oxide is used together with a back-plane, the transistor performance can be tuned thanks to back-biasing. Indeed, the coupling between the back bias ($V_B$) and the channel through a 10 nm thin BOX is larger than in a 32 nm bulk technology. The $g = |dV_{th}/dV_B|$ factor in thin BOX devices is in between 70 and 170 mV/V, depending of back-plane doping and bias. In the worst case (w/o back-plane), a 10% on-state current ($I_{ON}$) increase or half a decade off-state current ($I_{OFF}$) reduction at $V_{DD} = 0.9$ V and $|V_B| = 0.3$ V (Fig. 7) can be achieved thanks to standard power management techniques. It must be pointed out that this back bias range of 0.3 V is the typical one used on bulk circuits. However, contrarily to bulk, the presence of the BOX in FDSOI permits to get rid of the junction leakages (except the possible junction leakage between back-planes). Consequently, a back bias over 0.3 V is certainly possible on FDSOI, which would lead to wider $I_{ON}$–$I_{OFF}$ tuning capabilities.

That way, the back bias can also be exploited in order to achieve multiple $V_{th}$ with a single midgap gate [12]. Figure 7 shows that at $|V_B| = V_{DD} = 0.9$ V, the $I_{OFF}$ can be shifted by 3.7 decade (in both directions) and the $I_{ON}$ by 70%. These values are perfectly compatible with multi-$V_{th}$ specifications.

Finally, it is worth noticing that the Forward Back Bias (FBB) effect is even higher at low $V_{DD}$. This can be explained by the fact that the $I_{ON}$ sensibility to $V_B$ is higher when low gate overdrives are considered.

## 4.2 Implementation of Transport Boosters

The previously mentioned FDSOI performance can be further improved by the use of on-state current boosters in order to target high performance applications. It should be noted that local stressors are often more effective on FDSOI than on bulk at a given geometry because of the mechanical properties of the buried $SiO_2$, less stiff than Si [13]. The term of "local stressors" refers to techniques like tensile or compressive Contact Etch Stop Layers (t- or c-CESL), embedded SiGe

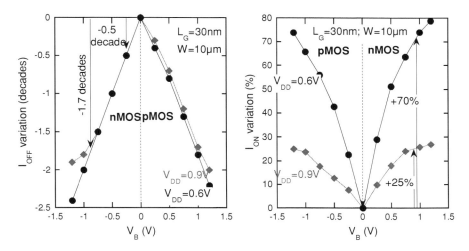

**Fig. 7** Off-state and on-state current sensitivity to the back bias for n and pMOSFETs with a 10 nm buried oxide thickness

source/drain (e-SiGe S/D) or Stress Memorization Techniques (SMT), for which the strain is localized at the edges of the channel and laterally transferred. On the contrary, other techniques like strained SOI (sSOI), (110), 45° rotated substrates or SiGe channel are called "global" or "wafer-level" boosters.

We have demonstrated that a tensile strain can be induced in the transistor channel both in the longitudinal (i.e. the source/drain one) and in the transverse directions (in case of mesa or STI with a local divot) by using a Physical Vapor Deposited TiN gate, since this material has a high compressive intrinsic stress. Highly non uniform and anisotropic strain configuration is thus obtained for transistors with aggressive dimensions. Figure 8 present the assessment of the impact of different boosters on the FDSOI architecture at the 65 or 45 nm design pitches. We can notice that strained-SOI (sSOI) is one of the most promising stressor for nMOSFETs, inducing $I_{ON}$ improvements of 25% for wide devices that increases up to 50% for $W = 50$ nm narrow transistors [14, 15]. Although the real efficiency of sSOI is still to be demonstrated for 22 nm related active area dimensions, these preliminary results let us predict a better scalability for sSOI than for t-CESL or SMT. Furthermore, the sSOI compatibility and quasi-additivity with t-CESL [15] is already proven on nMOSFETs.

Concerning pMOSFETs, the use of sSOI wafers has no major effect on the performance, but it turns out that sSOI is compatible with other pMOSFET boosters, such as rotated substrates [14], elevated SiGe source/drain [16], SiGe channels [17] and (110) substrates [18]. Indeed, for pMOSFETs, several options can be implemented to enhance the on-state current. The simplest one is to use 45° rotated substrates, which translates in a 11% performance boost [13]. Elevated SiGe source/drain electrodes lead to a 18% improvement thanks to access resistance reduction (up to 37% if a strain can also be generated into the channel) [16]. Here again, the scalability of global boosters is certainly better than that of local ones (c-CESL and elevated SiGe source/drain). Finally, the use of (110) substrates with a standard ⟨110⟩ channel direction enables up to 30% $I_{ON}$ improvement [19]. This latter configuration could be combined with SiGe channels and to other local techniques such as c-CESL and e-SiGe stressors.

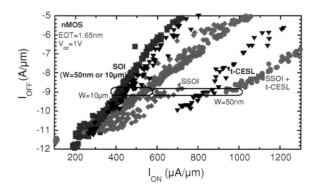

**Fig. 8** $I_{ON}$–$I_{OFF}$ improvements thanks to the implementation of performance boosters on nMOSFETs

## 5 Conclusions

Fully-Depleted SOI technology has acquired during the last few years a large technological maturity. It is thus nowadays a serious alternative to bulk conventional technologies for the 22 nm node and beyond, since it brings major breakthroughs in terms of electrostatic control (and thus scalability) and variability. We have shown in particular in this chapter that this technology is likely to be scaled down to the 10 nm range and that several performance booster options can be implemented efficiently. Gate stack materials allowing the design of low to high threshold voltage transistors are identified and have been demonstrated. Using an ultrathin buried oxide together with an implanted back-plane, additional flexibility is brought in terms of threshold voltage adjustment, and conventional power management techniques based on back-biasing remain efficient even in very aggressively scaled devices.

**Acknowledgments** This work was partially carried out in the frame of the LETI/ST/IBM joint program. It has been partly funded by the French Ministry of Industry, Economy and Finance through the MEDEA Decisif project and by the OSEO Nanosmart program. The authors thank the LETI facilities for device processing.

## References

1. International Technology Roadmap for Semiconductors (2009) Edition
2. Suzuki, K., Tanaka, T., Tosaka, Y., et al.: Scaling theory for double-gate SOI MOSFETs. IEEE Trans. Electron. Devices **40**, 2326–2329 (1993)
3. Buj-Durfournet, C., Andrieu, F., Faynot, O., et al.: Counter-doping as a solution for multi threshold voltage on FDSOI MOSFETs with a single TiN/HfO$_2$ gate stack. In: Proceedings of the Solid State Device and Materials (SSDM) Conference (2009)
4. Weber, O., Faynot, O., Andrieu, F., et al.: High immunity to threshold voltage variability in undoped ultra-thin FDSOI MOSFETs and its physical understanding. In: Proceedings of the International Electron Devices Meeting (IEDM) Technology Digest, pp. 641–644 (2008)
5. Fenouillet-Beranger, C., Denorme, S., Perreau, P., et al.: FDSOI devices with thin BOX and ground plane integration for 32 nm node and below. In: Proceedings of the Solid-State Device Research Conference (ESSDERC), pp. 206–209 (2008)
6. Andrieu, F., Weber, O., Baudot, S., et al.: Fully depleted silicon-on-insulator with back bias and strain for low power and high performance applications. In: Proceedings of the International Conference on IC Design and Technology (ICICDT) (2010)
7. Andrieu, F., Weber, O., Mazurier, J., et al.: Low leakage and low variability ultra-thin body and buried oxide (UT2B) SOI technology for 20 nm low power CMOS and beyond. In: Proceedings of the International Symposium on VLSI Technology (2010)
8. Barral, V., Poiroux, T., Andrieu, F., et al.: Strained FDSOI CMOS technology scalability down to 2.5 nm film thickness and 18 nm gate length with a TiN/HfO$_2$ gate stack. In: Proceedings of the International Electron Device Meeting (IEDM) Technology Digest, pp. 61–64 (2007)
9. Scheiblin, P., Weber, O., Andrieu, F., et al.: Ultra-thin body and BOX SOI roadmap for low power and low VT-variability MOSFETs. In: Proceedings of the EUROSOI Conference (2008)

10. Vandooren, A., Thean, A.V.Y., Du, Y., et al.: Mixed-signal performance of sub-100 nm fully-depleted SOI devices with metal gate, high-K (HfO$_2$) dielectric and elevated source/drain extensions. In: Proceedings of the International Electron Device Meeting (IEDM) Technology Digest, pp. 978–981 (2003)
11. Cheng, K., Khakifirooz, A., Kukarny, P., et al.: Extremely thin SOI (ETSOI) CMOS with record low variability for low power system-on-chip applications. In: Proceedings of the International Electron Device Meeting (IEDM) Technology Digest, pp. 49–52 (2009)
12. Fenouillet-Beranger, C., Thomas, O., Perreau, P., et al.: Efficient multi-VT FDSOI technology with UTBOX for low power circuit design. In: Proceedings of the International Symposium on VLSI Technology (2010)
13. Fenouillet-Beranger, C., Pham Nguyen, L., Perreau, P., et al.: Ultra compact FDSOI transistors (including strain and orientation) processing and performance. ECS Trans. (2009)
14. Baudot, S., Andrieu, F., Faynot, O., et al.: Electrical and diffraction characterization of short and narrow MOSFETs on Fully Depleted strained Silicon-On-Insulator (sSOI). Solid State Electron. **54**, 861–869 (2010)
15. Andrieu, F., Fenouillet-Béranger, C., Weber, O., et al.: Ultrathin body and BOX SOI and sSOI for low power application at the 22 nm technology node and below. In: Proceedings of the Solid State Device and Material (SSDM) Conference (2009)
16. Baudot, S., Andrieu, F., Weber, O., et al.: Fully-Depleted Strained Silicon-On-Insulator p-MOSFETs with recessed and embedded silicon-germanium source/drain (to be published)
17. Andrieu, F., Ernst, T., Faynot, O., et al.: Co-integrated dual strained channel on fully depleted sSDOI CMOSFETs with HfO$_2$/TiN gate stack down to 15 nm gate length. In: Proceedings of the IEEE International SOI Conference, pp. 223–225 (2005)
18. Mizuno, T., Sugiyama, N., Tezuka, T., et al.: (110)-surface strained-SOI CMOS devices. IEEE Trans. Electron. Devices **52**, 367–374 (2005)
19. Signamarcheix, T., Andrieu, F., Biasse, B., et al.: Fully depleted silicon on insulator MOSFETs on (110) surface for hybrid orientation technologies. In: Proceedings of EUROSOI Conference (2010)

# Ultrathin n-Channel and p-Channel SOI MOSFETs

F. Gámiz, L. Donetti, C. Sampedro, A. Godoy, N. Rodríguez and F. Jiménez-Molinos

**Abstract** We review the electrostatic and transport properties of charge carriers in ultrathin single gate (SG) and double gate (DG) SOI transistors. Both electron and hole inversion layers are studied and the influence of silicon thickness and of different crystallographic orientations is evaluated. The origin of volume inversion effect and its consequences are investigated for both types of carrier and for the different surface orientations considered. Finally we discuss the importance of correctly modeling phonons in ultra-thin SOI structures by studying acoustic phonon confinement and its impact on carrier mobility.

## 1 Introduction

It is well known that Silicon-On-Insulator (SOI) technology has a number of advantages over bulk silicon [1]. Among its prominent features we can list: a better control of short channel effects, lower parasitic capacitance, greater tolerance to radiation; moreover all this is achieved while maintaining compatibility with existing silicon fabrication facilities. The full potential of SOI technology can be exploited when the thickness of the silicon layer is reduced to the order of ten or less nanometers; in this case another important feature of SOI devices is a higher immunity to dopant fluctuations because there is no need for a highly doped channel. From the simulation and modeling point of view, such ultra-thin SOI

---

F. Gámiz (✉), L. Donetti, C. Sampedro, A. Godoy, N. Rodríguez and F. Jiménez-Molinos
Nanoelectronic Research Group, Departamento de Electrónica, Universidad de Granada, Granada, Spain
e-mail: fgamiz@ugr.es

devices require the development of new scattering and transport models, because those used for bulk or for thick SOI devices where the two Si–SiO$_2$ interfaces can be de-coupled, must be modified. Novel effects such as volume inversion [2] and subband modulation [3] appear, and scattering mechanisms are also modified. In this chapter we will review the properties of electrons and holes in ultrathin SOI devices and show how they are affected by the silicon layer thickness and crystallographic orientation of the device.

## 2 Carrier Transport in Ultrathin Layers

In ultra-thin SOI layers, carrier behavior is affected by the presence of the two potential barriers at the Si–SiO$_2$ interfaces. Electron wavelength in the confinement direction is comparable to the spatial confinement length: quantum effects are produced by both geometric and electric field confinement. Therefore, quantum effects are not only as important as in bulk inversion layers, but they also depend on silicon layer thickness and can give rise to novel effects such as volume inversion [2]. This phenomenon occurs when the distance between the potential wells at each interface is small and inversion carriers are not confined to one of the interfaces but are spread throughout the whole channel. This affects not only the total carrier density in the channel, but also, and notably, their distribution and, as a consequence, their scattering rates. Quantum effects are highly dependent on the confinement effective mass of carriers; therefore, they are strongly affected by substrate crystal orientation. As a consequence, volume inversion effects change for different orientations and, clearly, depend on the type of carrier, i.e. electrons or holes. To model the silicon conduction band structure we employ the usual effective mass approximation with six ellipsoidal valleys with non-parabolicity correction [4] as in bulk inversion layers. As is well known, such an approach is not useful for holes because the valence band structure is strongly non-parabolic and anisotropic. We employed the six-band $k \cdot p$ model [5], which has been shown to reproduce accurately hole properties in bulk semiconductors and inversion layers [6]. Due to its computational cost, however, self-consistent solution schemes necessary in the case of multiple gates have been developed only recently [7–9]. The use of the effective mass model and the $k \cdot p$ model, both of which have been developed for bulk materials, are questionable at the ultra-thin limit, when the channel consists of only a few atomic layers. In this case, atomistic models, while potentially more accurate, are orders of magnitude slower than such "bulk" models, which have, however, been proven reliable down to structures just a few nanometers thick [10].

We compute carrier mobility, $\mu$, employing a single particle Monte Carlo method [11] or Kubo-Greenwood formula [12]. Electron and hole mobility in SOI devices are studied as a function of silicon layer thickness, and for different surface and channel directions. The results can be explained by taking into account the carrier effective mass in the transport direction and the magnitude of scattering

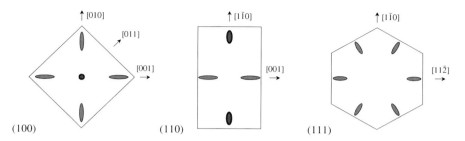

**Fig. 1** Schematic representation of electron valleys in two-dimensional inversion layers in (100), (110), and (111) oriented devices

**Table 1** Effective masses in silicon conduction band valleys for different surface orientations

| Surface orientation | $m_1$ ($m_0$) | | $m_2$ ($m_0$) | | $m_3$ ($m_0$) | $n_v$ |
|---|---|---|---|---|---|---|
| (100) | 0.19 | [010] | 0.19 | [001] | 0.916 | 2 |
|  | 0.19 | [010] [001] | 0.916 | [001] [010] | 0.19 | 4 |
| (110) | 0.19 | [001] | 0.553 | [1$\bar{1}$0] | 0.315 | 4 |
|  | 0.19 | [1$\bar{1}$0] | 0.916 | [001] | 0.19 | 2 |
| (111) | 0.19 | [1$\bar{1}$0] [10$\bar{1}$] [01$\bar{1}$] | 0.674 | [11$\bar{2}$] [1$\bar{2}$1] [$\bar{2}$11] | 0.258 | 6 |

rates in different cases. In the effective mass approximation used in electron simulations, the mass tensor of each valley is constant; however, owing to the relative orientation of such valleys in the device reference system, the quantization and transport masses vary according to the surface and channel orientation of the devices (see Fig. 1; Table 1). Moreover, the relative population of the differently oriented valleys changes as a function of the silicon thickness and the applied gate bias, so that the average effective mass can change if population redistribution occurs between valleys with different transport mass. On the other hand, since the silicon valence band is not parabolic and strongly anisotropic, it is not possible to define an "effective mass" trivially, in the same way it is done for the conduction band and with the same meaning. However, we will show later that it is possible to define an average effective mass, which is useful to understand the simulation results. Such an effective mass depends on the whole valence band-structure of two-dimensional hole inversion layers, which is not fixed by the $k \cdot p$ model alone, but depends on the confining potential. Therefore changes in the effective mass can be expected with silicon thickness and applied gate bias. Scattering rates also show strong variations as a function of silicon layer thickness. Phonon scattering is proportional to a form factor which depends on wave-function confinement: for ultra-thin layers it increases strongly as the silicon thickness decreases but at intermediate thicknesses it is suppressed (especially in double gate (DG) devices) with respect to bulk devices as a consequence of volume inversion. Surface roughness (SR) scattering shows a similar behavior, since it is affected by the average carrier distance from the interfaces and therefore it is directly related to the spatial distribution of carriers.

It is not only the rates of scattering mechanisms that depend on silicon thickness, $t_{Si}$, in SOI structures. The scattering models also have to be modified due to the presence of the two Si–SiO$_2$ interfaces close to each other. Surface roughness has to be modeled in an appropriate way [13, 14] taking into account the variations in channel thickness due to interface fluctuations. Phonon scattering in ultra-thin SOI structures is also affected by the modification of phonon modes due to the acoustic mismatch between different materials: in the following sections we will measure the effect of acoustic phonon confinement on electron and hole mobility.

Since in ultra-thin SOI devices channel doping is not needed to control short channel effects [15], the results shown in the following sections are obtained for undoped silicon layers, and charged impurity scattering is not taken into account for mobility calculations.

## 3 Electron Mobility on (100) Substrates

It is well known that in silicon the conduction band can be approximated by six ellipsoidal valleys with $m_l = 0.916 m_0$ and $m_t = 0.19 m_0$. In the case of Si-(100), two of the ellipsoids have confinement mass $m_3 = m_l$ (non-primed subbands) while the other four have $m_3 = m_t$ (primed subbands) as reported in Table 1. As $m_t$ is smaller than $m_l$, non-primed subbands have lower energy levels than primed subbands, and therefore they also have a larger occupation. However the higher degeneracy $n_v$ of primed valleys (four vs. two), produces a higher overall population when the energy difference is smaller, that is when quantization effects are weaker. Thus, for small values of inversion charge, a larger fraction of electrons occupy primed subbands for silicon layers that are sufficiently thick and a population inversion occurs for thinner silicon layers when most of the electrons belong to non-primed subbands. For high electron concentrations, non-primed valleys are always more populated than primed ones, but a strong increase in the non-primed subband population for thin silicon layers is still observed [16]. This fact is important because of the different conduction effective mass: for non-primed valleys, the conduction mass is small and isotropic ($m_1 = m_2 = m_t$), while for primed ones it is larger ($m_1 = m_l$, $m_2 = m_t$) and anisotropic, so that changes in the population of the valleys alter the electron average conduction mass. In particular, we observe a decrease of conduction mass for silicon layers thinner than 10 nm; such a decrease is more pronounced for low electron densities $N_{inv}$, and happens at thinner layers for high $N_{inv}$ [3, 16]. This is the so-called subband modulation effect. This redistribution of the inversion electrons as the silicon layer shrinks also produces a reduction in the intervalley scattering rate between non-equivalent

**Fig. 2** Phonon form factor for the fundamental subband as a function of silicon thickness; the *dotted line* represent the bulk value for reference

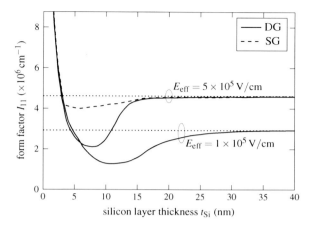

valleys (f scattering) due to the greater separation of the energy levels between primed and non-primed subbands.

Another important effect that appears in SOI-inversion layers as the silicon layer thickness $t_{Si}$ is reduced is a variation of the phonon-scattering rate, produced by differences in carrier confinement between SOI and bulk devices. Phonon scattering is related to wavefunction confinement through the form factor $I_{ij}$ which multiplies the scattering rates:

$$I_{ij} = \int |\psi_i(z)|^2 |\psi_j(z)|^2 dz, \qquad (1)$$

where $\psi_i(z)$ is the electron wavefunction envelope in the direction perpendicular to the interface in the $i$-th subband. When confinement is greater, the overlap integral of envelope wavefunctions is larger, so that the phonon scattering rate increases. Figure 2 shows the form factor for the ground subband as a function of silicon thickness for two different values of the transverse effective field $E_{eff}$, for single-gate SOI (SGSOI) and double gate SOI (DGSOI) devices. Here the effective field is defined as $E_{eff} = (Q_{inv}/2 + Q_{dep})/\varepsilon_{Si}$, where $Q_{inv}$ and $Q_{dep}$ are the inversion and depletion charge densities, and $\varepsilon_{Si}$ the dielectric constant of silicon. For thinner samples, the form factor is very large, due to the geometrical confinement of electrons in a very narrow channel. In this thickness range, $I_{11}$ shows weak variations with the effective field, and also the difference between SG and DG devices is small. As the silicon slab thickness increases, the form factor is reduced, until a minimum is reached in the region between 5 and 15 nm for $E_{eff} = 5 \times 10^5$ V/cm and between 5 and 25 nm for $E_{eff} = 1 \times 10^5$ V/cm. In this range we can observe a big difference between SG and DG devices; in the latter case the interaction between the two potential wells reduces the form factor in a larger measure. Then, for thicker samples, $I_{11}$ increases to approach the value obtained in bulk inversion layers. Consequently, between 5 and 25 nm for an effective field $E_{eff} = 10^5$ V/cm, and between 5 and 15 nm for $E_{eff} = 5 \times 10^5$ V/cm

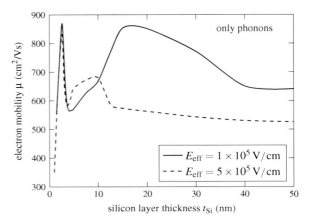

**Fig. 3** Phonon-limited electron mobility for DGSOI devices as a function of $t_{Si}$

the phonon scattering rate in the DGSOI inversion layer decreases, instead of increasing as expected. This is an important result, a direct consequence of the *volume inversion* effect.

The result of the previously described effects is a non-trivial behavior of phonon limited electron mobility as a function of $t_{Si}$ that can be observed in Fig. 3. For very large silicon layers, the two channels of a DGSOI device are separated and no interaction between them appears. This situation corresponds to two conventional inversion layers in parallel, separated by a large potential barrier. The behavior of electrons in each of these inversion layers is the same as that observed in a bulk silicon inversion layer. In a second region, whose limits depend on the electric field being considered, volume inversion occurs, because energy levels and wavefunctions vary significantly as a consequence of the interaction between the two channels, which, consequently, are not independent. In the third and last region ($t_{Si} < 4$ nm), mobility falls abruptly, because electron confinement is stronger due to geometrical constraint and, therefore, the form factor and the phonon scattering rate increase. The limits of this region do not depend on the transverse electric field. In fact, electron mobility in this region is hardly modified by $E_{eff}$. As can be seen in the mobility curves of Fig. 3, electron mobility increases abruptly in the range between 3 and 4 nm. This is a little surprising since, as stated above, the form factor and the phonon scattering rate strongly increase as $t_{Si}$ is reduced in such a range. However, the sharp increase in electron mobility can be understood by taking into account the subband modulation effect mentioned before: there exists a range of $t_{Si}$ where the decrease in the average conduction mass produces a mobility increase. For even thinner devices, the increase in phonon scattering is very large and dominates the residual decrease of effective conduction mass, and therefore electron mobility falls abruptly. SR scattering has a similar dependence to phonon scattering on silicon layer thickness because it depends strongly on the average carrier distance from the rough interfaces: the volume inversion effect contributes to low scattering rates in a $t_{Si}$ range and a strong increase is observed for the thinner devices. The mobility peak produced by the subband modulation effect is reduced by SR scattering but is still present, as we will see in the following section.

## 4 Orientation Effects on Electron Mobility

Changing the device surface and channel directions modifies the relative orientation of the ellipsoidal valleys with respect to confinement and transport directions. The first effect is that quantization effective mass $m_3$ changes, according to the procedure developed in Ref. [17]; the resulting values are reported in Table 1. Then, the restricted two-dimensional dispersion in the transport plane is elliptical with effective masses $m_1$ and $m_2$, also reported in Table 1, along with the corresponding direction of the ellipse axis. For the (110) surface orientation, two nonequivalent sets of valleys are present, as in the (100) case: two bulk energy ellipsoids have an effective mass perpendicular to the interface $m_3 = m_t$, while the other four have higher quantization masses. In the case of the (111) orientation, the six ellipsoids have the same quantization mass, and therefore only one set of subbands is obtained. Due to the different quantization masses, the electron distribution, subband energy levels, wavefunctions, form factors and scattering rates depend on the surface orientation. In addition, as observed in Fig. 1, the effective masses of the constant-energy ellipses associated with motion in the parallel direction are, in general, anisotropic. As a consequence, one can expect to see anisotropic conduction, i.e. for a fixed quantization direction, conduction, and therefore mobility, can depend on the direction of the drift electric field.

Figure 4 shows the inversion carrier distribution across the channel for devices with $t_{Si} = 5$ and 12 nm, with (100), (110), and (111) surface orientations. Electron distribution for (111) and (110) orientations is similar, but this is quite different from the distribution for the (100) case. This fact can be explained by looking at

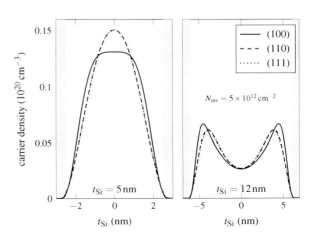

**Fig. 4** Electron density across the channel for devices with different $t_{Si}$ and different surface orientations

Table 1. The quantization masses for (110) and (111) orientations are comparable, but very different to that for (100). The higher quantization mass in Si-(100) produces lower energy subbands in the two valleys with a longitudinal mass perpendicular to the interface. For thin silicon layers or high inversion charge concentrations (i.e. when quantum effects are more important) these two subbands are more populated than the others and the electrons are closer to the Si–SiO$_2$ interfaces. On the other hand, in the (110) and (111) cases, the electrons are, on average, farther from the interfaces, and this also means a lower gate-channel capacitance and therefore a lower inversion charge concentration (and a lower drain current) for a given gate bias.

Figure 5a and b show the mobility obtained for devices with $t_{Si} = 9$ and 4 nm, respectively, taking into account phonon and SR scattering mechanisms. The first result we see is the strong dependence of electron mobility on surface orientation. The highest mobility values are obtained for the (100) orientation, corresponding to lower conduction effective mass, as shown in Table 1. The other two orientations have larger conduction effective masses and therefore lower mobility values. As observed experimentally [18], electron mobility can be degraded by a factor of two for the $(110)/[1\bar{1}0]$ orientation. However, this degradation is much less if a different channel direction (110)/[001] is selected for the same surface orientation (110). This is another noteworthy result: for (110) surface orientation, mobility shows a strong anisotropy in the transport plane. Indeed, Table 1 tells us that subbands with lower energy values (larger $m_3$) correspond to the four ellipses with major axis in the $[1\bar{1}0]$ direction. Therefore, these subbands become more populated and dominate transport. If we apply a drift electric field in the [001] direction, these subbands have an effective mass in the drift direction equal to $m_{drift} = m_1 = 0.19m_0$. On the other hand, if we apply a drift electric field in the

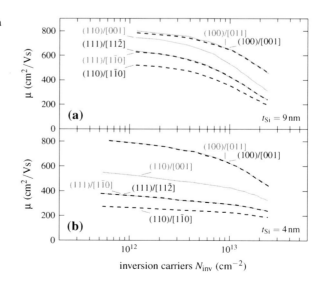

**Fig. 5** Electron mobility as a function of inversion carrier density for different device orientations

[1$\bar{1}$0] direction, the conduction effective mass is $m_{\text{drift}} = m_2 = 0.553 m_0$, almost three times higher than in the [001] direction. This produces the big difference in mobility shown in Figs. 5 and 6 for these two channel directions. In addition, the anisotropy becomes stronger as the silicon thickness decreases. Such anisotropy of electron mobility is not observed when we consider other surface orientations. In the case of the (100) surface orientation, the subbands with lower energy correspond to the two bulk ellipsoids with $m_3 = m_1$. These two valleys show the same conduction mass regardless of the drift electric field orientation. On the other hand, the other four valleys (corresponding to $m_3 = m_t$) present a different effective mass in the [010] and [001] directions. Only when the contribution to conduction of these four high-energy valleys becomes significant, do we observe a dependence of mobility on channel orientation. For example, a slightly anisotropic behavior can be seen in Fig. 5a: a slight dependence of the mobility curves for [001] and [011] channel directions at low inversion charge concentrations is observed for the (100) surface orientation. This effect does not appear in Fig. 5b, where the silicon thickness is $t_{\text{Si}} = 4$ nm.

In the (111) case, Table 1 tells us that each valley has an anisotropic mass tensor: the lack of overall transport anisotropy is due to the fact that the ellipses are symmetrically placed. By comparing Fig. 5a and b we can see that for smaller $t_{\text{Si}}$ the dependence of $\mu$ on electron density $N_{\text{inv}}$ is weaker, especially for the devices with (110) and (111) surface orientations. We have already seen in the previous section that for such thin Si layers the carrier distribution across the channel is essentially determined by spatial confinement and the quantization mass, and only slightly depends on the potential profile.

The behavior of the electron mobility corresponding to different surface and channel orientations as a function of silicon thickness is shown in Fig. 6. As already observed in the previous section, for each orientation a range of silicon thickness exists where electron mobility is larger than the bulk value due to the

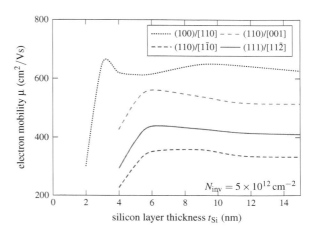

**Fig. 6** Electron mobility as a function of $t_{\text{Si}}$ for different device orientations

volume inversion effect. However, the silicon thickness interval where electron mobility is improved depends on the particular surface/channel combination.

## 5 Hole Mobility

To compute valence band-structure properties in ultrathin SOI devices we employ a fully self-consistent solver for the $k \cdot p$ and Poisson equations [7]. The 6-band $k \cdot p$ Hamiltonian is discretized along the confinement direction, with the appropriate rotation on $k$ space to take into account substrate orientation. The discretized Hamiltonian depends on the wave-vector $k$ in the transport plane (perpendicular to confinement direction); its eigenvalues for different values of $k$ give the energy levels and dispersion relationships $E_i(k)$ of hole subbands, while the eigenvectors are the corresponding six-component wavefunctions. The Poisson equation (discretized on the same grid in the confinement direction) is then solved taking as input the charge distribution computed using the previously obtained energy levels and wavefunctions. This whole procedure is repeated until self-consistency is reached, that is, until the potential profile obtained by solving Poisson equations and that used in the $k \cdot p$ equation coincide within a given tolerance. We manage to achieve sufficient accuracy in a reasonable computational time by employing a specially designed mesh in the $k$ plane. Momentum-space integration for the calculation of carrier concentration is performed by summing over the triangles defined by the above-mentioned mesh; inside each of these, the energy of the hole subbands is linearly interpolated so that the integration is trivial (see [7] for details). Hole mobility is then computed employing the Kubo-Greenwood formula [6]: scattering rates are computed as integrals in the $k$ plane calculated as explained previously for the carrier concentration [7]. Surface orientations that differ from the usual (100) one are particularly interesting in the case of holes, because it has been found that, unlike what happens with electrons, they show higher mobility [19]. Therefore, in the following, we simulate DGSOI devices for different surface and channel orientations to determine whether the performance advantages of alternative orientations persists for SOI devices and, especially, for ultra-thin silicon layers.

For all the silicon thicknesses considered, surface orientation has a similar impact on inversion charge: carrier concentration $P_{\text{inv}}$ slightly increases, moving from (001) to (111) and (011) orientations, with differences between 5% and 10%. The effect on threshold voltage is also quite small. However, when surface orientation changes, the charge distribution profile is strongly affected, as can be seen in Fig. 7, where we represent carrier density across the channel for devices with $t_{\text{Si}} = 5$ and 12 nm structures. The average distance of the carrier from the interface is smallest in the (011) case, larger in the (111) case and, largest of all in the (001) case. At $t_{\text{Si}} = 12$ nm, this translates into different widths for the two peaks of the carrier distribution. On the other hand, at $t_{\text{Si}} = 5$ nm, there are substantial differences in the shape of the distribution: one or two peaks may be present

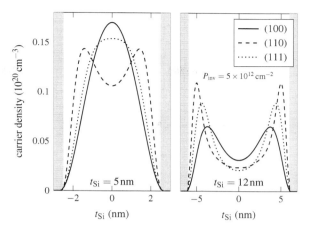

**Fig. 7** Hole density across the channel for devices with different $t_{Si}$ and different surface orientations

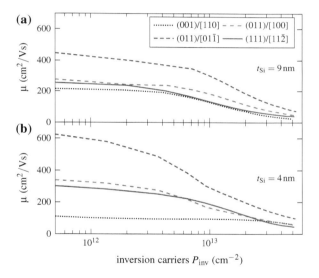

**Fig. 8** Hole mobility as a function of inversion carrier density for different device orientations

depending on surface orientation and gate bias. The dependence of the average distance on gate bias is stronger for (011) devices; in this case two peaks are clearly present at high bias, even for such a thin channel. This fact reveals that volume inversion could differently affect devices built on different crystallographic orientations.

Turning to mobility, the results shown in Fig. 8 qualitatively reproduce those obtained for bulk MOSFETs, although mobility values are different. As in the case of electrons, mobility is almost isotropic for (100) and (111) surface orientations and therefore only one curve is plotted in Fig. 8 for each of these orientations, while it is strongly anisotropic for the (110) surface. However, the relative performance of different orientations is the opposite: the usual (100) orientation

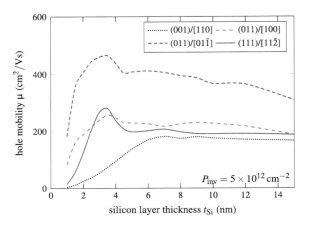

**Fig. 9** Hole mobility as a function of $t_{Si}$ for different device orientations

presents the smallest $\mu$, followed by (111) devices, while the highest values are obtained in the (110) case, with much larger values for the $[1\bar{1}0]$ than the [001] channel direction. For thinner silicon layers we can observe a qualitative difference: $\mu$ has a weaker dependence on $P_{inv}$ in the case of (001) devices. This fact implies that the improvement factors of alternative orientations are greater at low $P_{inv}$ density and decrease (or even vanish) for larger values of $P_{inv}$. The weak mobility dependence on inversion charge density can be attributed to a weaker than usual effect on the scattering rates. Generally, with a higher confining field, the phonon scattering form factor increases, due to a stronger wavefunction confinement, and SR scattering increases since carriers are closer to the interfaces. However, for thin channels, quantum confinement is essentially caused by geometric constraints and the potential well profile has a weak influence on wavefunction shape; as a consequence, scattering rates too are only slightly affected by the applied bias. As an example, we can see in Fig. 7 that hole distribution peaks around the channel center in the case of the (001) orientation, so that bias has a weak effect on charge distribution.

The dependence of hole mobility on $t_{Si}$ is shown in Fig. 9, for different values of $P_{inv}$. As in the case of electrons, a range of silicon layer thicknesses exists where $\mu$ is larger than in bulk inversion layers. Such a range, showing the effects of volume inversion, exists for every surface and channel orientation considered, but its position and width and the mobility increment depend on inversion charge density and on the orientation. Below the volume inversion range, for very thin channels, $\mu$ is always strongly degraded; however, it is important to emphasize that there are large quantitative differences between curves corresponding to different surface and channel orientations. The most important observation is that hole mobility is severely degraded for (001) devices below 6 nm, while this happens at smaller values of $t_{Si}$ for (011) and (111) devices. Therefore, the mobility enhancement of alternative orientations for ultra-thin devices not only persists; it actually increases and can be quite large, especially at small values of $P_{inv}$.

Restricting our attention to the first subband which, in such thin devices, contains the majority of carriers, we may wonder if the increase in the enhancement factor of alternative orientations is due to scattering or to bandstructure modifications. The phonon scattering form factors computed for the different devices have a very similar behavior as a function of $t_{Si}$ for such thin layers; actually, those relative to the (011) orientation are the largest, followed by (111) while in the (001) case they are the smallest. Also SR scattering must be higher for (011) and (111) devices since in these cases carriers are, on average, closer to the Si–SiO$_2$ interfaces (see Fig. 7).

Let us now turn to bandstructure. Its effects are usually represented by the effective transport mass $m^i_\alpha$, related to the second derivative of subband dispersion as:

$$\frac{1}{m^i_\alpha} = \frac{1}{\hbar^2} \frac{\partial^2 E_i(k)}{\partial k_\alpha^2}, \quad (2)$$

where $E_i(k)$ is the energy of the $i$-th subband as a function of momentum $k$ and $k_\alpha$ is the momentum component in the transport direction. If the dispersion is parabolic, $m^i_\alpha$ is a constant for each subband $i$ and direction $\alpha$; however, this is not the case for holes, where the energy dispersion is strongly anisotropic and non-parabolic. One could take the value at $k = 0$ as a representative value, however it can be shown that $m^i_\alpha$ present strong variations as a function of $k$. Therefore, a better alternative is to define the average effective transport mass $M^i_\alpha$ weighting the average with the occupation probability, obtaining

$$\frac{1}{M^i_\alpha} = \left\langle \frac{1}{m^i_\alpha} \right\rangle_i = \int \frac{f(E_i(k))}{m^i_\alpha} dk \bigg/ \int f(E_i(k)) dk, \quad (3)$$

where $f(E)$ is the Fermi-Dirac distribution function. Figure 10 shows $M^1_\alpha$ as functions of $t_{Si}$; we can see a strong decrease at small $t_{Si}$ for the (110) surface orientation, especially for the $[1\bar{1}0]$ direction. This fact compensates, to a certain extent, for the large increase in scattering rates (both phonon and SR) occurring at very small thicknesses, giving rise to the mobility behavior shown in Fig. 9.

## 6 Confined Phonons

We have already seen that in SOI structures, especially thinner ones, the conduction and valence band barriers between the silicon and oxide layers modify the confinement properties of electron and hole inversion layers. In a similar way, acoustic phonons are affected by the acoustic mismatch between the semiconductor and insulator layers. We considered the elastic model of acoustic phonons as elastic waves in a continuous medium (see, for example Ref [20] for a detailed

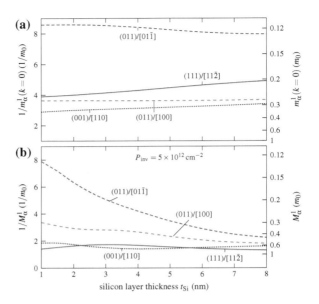

**Fig. 10** Effective mass at $k = 0$ (**a**), and average effective mass (**b**) as a function of $t_{Si}$

discussion): in a three-layer structure we complement the wave equation with continuity of the displacement and stress tensor components [21]. As external boundary conditions (on the outside surfaces of the oxide layers), we consider two different sets of boundary conditions (BC): fixed BC (vanishing displacement) and free BC (vanishing perpendicular component of stress tensor). Clearly, neither condition accurately models a real device where gates are present outside the oxide layers and other elements, including complex gate stacks, can be used. However, the two BCs considered represent the extreme situations, so we can use them to delimit the results we would obtain for any specific case. By imposing either set of BC, quantized phonon modes are obtained: for any value of the phonon momentum in the transport plane $q$, the perpendicular momentum $q_z$ (parallel to confinement direction) is quantized, and only a discrete set of values of phonon energy (frequency) are permitted. Confined phonon dispersion relationships are therefore represented by a set of branches, as we can see in Fig. 11.

Once phonon dispersion in the three-layer structure has been computed, we can evaluate the interaction Hamiltonian and the scattering rates for both electrons and holes. These are computed employing the *Fermi Golden Rule* and integrating over the carrier final state, taking into account the contribution of phonon absorption and emission and different phonon branches: in the case of holes, we use the tabulated values of hole subband energy dispersion obtained with the $k \cdot p$ and Poisson solvers. For both electrons and holes, the resulting scattering rates obtained with fixed BC are larger than those obtained with the bulk phonon model except at very small carrier kinetic energy; those obtained with free BC are larger than those corresponding to fixed BC and always larger than the bulk ones [21, 22].

**Fig. 11** Confined phonon modes obtained with $t_{Si} = 3$ nm and 1 nm for both oxide thicknesses, for free and fixed BC

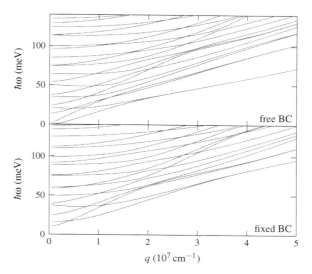

**Fig. 12** Electron mobility computed with different acoustic phonon models

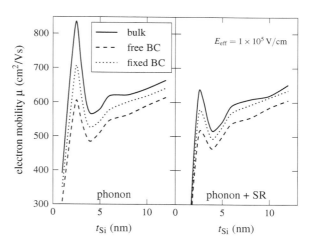

Electron mobility is shown in Fig. 12a and b: for the former only phonon scattering is considered, while for the latter SR scattering is also taken into account; the three curves in each figure correspond to the three different models of acoustic phonons: the bulk model, the confined model with free BC and the confined model with fixed BC. We can see that confined phonons always produce lower mobility than the bulk phonons with either set of BC. When only phonon scattering is considered we can still observe the peak due to subband modulation with $t_{Si}$ between 2 and 4 nm but the corresponding value is 15–30% lower depending on the BC. When SR scattering is added, mobility is suppressed, especially for thin silicon layers: the effect of phonon confinement is clearly

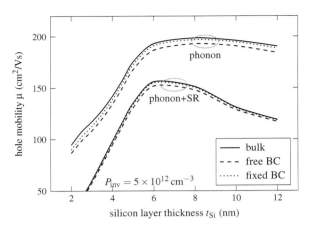

**Fig. 13** Hole mobility computed with different acoustic phonon models

masked. However, especially at low field values, the confinement still affects mobility in a clearly observable manner.

In the case of holes, the computed mobility is shown in Fig. 13. When only phonon scattering is taken into account, we observe a small effect of phonon confinement, with free BC producing the smallest mobility. However, if SR scattering is also taken into account, hole mobility is scarcely affected when the confined phonon model is used instead of the bulk model.

**Acknowledgments** The work of L.D. is done as part of the program *Ramón y Cajal* of the Ministerio de Ciencia e Innovación (M.C.I.) of Spain. Financial support from M.C.I. (contracts TEC2008-06758-C02-01 and FIS2008-05805), Junta de Andalucía (project TIC-P06-1899), EU EUROSOI + Thematic Network (FP7-CA-216373) and EU NANOSIL Network of Excellence (FP7-NOE-216171) is also acknowledged.

# References

1. Cristoloveanu, S., Li, S.S.: Electrical Characterization of Silicon on Insulator Materials And Devices. Kluwer, Boston (1995)
2. Balestra, F., Cristolovenau, S., Benachir, M., Brini, J., Elewa, T.: Double-gate silicon-on-insulator transistor with volume inversion: A new device with greatly enhanced performance. IEEE Electron Device Lett. **8**, 410 (1987)
3. Gamiz, F., Fischetti, M.V.: Monte Carlo simulation of double-gate silicon-on-insulator inversion layers: the role of volume inversion. J. Appl. Phys. **89**, 5487 (2001)
4. Fischetti, M.V., Laux, S.E.: Monte Carlo study of electron transport in inversion layers. Phys. Rev. B **48**, 2244 (1993)
5. Luttinger, J.M., Kohn, W.: Motion of electrons and holes in perturbed periodic fields. Phys. Rev. **97**, 869 (1955)
6. Fischetti, M.V., Ren, Z., Solomon, P.M., Yang, M., Rim, K.: Six-band k·p calculation of the hole mobility in silicon inversion layers: dependence con surface orientation, strain and silicon thickness. J. Appl. Phys. **94**, 1079 (2003)
7. Donetti, L., Gamiz, F., Rodriguez, N.: Simulation of hole mobility in two-dimensional systems. Semicond. Sci. Technol. **24**, 035016 (2009)

8. Pham, A., Jungemann, C., Meinerzhagen, B.: Physics-based modeling of hole inversion-layer mobility in strained-SiGe-on-insulator. IEEE Trans. Electron Devices **54**, 2174 (2007)
9. Zhang, Y., Kim, J., Fischetti, M.V.: Self-consistent calculation for valence subband structure and hole mobility in p-channel inversion layers. J. Comput. Electron. **7**, 176 (2008)
10. Rideau, D., Feraille, M., Michaillat, M., Niquet, Y.M., Tavernier, C., Jaouen, H.: On the validity of the effective mass approximation and the Luttinger k·p model in fully depleted SOI MOSFETs. Solid-State Electron. **53**, 452 (2009)
11. Jacoboni, C., Reggiani, L.: The Monte Carlo method for the solution of charge transport in semiconductors with applications to covalent materials. Rev. Mod. Phys. **55**, 645 (1983)
12. Fischetti, M.V.: Long-range Coulomb interactions in small Si devices. Part II. Effective electron mobility in thin-oxide structures. J. Appl. Phys. **89**, 1232 (2001)
13. Gamiz, F., Roldan, J.B., Cartujo-Cassinello, P., Lopez-Villanueva, J.A., Cartujo, P.: Role of surface-roughness scattering in double gate silicon-on-insulator inversion layer. J. Appl. Phys. **89**, 1764 (2001)
14. Donetti, L., Gamiz, F., Rodriguez, N., Godoy, A., Sampedro, C.: The effect of surface roughness scattering on hole mobility in double gate silicon-on-insulator devices. J. Appl. Phys. **106**, 023705 (2009)
15. Celler, G.C., Cristoloveanu, S.: Frontiers of silicon-on-insulator. J. Appl. Phys. **93**, 4955 (2003)
16. Gamiz, F., Godoy, A., Donetti, L., Sampedro, C., Roldan, J.B., Ruiz, F., Tienda, I., Rodriguez, N., Jimenez-Molinos, F.: Monte Carlo simulation of nanoelectronic devices. J. Comput. Electron. **8**, 174–191 (2009)
17. Rahman, A., Lundstrom, M.S., Ghosh, A.W.: Generalized effective-mass approach for n-type metal-oxide-semiconductor field effect transistors on arbitrarily oriented wafers. J. Appl. Phys. **97**, 053702 (2005)
18. Yang, M., Gusev, E.P., Ieong, M., Gluschenkov, O., Boyd, D.C., Chan, K.K., et al.: Performance dependence of CMOS on silicon substrate orientation for ultrathin oxynitride and HfO2 gate dielectrics. IEEE Electron Devices Lett. **24**, 339 (2003)
19. Irie, H., Kita, K., Kyuno, K., Toriumi, A.: In-plane mobility anisotropy and universallity under uni-axial strains in n- and p-MOS inversion layers on (100), (110), and (111) Si. In: IEDM '04, p. 225 (2004)
20. Bannov, N., Mitin, V., Stroscio, M.: Confined acoustic phonons in a free-standing quantum well and their interaction with electrons. Phys. Stat. Sol. (B) **183**, 131 (1994)
21. Donetti, L., Gámiz, F., Roldán, J.B., Godoy, A.: Acoustic phonon confinement in silicon nanolayers: effect on electron mobility. J. Appl. Phys. **100**, 013701 (2006)
22. Donetti, L., Gamiz, F., Rodriguez, N., Godoy, A.: Hole mobility in ultrathin double-gate SOI devices: the effect of acoustic phonon confinement. IEEE Electron Device Lett. **30**, 1338–1340 (2009)

# Junctionless Transistors: Physics and Properties

J. P. Colinge, C. W. Lee, N. Dehdashti Akhavan, R. Yan, I. Ferain, P. Razavi, A. Kranti and R. Yu

**Abstract** Junctionless transistors are variable resistors controlled by a gate electrode. The silicon channel is a heavily doped nanowire that can be fully depleted to turn the device off. The electrical characteristics are identical to those of normal MOSFETs, but the physics is quite different. This paper compares the conduction mechanisms in three types of MOS devices: inversion-mode, accumulation-mode and junctionless MOSFET.

## 1 Introduction

All existing transistors are based on the use of semiconductor junctions. Because of the laws of diffusion and the statistical nature of the distribution of the doping atoms in the semiconductor, the formation of ultrashallow junctions with high doping concentration gradients has become an increasingly difficult challenge for the semiconductor industry. Junctionless transistors (also called gated resistor) have no junctions and no doping concentration gradients. These devices have full CMOS functionality and are made using silicon nanowires.

The key to fabricating a junctionless gated resistor is the formation of a semiconductor layer that is thin and narrow enough to allow for full depletion of carriers when the device is turned off. The semiconductor also needs to be heavily doped to allow for a decent amount of current flow when the device is turned on.

J. P. Colinge (✉), C. W. Lee, N. Dehdashti Akhavan, R. Yan, I. Ferain,
P. Razavi, A. Kranti and R. Yu
Tyndall National Institute, University College Cork, Cork, Republic of Ireland
e-mail: jean-pierre.colinge@tyndall.ie

Putting these two constraints together imposes the use of nanoscale dimensions and high doping concentrations [1, 2].

## 2 Conduction Mechanisms

MOSFETs (including accumulation-mode FETs) are normally off devices, as the drain junction is reverse biased and blocks any current flow if no channel is created between source and drain. To turn the device on, the gate voltage is increased in order to create an inversion channel. The junctionless transistor, on the other hand, is basically a normally on device where the workfunction difference between the gate electrode and the silicon nanowire shifts the flatband voltage and the threshold voltage to positive values. When the device is turned on and in flatband conditions, it basically behaves as a resistor and the electric field perpendicular to current flow is basically equal to zero in the "bulk" channel.

Figure 1 shows the drain current as a function of gate voltage in the three types of SOI MOSFETs: inversion-mode "$N^+PN^+$", accumulation-mode "$N^+NN^+$" and heavily doped junctionless "$N^+N^+N^+$" transistor.

- Below threshold the inversion-mode device is depleted (either fully or partially) and the flat-and voltage is situated below the threshold voltage, at a gate voltage at which the device is off (Fig. 1a). Below flatband the body is p-type neutral. Above threshold, the body of the channel is depleted and a surface inversion layer is formed.
- Below threshold accumulation-mode devices are fully depleted. Threshold is reached when the gate voltage is increased in such a way that a portion of the channel region is no longer depleted. At that point, the channel region is technically partially depleted. As gate voltage is further increased, flat-band is

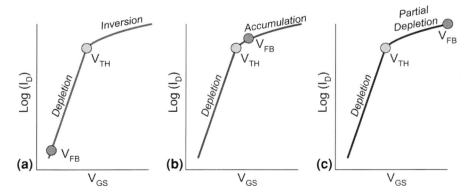

**Fig. 1** Drain current (log scale) as a function of gate voltage in **a** an inversion-mode MOSFET; **b** an accumulation-mode MOSFET and **c** a heavily-doped junctionless transistor

# Junctionless Transistors: Physics and Properties

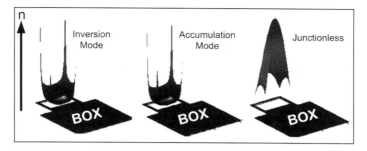

**Fig. 2** Electron concentration profile above threshold

reached: the channel region is now neutral (i.e. no longer depleted, even partially). Further increasing the gate voltage creates a surface accumulation channel (Fig. 1b).

- The heavily doped junctionless transistor is fully depleted below threshold. As gate voltage is increased, the electron concentration in the channel increases, and threshold is reached when the peak electron concentration in the channel reaches the doping concentration $N_D$. Further increasing the gate voltage increases the "diameter" of the region where $n = N_D$, until the entire cross-section of the device becomes neutral (i.e. no longer depleted, even partially), at which point flatband is reached (Fig. 1c). It is possible to further increase the gate voltage to create accumulation channels, but this is probably not desirable, as the high doping concentration in the channel already insures a large current drive.

Figure 2 compares the electron profile above threshold in an inversion-mode pi-gate MOSFET, an accumulation-mode pi-gate transistor and a junctionless pi-gate device in the on state (above threshold). In the inversion-mode the electrons are confined in inversion layers at the top of the device and along its sidewalls, with marked peaks at the corners. The cross-section of the silicon channel is 10 nm × 10 nm. If confinement effects are neglected, there is no significant volume inversion. The profile carrier concentration in the accumulation-mode device is remarkably similar to that in the inversion-mode device, and the majority of the electrons are confined in inversion layers at the top of the device and along its sidewalls, again with marked peaks at the corners. The electron concentration in the center of the device is equal to the doping concentration ($N_D = 10^{17}$ cm$^{-3}$), such that a small body current is added to the current in the accumulation layers.

Below threshold, the junctionless channel is depleted of electrons, and the current varies exponentially with gate voltage (Fig. 3a). At threshold, a neutral silicon filament forms between source and drain (Fig. 3b). The cross-section of filament increases when gate voltage is increased (Fig. 3c) until depletion disappears and the device is in flatband condition (Fig. 3d).

When the device is turned off (Fig. 3a), the effective channel length, hereby defined as the distance between the non-depleted source and drain regions varies

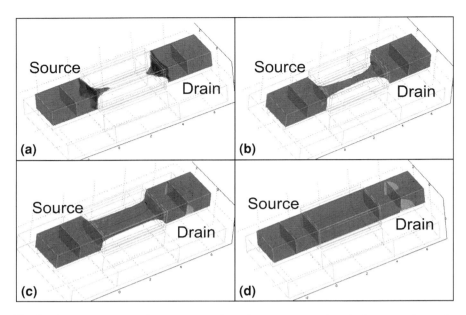

**Fig. 3** Electron concentration contour plots in an $n$-type junctionless transistor for $V_{DS} = 50$ mV. **a** $V_G < V_{TH}$; **b** $V_G = V_{TH}$; **c** $V_G > V_{TH}$; **d** $V_G = V_{FB} \gg V_{TH}$

from a distance less than the physical gate length in the center of the nanowire to a distance larger than the physical gate length near the periphery of the nanowire. When the gate voltage is negative enough, the distance between the non-depleted source and drain regions can be larger than the physical gate length across the entire section of the device. This has a significant impact on short-channel properties. In a "regular", inversion-mode trigate device, assuming the distance between the source and drain junctions is exactly equal to the physical gate length (Fig. 4a), the presence of PN$^+$ junctions creates a reduction of the effective gate length, resulting in a short-channel effect (SCE). This effect has been quantified by Skotnicki et al. [3] using the voltage-doping transformation model (VDT). The VDT can be used to translate the effects of shrinking device parameters such as gate length or drain voltage into electrical parameters. In the particular case of the SCE and the drain-induced barrier lowering (DIBL), the following expressions can be derived from the VDT model [4].

$$\text{SCE} = 0.64 \frac{\varepsilon_{Si}}{\varepsilon_{ox}} \left[ 1 + \frac{x_j^2}{L_{el}^2} \right] \frac{t_{ox}}{L_{el}} \frac{t_{dep}}{L_{el}} V_{bi} \equiv 0.64 \frac{\varepsilon_{Si}}{\varepsilon_{ox}} EI \, V_{bi} \qquad (1)$$

and

$$\text{DIBL} = 0.80 \frac{\varepsilon_{Si}}{\varepsilon_{ox}} \left[ 1 + \frac{x_j^2}{L_{el}^2} \right] \frac{t_{ox}}{L_{el}} \frac{t_{dep}}{L_{el}} V_{DS} \equiv 0.80 \frac{\varepsilon_{Si}}{\varepsilon_{ox}} EI \, V_{DS} \qquad (2)$$

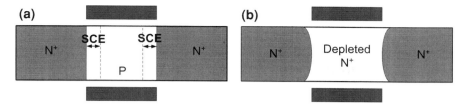

**Fig. 4** Illustration of effective channel length in an inversion-mode device (**a**) and a junctionless transistor (**b**)

where $L_{el}$ is the electrical (effective) channel length, $V_{bi}$ is the source or drain built-in potential, tox is the gate oxide thickness, $x_j$ is the source and drain junction depth and tdep is the penetration depth of the gate field in the channel region, which is equal to the depth of the depletion region underneath the gate in a bulk MOSFET. The parameter EI is called the "Electrostatic Integrity" factor. It depends on the device geometry and is a measure of the way the electric field lined from the drain influence the channel region, thereby causing SCE and DIBL effects. Based on the above expressions, the threshold voltage of a MOSFET with a given channel length Lel can be calculated using the following relationship:

$$V_{TH} = V_{TH\infty} - \text{SCE} - \text{DIBL} \qquad (3)$$

where $V_{TH\infty}$ is the threshold voltage of a long-channel device. The decrease of threshold voltage with decreased gate length is a well-known short-channel effect called the "threshold voltage roll-off". The SCE is illustrated in Fig. 4a for an inversion-mode transistor. In a junctionless device in the off state, the electrostatic squeezing effect causes the distance between the non-depleted source and drain regions to be larger than the physical gate length (Fig. 4b). This is a beneficial factor that reduces short-channel effects [2] and can possibly reduce source-to-drain direct tunneling in very short-channel devices.

Unlike accumulation-mode and inversion-mode devices the channel of junctionless transistors is in the bulk of the nanowire (i.e. it is not a surface channel). As a result, carriers in the channel are exposed to a low electric field in the directions to current flow. This strongly reduces the degradation of mobility when gate voltage is increased in the on state [5]. Figure 5 shows the position of the channel in both subthreshold operation and above threshold. In an inversion-mode device, subthreshold conduction mainly takes place in the top corners of the device (Fig. 5a). Above threshold, surface channels are formed on three sides of the nanowire, with carrier concentration peaks in the corners Fig. 5d). In an accumulation-mode transistor, the subthreshold current flows through the bulk of the device, near the center or the back of the nanowire (Fig. 5b). When the device is turned on, a small current flows through the body of the nanowire, but this current typically amounts for les than ten percents of the overall current drive. Most of the current flows in surface and corner accumulation channels, like in an inversion-mode device (Fig. 5e). In the junctionless device the subthreshold current flows in

**Fig. 5** Location of conduction path in the different devices. Subthreshold conduction path in **a** inversion-mode, **b** accumulation-mode and **c** junctionless devices. Conduction channels above threshold in **d** inversion-mode, **e** accumulation-mode and **f** junctionless devices

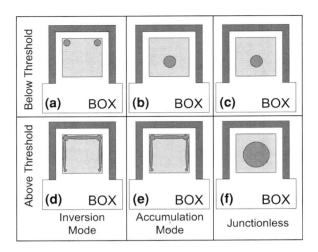

the center of the nanowire, as in the accumulation-mode device (Fig. 5c). When threshold is reached, the channel leaves full depletion and a neutral ("undepleted") channel forms between source and drain, in the center of the device (Fig. 5f). Thus, the junctionless transistor is partially depleted when it is turned on.

Above threshold, channel saturation and pinchoff occurs like in a regular MOSFET or in a JFET (Fig. 5). As a result of pinchoff, the output characteristics of junctionless transistors are virtually identical to those of conventional MOSFETs [1].

No analytical mode has been developed for the junctionless device as yet, but, in first approximation it is probably safe to use the model developed for calculating the body current of accumulation-mode transistors [6], which yields:

$$I_{Dsat} \approx \frac{1}{2} \frac{q\mu N_D W_{si} T_{si}}{L} V_{Dsat}^2 \qquad (4)$$

with

$$V_{Dsat} = V_G - V_{FB} - \left(\frac{qN_D T_{si}}{2\varepsilon_{si}} + \frac{qN_D T_{si}}{C_{ox}}\right) \qquad (5)$$

This is to be compared with the general expression for the current in an inversion-mode trigate device:

$$I_{Dsat} \approx \frac{1}{2} \frac{\mu C_{ox} W_{eff}}{L} V_{Dsat}^2 \text{ with } W_{eff} = 2T_{si} + W_{si} \qquad (6)$$

The current drive of the inversion-mode device is directly proportional to the gate oxide capacitance, but so is the capacitance of the gate electrode. As a result, the intrinsic delay time of the device ($CV/I$) is independent on the gate oxide thickness. It can only be improved by enhancing the mobility and decreasing gate length. In the junctionless device the current is not directly related to the gate

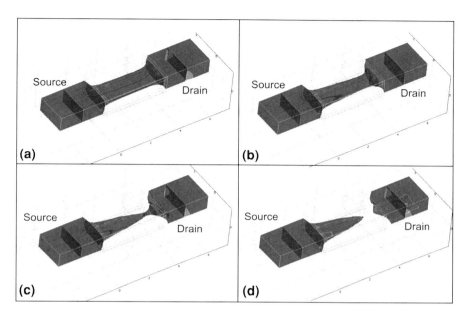

**Fig. 6** Electron concentration contour plots in an *n*-type junctionless transistor. **a** $V_D = 50$ mV; **b** $V_D = 200$ mV; **c** $V_D = 400$ mV; **d** $V_D = 600$ mV

oxide thickness. It is, instead, proportional to the doping concentration in the nanowire—and the doping concentrations that are used are at the same level as in the source and drain extensions. This can give the junctionless device a speed advantage over regular devices [1] (Fig. 6).

## 3 Threshold Voltage

The threshold voltage of junctionless devices depends on silicon film thickness, width of the nanowire, doping concentration and gate oxide thickness (Fig. 7). One can easily obtain different threshold voltages by varying the width of the nanowires, if doping concentration is kept constant, which may be useful for producing devices with multiple values of $V_{TH}$ on a chip. Figure 7 shows the variation of threshold voltage in a long-channel junctionless device as a function of silicon width ($W_{si}$) and thickness ($T_{si}$). The EOT is 0.5 nm and the doping concentration is $2 \times 10^{19}$ cm$^{-3}$. Threshold voltage can be varied from 1 to $-0.2$ V by varying $W_{si}$ from 5 to 20 nm and $T_{si}$ from 5 to 15 nm.

It is also important to evaluate the sensitivity of $V_{TH}$ variations with fabrication parameters. Figure 8 shows the results of such an analysis. If one of the dimensions ($T_{si}$ or $W_{si}$) is small enough, the variations of the other dimension do not impact too much the threshold voltage. For example, if the silicon thickness is

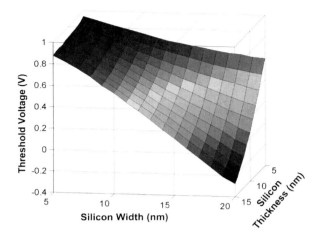

**Fig. 7** Long-channel $V_{TH}$ defined as peak in $dg_m/dV_G$ for $T_{ox} = 0.5$ nm and $N_D = 2 \times 10^{19}$ cm$^{-3}$

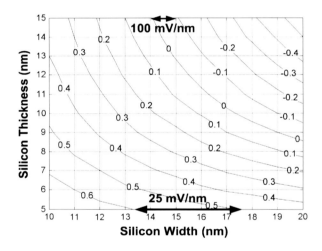

**Fig. 8** Contour plot of threshold voltage in an $n$-channel junctionless device with $T_{ox} = 1$ nm and $N_D = 2 \times 10^{19}$ cm$^{-3}$, as a function of nanowire thickness and width

5 nm, the variation $\Delta V_{TH}/\Delta W_{si}$ is equal to 25 mV/nm. At the same time, the variation of threshold voltage $\Delta V_{TH}/\Delta T_{si}$ is equal to 100 mV/nm. Since thin-film SOI wafers with a $\sigma T_{Si} < 0.2$ nm can nowadays be produced [7], threshold voltage variations on the order of $\sigma V_{TH} < 35$ mV can be expected at wafer level, provided a lithography width control of 0.5 nm. The use of a thinner EOT decreases the sensitivity of $V_{TH}$ on dimensions.

Doping fluctuations are a serious problem in nanoscale devices. Even in the so-called "undoped" channels the doping concentration is not equal to zero but to a value of a few $10^{15}$ cm$^{-3}$. This means that there a chance of approximately one in a thousand to find a (boron) doping atom in a device with a channel volume of $10 \times 10 \times 10$ nm$^3$. In the corresponding junctionless device with a doping concentration $N_D$ of $5 \times 10^{19}$ cm$^{-3}$, there will be an average 50 doping atoms per

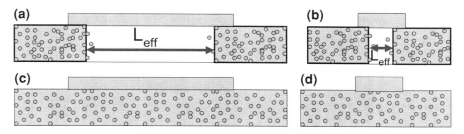

**Fig. 9** Scattering of source and drain doping impurities in the channel of **a** a long-channel and **b** a short-channel inversion-mode MOSFETs; **c** long-channel and **d** short-channel junctionless devices

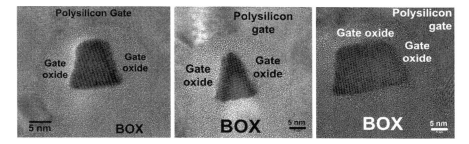

**Fig. 10** TEM cross section of a junctionless transistors with increasing width from left to right

channel, which makes the device much more robust against doping fluctuation problems.

Another problem associated with the statistical distribution of doping impurities is the variation of effective channel length, $L_{eff}$, defined as the distance between the source junction and the drain junction [8]. This is illustrated in Fig. 9, the statistical nature of the doping atom distribution at the source and drain junctions causes the effective channel length to fluctuate from device to device. These fluctuations are inherent to the ion implant and diffusion processes. Furthermore, dopants from the source and drain can scatter in the channel region and influence the threshold voltage. In the junctionless device, there is no gradient of doping concentration between source, channel and drain. The effective channel length can no longer be defined as the distance between two junctions. The effective gate length is basically equal to the physical gate length, although it may be somewhat longer when the device is turned off (Fig. 4b).

## 4 Experimental Results

Junctionless gated resistor devices were made using standard SOI wafers. The SOI layer was thinned down to 15 nm and patterned into nanoribbons using e-beam

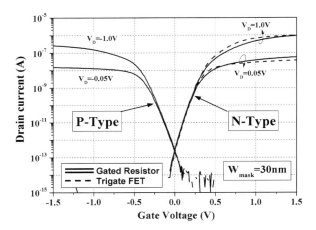

**Fig. 11** Drain current versus gate voltage for N- and P-type junctionless gated resistors and a trigate inversion-mode, N-channel MOSFET. $N_D = 1 \times 10^{19}$ cm$^{-3}$ in the junctionless device. $L = 1$ μm

lithography. Using a combination of plasma lateral overetch and gate oxidation, the thickness of the nanowires was reduced to nm, and their width was reduced to dimensions as small as 5 nm. Ion implantation was used to dope the devices uniformly N$^+$ or P$^+$ with a concentration of $1 \times 10^{19}$–$5 \times 10^{19}$ cm$^{-3}$, which is a typical LDD doping concentration, to realize N-channel and P-channel devices, respectively.

Figure 10 shows the TEM cross-section of a several devices with different widths. Due to processing parameters the cross-section of the devices is not a rectangle, but rather a trapezium or even a triangle. Devices with a silicon thickness of approximately 10 nm and a width ranging from 5 to 30 nm were made. Junctionless transistors have excellent on–off switching behavior and an on/off ratio larger than $10^5$ for $V_{DD}$=0.5 V (Fig. 11). The off current could not be measured as it is lower than 1 fA, which is the sensitivity limit of the measuring apparatus.

The subthreshold slope at room temperature is 64 mV/decade, and it remains very close to the "ideal" value of (kT/q) ln(10) over the temperature range 225–475 K. Figure 12 shows the evolution of subthreshold slope, $SS$ (in mV/decade) at $V_{DS}$=50 mV in a N-channel inversion-mode trigate MOSFET and a junctionless gated resistor. The subthreshold slope of both devices follows the following law: $SS = n \, (kT/q) \, ln(10)$ with $n = 1.066$, which is very close to the lowest theoretical limit ($n = 1$).

Equation 4 predicts an increase of drain current with doping concentration. This is indeed observed (Fig. 13), but it should be noted that the current increase is due not only to the reduction of channel resistivity with doping, but also to the improved conduction in the source and drain, which have the same doping concentration as the channel. In the measured devices, the length of the source, the drain and the channel region are all equal to 1 μm.

**Fig. 12** Measured subthreshold slope versus temperature in a junctionless gated resistor and an inversion-mode trigate MOSFET

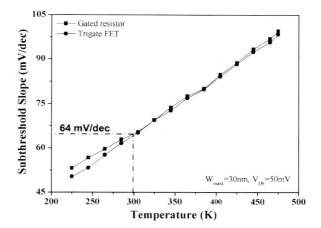

**Fig. 13** Current versus gate voltage in junctionless devices with two different doping concentrations

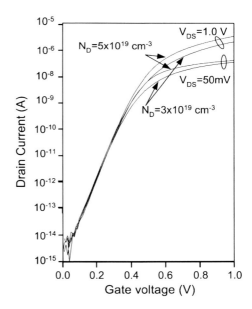

## 5 Mobility Considerations

One might be concerned by the effect of the high channel doping concentration of junctionless gated resistors on carrier mobility. It is well known that ionized impurity scattering degrades carrier mobility. The electron mobility in silicon is shown in Fig. 14 as a function of donor atom concentration, $N_D$ [9]. The mobility drops from 1,400 cm$^2$/Vs in lightly doped silicon to 80 cm$^2$/Vs for $N_D=10^{19}$ cm$^{-3}$.

**Fig. 14** Electron mobility in silicon as a function of donor doping atom concentration and as a function of electric field in the channel. The latter curve shows the mobility/field for several key technology nodes

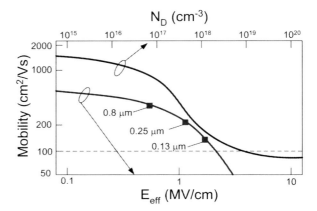

Interestingly, mobility does not significantly degrade any further as the doping concentration is increased beyond $10^{19}$ cm$^{-3}$. A similar behavior is observed for holes in P-type doped silicon [9].

Channel mobility in inversion-mode devices is affected by the (vertical) electric field in the channel, $E_{eff}$. Since $E_{eff}$ increases when the effective oxide thickness, EOT, is reduced, surface channel mobility has steadily decreased in successive technology nodes [10] and would now be well below 100 cm$^2$/Vs at the 45-nm node, if it wasn't for the introduction of strained silicon technology: dealing with mobilities of 80 cm$^2$/Vs in junctionless gated resistors does not sound like a completely outrageous idea when this is taken in consideration. It is also important to note that the conduction channel in junctionless devices is in the center of the device, and that the electric field perpendicular to the current flow, $E_{eff}$, is very small. As a result, mobility in junctionless devices is not expected to decrease as EOT is decreased [3].

Mobility in junctionless devices is largely dominated by ionized impurity scattering, and acoustic phonons seem to have little effect on mobility. Figure 15 shows the mobility, measured from the peak of transconductance as a function of gate voltage, in trigate SOI MOSFETs and in junctionless transistors. The inversion-mode trigate devices have either an undoped channel ($N_A \approx 5 \times 10^{15}$ cm$^{-3}$) or a doped channel ($N_A \approx 5 \times 10^{17}$ cm$^{-3}$). In the undoped devices the peak mobility is 350 cm$^2$/Vs at room temperature, but drops by 36% as temperature is increased to 200°C. The doped devices have a lower room-temperature mobility (220 cm$^2$/Vs) which also drops by approximately 36% as temperature is increased to 200°C. The heavily doped ($N_D \approx 2 \times 10^{19}$ cm$^{-3}$) junctionless devices have a much lower room-temperature mobility: 80 cm$^2$/Vs. However, the mobility decreases <7% as temperature is increased to 200°C. This clearly illustrated the fact that mobility is limited by ionized impurity scattering and is relatively insensitive to phonon scattering in heavily doped junctionless transistors. A similar trend is observed in p-channel devices (Fig. 15).

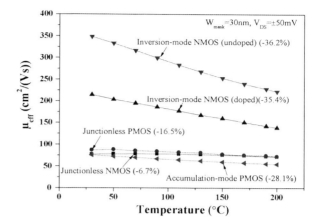

**Fig. 15** Mobility in heavily doped junctionless transistors and in regular trigate MOSFETs with doped and undoped channels, as a function of temperature

# 6 Conclusion

Junctionless transistors are variable resistors controlled by a gate electrode. The silicon channel is a heavily doped nanowire that can be fully depleted to turn the device off. The electrical characteristics are identical to those of normal MOSFETs, but the physics is quite different. Three types of MOS devices: inversion-mode, accumulation-mode and junctionless MOSFETs are compared. The dependence of threshold voltage on device geometry and doping fluctuation effects in junctionless transistors is discussed. An analysis of mobility reduction effects is presented.

# References

1. Colinge, J.P., Lee, C.W., Afzalian, A., Dehdashti Akhavan, N., et al.: Nanowire transistors without junctions. Nat. Nanotechnol. **5**, 225 (2010)
2. Lee, C.W., Afzalian, A., Dehdashti Akhavan, N., Yan, R., et al.: Junctionless multigate field-effect transistor. Appl. Phys. Lett. **94**, 053511 (2009)
3. Skotnicki, T., Merckel, G., Pedron, T.: The voltage-doping transformation: a new approach to the modeling of MOSFET short-channel effects. IEEE Electron Device Lett. **9**, 109 (1988)
4. Skotnicki, T.: Heading for decananometer CMOS—is navigation among icebergs still a viable strategy? In: Proceedings of the 30th European Solid-State Device Research Conference (ESSDERC) (2000)
5. Colinge, J.P., Lee, C.W., Ferain, I., Dehdashti Akhavan, N., et al.: Reduced electric field in junctionless transistors. Appl. Phys. Lett. **96**, 073510 (2010)
6. Colinge, J.P.: Conduction mechanisms in thin-film, accumulation-mode p-channel SOI MOSFETs. IEEE Trans. Electron Devices **37**, 718 (1990)
7. Weber, O., Faynot, O., Andrieu, F., Buj-Dufournet, C., et al.: High Immunity to Threshold Voltage Variability in Undoped Ultra-Thin FDSOI MOSFETs and its Physical Understanding. Technical Digest of IEDM 245 (2008)

8. Xiong, S., Bokor, J.: Sensitivity of Double-Gate and FinFET Devices to Process Variations. IEEE Trans. Electron Devices **50**, 2255 (2003)
9. Jacoboni, C., Canali, C., Ottaviani, G., Quaranta, A.A.: A review of some charge transport properties of silicon. Solid State Electron **20**, 77 (1977)
10. Thompson, S.E., Armstrong, M., Auth, C., Buchler, M., et al.: A 90-nm logic technology featuring strained-silicon. IEEE Trans. Electron Devices **50**, 1790 (2004)

# Gate Modulated Resonant Tunneling Transistor (RT-FET): Performance Investigation of a Steep Slope, High On-Current Device Through 3D Non-Equilibrium Green Function Simulations

Aryan Afzalian, Jean-Pierre Colinge and Denis Flandre

**Abstract** Performances of a new concept of nanoscale MOSFET, the gate modulated resonant tunneling (RT)-FET, are investigated through 3D non-equilibrium green function simulations enlightening the main physical mechanisms. Modulation by gate voltage of resonant tunneling states induced by channel and additional tunnel barrier(s) enables very low RT-limited $I_{off}$ current together with high thermionic $I_{on}$ current. A region of subthreshold slope values as low as 45 mV/dec is achieved just below threshold, enabling a fast transition between off and on regimes. High $I_{on}/I_{off}$ current ratios with low voltage operation and good delay characteristics are predicted. The 10 nm Si RT nanowire investigated here could operate with a supply voltage as low as 0.5 V, $I_{on}/I_{off} > 10^4$ (an order of magnitude improvement compared to a classical nanowire) and low leakage.

## 1 Introduction

In a MOSFET the lower limit for the subthreshold slope (SS) is $\ln(10) \times kT/q$ (kT/q in short) or 59.6 mV/dec at 300 K. This sets a practical limit to reduction of supply voltage and power consumption of a circuit. Achieving steeper SS transistors has become vital for further CMOS downscaling. Accordingly, different

A. Afzalian (✉) and D. Flandre
ICTEAM Institute, Université Catholique de Louvain, Place du Levant 3,
1348 Louvain-La-Neuve, Belgium
e-mail: aryan.afzalian@uclouvain.be

A. Afzalian and J.-P. Colinge
Tyndall National Institute, University College Cork, Lee Maltings, Prospect Row,
Cork, Ireland

concepts have been proposed such as Tunnel-FETs (TFETs) [1–3], I-MOS [4, 5], Suspended-Gate FET [6] and Ferro-electric FET [7, 8]. All these devices still present important challenges before they could be integrated and achieve performances that match those of a MOSFET. For instance, TFETs suffer from very low on-current level in Si [2, 3, 9] so that alternative channel materials and heterojunctions would be required in order to compete with classical Si-MOSFETs [10–12]. We have recently shown the possibility of achieving sub-kT/q SS near threshold together with high on-current by using a "boosted" Si CMOS transistor modified to comprise tunnel barriers near the gate edges [13, 14] (Fig. 1). In this chapter, the physics and the performance limits of this new device, the resonant tunneling (RT)-FET, are thoroughly investigated through quantum simulations in silicon nanowires.

In Sect. 2, the simulated structure and the simulation algorithm are first presented. Then the invariance of the density of state (DoS) in the channel, and therefore of the transmission, in subthreshold regime that leads to the carrier and subthreshold limitation of classical transistors is reviewed. Section 3 investigates the physics and performance limits of RT-FETs. First, the phenomenon we called the modulated barrier resonant tunneling (MBRT) effect and its influence on transistor characteristics is explained. Then the performances optimization of RT-FETs is investigated versus physical parameters and compared to classical transistor in ultra scaled constricted Si nanowires. Due to its particular physics, the new device can achieve very low $I_{off}$ together with high $I_{on}$ current. A region of SS values as low as 45 mV/dec is achieved just below threshold, enabling a fast transition between off and on regimes. High $I_{on}/I_{off}$ current ratios with low voltage operation and good delay characteristics are predicted.

## 2 Simulation Algorithm and Subthreshold Slope Basics

SOI nanowire transistors with a channel length, $L$, of 10 nm were simulated in the ballistic limit using a self-consistent three-dimensional non-equilibrium Green function (NEGF) quantum simulator [15]. Figure 1 shows a schematic device representation and gives the parameters of the SOI Gate-all-Around n-channel nanowire with dielectric constrictions of the Source (S)/Drain (D) extensions (local reduction of length $L_C$ and thickness $\Delta t_{si}$) located at a distance $L_{out}$ from the S and D/channel junctions. Such constrictions result in tunnel barriers (TBs) between a few meV to several hundreds of meV simply by tuning $\Delta t_{si}$ [15]. We also note that the effect discussed here is not strictly limited to constriction tunnel barriers. Our simulation results have shown similar effects by replacing the constrictions area by other usual means of creating barriers (a small P+ pocket in the N+ source and drain, surface charge defects through grain boundaries [16] (in this case a delta shaped barrier is created), vertical devices and heterojunctions, Schottky barrier using dopant segregation [14], oxide charges,...). The gates can overlap over the S–D extensions side by a length $L_{ov}$.

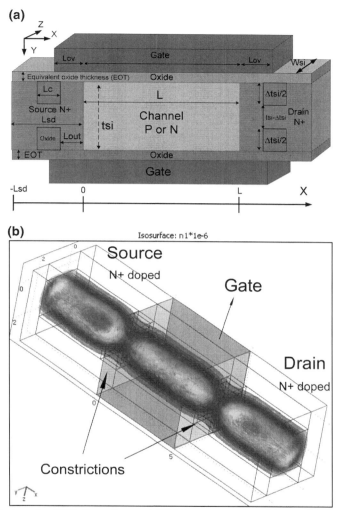

**Fig. 1** a Rectangular [100] Gate-all-around SOI N-type nanowire with constrictions (for figure clarity the lateral gates and gate oxides are not shown). $t_{si} = w_{si} = 2$ nm. Channel: $L = 10$ nm, doping $N^- = 10^{15}$ cm$^{-3}$. Source/Drain extensions: $L_{sd} = 5$ nm, doping $N^+ = 10^{20}$ cm$^{-3}$ Oxide: EOT = 0.5 nm. T = 300 K. **b** 3D plot of the simulated electron concentration in the device with $\Delta t_{si} = 0.5$ nm, $L_C = 1$ nm, $L_{out} = L_{ov} = 0$. Due to the constrictions, tunnel barriers of a few 100 meV are created at the gate edges

This simulator was specifically optimized for the efficient simulation of nanowires with tunnel barriers and non-constant cross-section. It is based on the use of a fast coupled mode space (FCMS) implementation of the NEGF, adaptive energy and non-uniform mesh algorithms. The Hamiltonian is written in the effective mass formalism but it includes a non-parabolic correction for transport effective mass and subband levels that is sufficiently accurate for the scope of this study [15].

## 2.1 Carrier Limitation in a Classical Transistor

In a MOSFET the lower limit for the subthreshold slope (SS) is $\ln(10) \times kT/q$ ($kT/q$ in short) or 59.6 mV/dec at 300 K.

$$SS \equiv \frac{\partial V_G}{\partial(\log_{10} I_D)} = \frac{\partial V_G}{\partial \varphi_S} \times \frac{\partial \varphi_S}{\partial(\log_{10} I_D)} \qquad (1)$$

This limit arises from the combination of two facts:

(1) The gate channel coupling limitation, i.e. it is at best possible to induce a surface potential, $\varphi_S$ variation in the channel (the potential responsible for the conduction band potential variation $\Delta E_{TCB}$, a N-MOSFET is considered here but the principle is the same for a P-MOSFET in the valence band) as large as the gate potential variation, i.e. $\frac{\partial V_G}{\partial \varphi_S} = n \geq 1$.
(2) The carrier limitation: the maximum current change available due to the change of channel barrier $\Delta E_{TCB} = -q\Delta\varphi s$ is limited by the Fermi–Dirac distribution which in subthreshold ($E_{TCB} > E_{FS}$ the source Fermi level) gives:

$$\Delta Id = Id_0 \cdot e^{-q\Delta \varphi s/kT} \qquad (2)$$

This exponential increase in current is related to the increase of the number of electrons from the source that can cross the channel and reach the drain when the channel barrier is lowered. Note that in (2), one only sees a variation due to the Fermi–Dirac distribution variation but no influence of the DoS variation with energy. This is because the transmission probability, T, of electrons from source to drain is in fact limited by the DoS in the channel and not that in the source: the DoS in the channel is usually much smaller than the DoS in the source and therefore through the exclusion principle, electrons in the source can only cross the channel through the available path of transmission, proportional to the DoS in the channel. When the channel barrier is lowered, DoS and transmission for energies in the vicinity of the top of the channel barrier $E_{TCB}$, i.e. those that will contribute to the current, are therefore virtually unchanged (Fig. 2). As a result, the current is proportional to the product of an equivalent constant average transmission around TCB times the exponential function, which yields (2) (Fig. 2b). Note than in an ultra scaled nanowire, due to the irregular 1D-DoS shape and its low value at source extension side, the variation of source DoS when changing $V_G$ can have a limited impact on transmission, especially for low value of channel barrier where Source and channel DoS can be of the same order of magnitude. Such a small

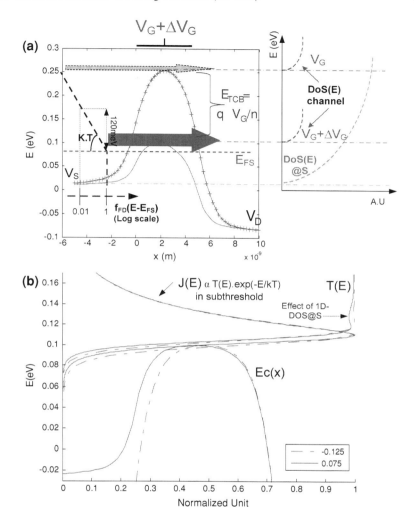

**Fig. 2 a** Schematic representation of the conduction band ($E_C$) and current flow in subthreshold regime of a N-MOS transistor. When $V_G$ is increased, the top of channel barrier (TCB) is lowered by a value $\Delta E_{TCB}$ and current increases exponentially. The DoS in the channel and therefore the transmission T stay mostly constant. **b** Simulation results for the 10 nm nanowire without barrier of Fig. 1: $E_C$ vs. normalized distance (($x - L_{sd}$)/($L + 2*L_{sd}$)), normalized current density J(E) and T (non normalized) versus energy (both curves have been rotated by 90°) below threshold at $V_G = -125$ and 75 mV. All the curves in **b** have been shifted down in energy so that $E_{TCB} = 0.1$ eV for comparison. Transmission and current are mostly unaffected by the relative motion of the source side of the channel barrier compared to TCB and the subthreshold slope is above the kT/q limit

fluctuation can be observed in Fig. 2b. However its impact on the current is negligible and if oscillations can appear in current or capacitance characteristics of ultra-scaled nanowires it is usually above threshold [17].

## 3 MBRT Effect in Constricted Si Nanowires

In Fig. 3, $I_D(V_G)$ and $SS(V_G)$ curves of RT-FETs with tunnel barriers, with increasing $L_C$ and overlap $L_{ov}$ are shown and compared to those of an identical classical "reference" nanowire transistor without barrier or overlap, TREF. We also show the current ratio $I_R = I_{DRTFET}/I_{DTREF}$ versus $V_G$. An overlap covering the constrictions is required in order to keep adequate electrostatic control of the gate over the tunnel barriers (TBs) [13, 14].

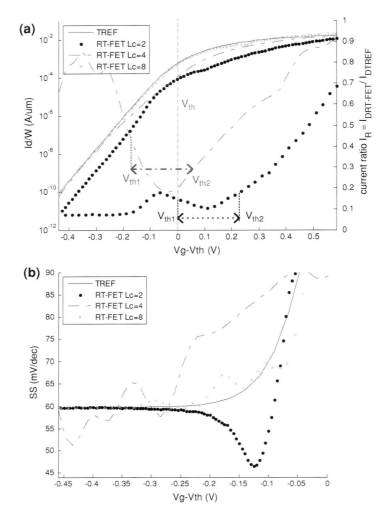

**Fig. 3** $I_D(V_G)$ and $SS(V_G)$ curves of TREF (1) and RT-FETs with, $L_C = L_{ov} = 2$ nm (2), $L_C = L_{ov} = 4$ nm (3) and $L_C = L_{ov} = 8$ nm (4). $V_d = 1$ V. $L = 10$ nm. The current ratio, $I_R(V_G)$ is also shown for case (2) and (3). $TB_S = 0.2$, $TB_D = 0.36$ eV. $L_{out} = 1$ nm

## 3.1 Concept

The device with $L_C = 2$ nm exhibits a 45 mV/dec steep SS just below threshold. This results from the gate modulation of resonant tunneling states induced by a differential motion of channel and additional tunnel barriers that modifies the longitudinal confinement in the device [14].

Figure 4a shows the band structure ($E_C$) as a function of position x and the transmission ($T$) as a function of energy of the RT-FET with $L_{out} = 1$ and $L_{ov} = 2$ nm before the steep SS region (SS = 60 mV/dec) at $V_G = -0.325$ V and at $V_G = -0.225$ V. Figure 4b shows the same in the steep SS region at $V_G = -0.125$ V, where $SS$ is 58 mV/dec and at $V_G = 0.075$ V where SS = 45 mV/dec (Fig. 3). Figure 4b can be compared to Fig. 2b obtained for the case of *TREF*. The transmission in the device with constrictions is quite different from that without constrictions and features very sharp peaks. This is the signature of resonant tunneling states. When the height of the TBs becomes typically comparable to that of the channel barrier, the TBs can induce resonant energy states in the quantum wells formed between tunnel-channel and/or tunnel-tunnel barriers. In these wells, electrons have discrete energy levels in the transport (x) direction for energy levels located in the conduction band, while for energies sufficiently above the wells, the electrons are essentially free. This decreases the DoS in the channel and therefore the current, compared to the reference transistor. Before the steep SS region, however, as illustrated in Fig. 4a, these wells are totally controlled by the gate potential. Increasing $V_G$ only shifts down the wells in energy but the relative shape of the well are preserved. As a consequence, the transmission close to the top of the channel barrier, however reduced due to the resonant tunneling effect, does not change when increasing $V_G$. As in the case of TREF, when increasing $V_G$, the number of electrons available to cross the channel only increases due to the Fermi–Dirac statistics and the slope is the same as that of $T_{REF}$ in this region. As a consequence, the current ratio is low (7%) but stays constant in this region. TBs can however locally move in a different way than the top of the channel barrier through 3D electrostatics effects at the gate edges and mixed influence between gate/source and gate/drain voltages. In Fig. 4b, one sees that, just below threshold, the distance in energy between source and channel barrier is sufficiently low so that the motion of $E_C$ with $V_G$ at the source gate edge is slowed down. This changes the relative shape of the wells with $V_G$ (i.e. creates a non-translational invariant movement) and induces changes in energy levels as well as their broadening ($\gamma$), which influence the DoS, transmission, current, and SS. Different motions can result in the change of the transmission and increase of the current. A combination of these can be seen in Fig. 4: (a) the depth of the potential well between TB$_S$ and TCB is reduced when $V_G$ is increased. This increases the energy of levels in this well and transfers them closer or above the top of the channel barrier effectively increasing the number of levels that can drive the current; (b) in the vicinity of its top, the shape of the tunnel barrier is changed such that its transparency is decreased. This reduces the coupling between source and channel

**Fig. 4** $E_C$ versus normalized distance $((x - L_{sd})/(L + 2*L_{sd}))$, and transmission T (rotated by 90° and non normalized) versus energy for TBT with $L_{out} = 1$ and $L_{ov} = 2$ nm **a** at $V_G = -325$ and $-225$ mV and **b** at $V_G = -125$ and 75 mV. All the curves have been shifted down in energy so that $E_{TCB} = 0.1$ eV for comparison. The deformation of the quantum well formed between $TB_S$ and channel barrier in **b** has increased the transmission and allowed for the steep subthreshold slope region observed in Fig. 3

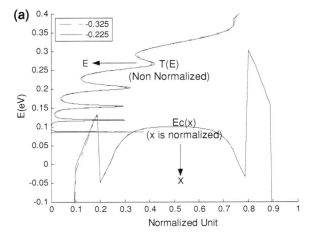

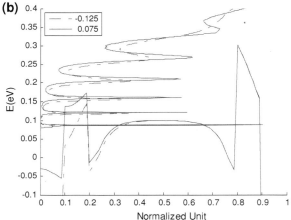

at these energies, effectively reducing the broadening of the energy levels close to the top of the barrier and shifting these levels down in energy. In this case the current can be increased through an increase of the transmission or through a more favorable probability of the Fermi–Dirac distribution related to the lower energy of the shifted levels. In the case of Fig. 4b, the combination of the two motions keeps the two lowest levels at about same energy but enhances the transmission. A reduction of *SS* below $kT/q$ is therefore achieved by this means even if the transparency of the TBs decreases in comparison to the channel barrier. This is because the electron concentration (and therefore $I_D$) increases faster than the Fermi–Dirac distribution of carriers due to the strong non-linear change of DoS shape with $V_G$. This change in DoS distribution modifies the transmission (Fig. 4b). This is not the case in the device without tunnel barrier, where the electrons are free of any confinement effects in the transport direction (x), (Fig. 2b). In this case, SS does not reach values below $kT/q$ (Fig. 3b).

**Fig. 5** $E_C$ versus normalized distance $((x - L_{sd})/(L + 2*L_{sd}))$, normalized current density J(E) and T (non normalized) versus energy for TBT with $L_{out} = 1$ and $L_{ov} = 2$ nm **a** below threshold at $V_G = 75$ mV and **b** above $V_{th2}$ at $V_G = 675$ mV. The position of the source Fermi level, $E_{FS}$, is also shown for comparison

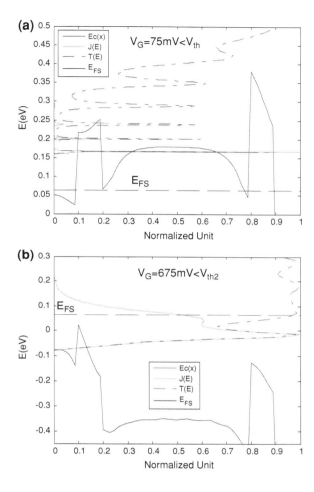

Finally, transistors with TBs can have two different thresholds. The first, $V_{th1}$ (=0.19 V in Fig. 3a), related to the resonant tunneling (RT)-current, happens like in a standard transistor when TCB passes below the source Fermi level $E_{FS}$. The second, $V_{th2}$ (=0.43 V in Fig. 3a), related to the thermionic current above the well, happens when the TBs pass below $E_{FS}$. A transistor dominated by the RT-current below its threshold voltage, $V_{th}$, as that with $L_C = L_{ov} = 2$ nm studied here (Fig. 5a), can achieve low off-current and steep slope region owing to the RTE. Its actual $V_{th}$ will be equal to $V_{th1}$. However for $V_G >= V_{th2}$, an important additional thermionic current will start flowing enabling further improvement of the slope and current ratio and hence very high on-current (Figs. 3a, 5b). By increasing $V_G$ above $V_{th2}$, an increasing part of the current is flowing above the well and therefore not or only slightly reduced by the resonant tunneling effect (RTE). The transistor is recovering the on-current level of TREF.

## 3.2 Performance Investigation

We compare the different devices in term of $I_{on}/I_{off}$ ratio and delay vs. supply voltage $V_{dd}$ (=$V_{on} - V_{off}$) for two different threshold voltages, i.e. 0.16 and 0.1 V, changing the gate workfunction to have same $V_{th}$ for all devices. The delay is estimated by $(Q_{on} - Q_{off})/I_{on}$, $Q_{on}$ and $Q_{off}$ being the total charge in the device at $V_G = V_{on}$ and $V_{off}$ respectively (Fig. 7). Depending on the tunnel barriers width, $L_C$, the current can be carried by resonant tunneling states in the well (RT-current) (Fig. 5a) and/or free or quasi free states above the well ("thermionic" like current) (Fig. 6, both currents flow in case of RT-FET of Fig. 6a). A transistor already dominated by the thermionic current below threshold, i.e. with $L_C$ too large, will have characteristics very similar to TREF but with a shifted threshold voltage, i.e. $V_{th} = V_{th2}$ (e.g. transistor with $L_C = 8$ nm). Its performances are not enhanced

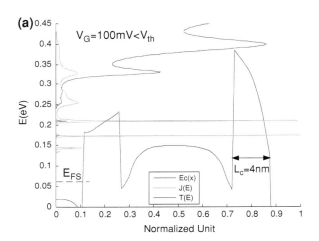
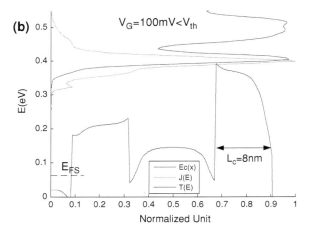

**Fig. 6** $E_C$ versus normalized distance $((x - L_{sd})/(L + 2*L_{sd}))$, normalized current density J(E) and T (non normalized) versus energy below threshold at $V_G = 100$ mV for TBT, **a** with $L_{out} = 1$ and $L_C = L_{ov} = 4$ nm and **b** with $L_{out} = 1$ and $L_C = L_{ov} = 8$ nm. The position of the source Fermi level, $E_{FS}$, is also shown for comparison

# Gate Modulated Resonant Tunneling Transistor (RT-FET)

compared to that of TREF and therefore not further investigated here. As said above, a transistor dominated by the RT-current below its threshold voltage, $V_{th}$, can achieve low off-current and steep slope region owing to the RT effect. Its actual $V_{th}$ will be equal to $V_{th1}$. The RT-FET with $L_C = 2$ nm has the lowest $I_{off}$ and enhanced $I_{on}/I_{off}$ ratio compared to TREF for both sets of $V_{th}$. However for $V_G \geq V_{th2}$, an important additional thermionic current will start flowing enabling further improvement of the slope and current ratio and enabling very high on-current (e.g. transistor with $L_C = 2$ nm, Fig. 3). The transistor is recovering the

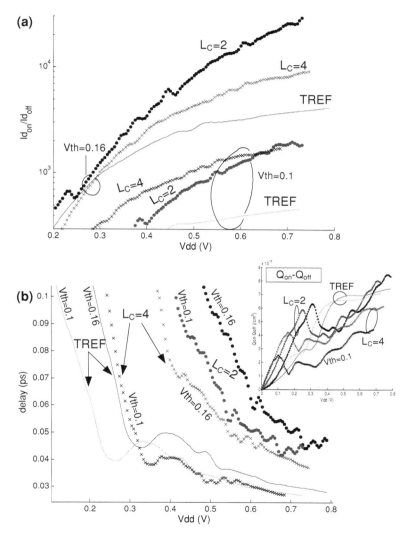

**Fig. 7** a $I_{on}/I_{off}$ ratio, b delay, and charge $Q_{on} - Q_{off}$ (inset) versus $V_{dd}$, of TREF and RT-FETs with $L_C = 2$ nm and $L_C = 4$ nm for $V_{th}$ set to 0.16 and 0.1 V

on-current level of TREF and its delay rapidly decreases in this range. Transistors having intermediate $L_C$ (i.e. $L_C = 4$ nm in Fig. 3) will have both thermionic and RT-current flowing in subthreshold (Fig. 6a). They will present a regime between $V_{th1}$ and $V_{th2}$ where their slope and current characteristics are in between subthreshold and above threshold. Their effective threshold voltage, $V_{th}$, will also be in between $V_{th1}$ and $V_{th2}$. Although their off-current compared to RT-FET with $L_C = 2$ nm is not as low, they feature very sharp turn on just above threshold which ensure good delay characteristics and comparatively good $I_{off}$ performances for very low threshold voltages (i.e. 0.1 V and below). This makes them interesting devices for ultra low voltage especially at ultra low threshold for high speed application where lower $I_{on}/I_{off}$ ratio can be trade-off for low delay and high $I_{on}$. Note also that the capacitance, or charge variation, increase of the RT-FETs due to the overlaps, is more or less compensated by a charge reduction due to the DoS reduction in the TBs and the quantum well due to RT effect. Also in this case extra peaks in the capacitance and charge curves related to the 0D DOS can be observed for TBT with $L_C = 2$ and 4 nm (inset of Fig. 7b). Finally performances of RT-FETs in terms of $I_{on}$ and delays for a given $V_{dd}$ depend on the distance between $V_{th}$ and $V_{th2}$ which is also influenced by barrier height. The curves obtained here for $L_C = 2$ nm were optimized for $V_{dd} = 0.7$ V, but better delays can be achieved for lower $V_{dd}$ by reducing barrier height.

## 4 Conclusions

Performances of a new concept of nanoscale MOSFET, the gate modulated resonant tunneling (RT)-FET, have been investigated through 3D non-equilibrium Green function simulations enlightening the main physical mechanisms.

A well optimized RT-FET, with tunnel barriers width, $L_C$, in the order of a few nanometers, features low off current and SS below the kT/q limit together with high on-current. We show that the improvement in SS results from the modulation (when the gate voltage is varied) of resonant tunneling states induced by a differential motion of channel and additional tunnel barrier(s) that modifies the longitudinal confinement in the device. This leads to an increase of the DoS in the channel and therefore an increase of the transmission with $V_G$. We have also shown that RT-FETs feature a second threshold voltage above threshold, related to a change of regime between resonant-tunneling current in the well to thermionic current above the well, allowing to filter more efficiently the off than the on-current. For gate voltage above this threshold, an important additional thermionic current will start flowing enabling further improvement of the slope and current ratio and hence very high on-current. The transistor is recovering the on-current level of a transistor without tunnel barrier.

High $I_{on}/I_{off}$ current ratios with low voltage operation and good delay characteristics are therefore predicted. The 10 nm Si RT nanowire with $L_C$ of a few nanometers investigated here could operate with a supply voltage as low as 0.5 V,

$I_{on}/I_{off} > 10^4$ (an order of magnitude improvement compared to a classical nanowire) and low leakage.

**Acknowledgments** This material is based upon works supported by FNRS Belgium and by Science Foundation Ireland under Grant 05/IN/I888. This work was supported in part by the European Community (EC) Seventh Framework Program through the Networks of Excellence NANOSIL and EUROSOI+ under Contracts 216171 and 216373.

# References

1. Esaki, L.: New phenomenon in germanium narrow junction. Phys. Rev. http://prola.aps.org/pagegif/PR/v109/i2/p603_1/p603 (1958)
2. Appenzeller, J., et al.: Band-to-band tunneling in carbon nanotube field-effect transistors. Phys. Rev. Lett. (2004). doi:10.1103/PhysRevLett.93.196805
3. Zhang, Q., et al.: Low-subthreshold-swing tunnel transistors. IEEE Electron Dev. Lett. (2006). doi:10.1109/LED.2006.871855
4. Matloubian, M., Chen, C., Mao, B.-Y., Sundaresan, R., Pollack, G.P.: Modeling of the subthreshold characteristics of SOI MOSFET's with floating body. IEEE Trans. Electron Dev. (1990). doi:10.1109/16.57160
5. Gopalakrishnan, K., Griffin, P. B., Plummer, J. D.: I-MOS: A novel semiconductor device with subthreshold slope lower than kT/q. IEDM Tech. Dig. (2002). doi:10.1109/IEDM.2002.1175835
6. Abelé, N., Fritschi, R., Boucart, K., Casset, F., Ancey, P., Ionescu, A.M.: Suspended-Gate MOSFET: bringing new MEMS functionality into solid-state MOS transistor. IEDM Tech. Dig. (2005). doi:10.1109/IEDM.2005.1609384
7. Salahuddin, S., Datta, S.: Use of negative capacitance to provide voltage amplification for ultralow power nanoscale devices. Nano Lett. (2008). doi:10.1021/nl071804g
8. Salvatore, G., Bouvet, D., Ionescu, A.: Demonstration of subthreshold swing smaller than 60 mV/decade in Fe-FET with P(VDF-TrFE)/SiO2 gate stack. IEDM Tech. Dig. (2007). doi:10.1109/IEDM.2008.4796642
9. Choi, W.Y., et al.: Tunneling field-effect transistors with subthreshold swing (SS) less than 60 mV/dec. IEEE Electron Dev. Lett. (2007). doi:10.1109/LED.2007.901273
10. Khatami, Y., Banerjee, K.: Steep subthreshold slope n- and p-type tunnel-FET devices for low-power and energy-efficient digital circuits. IEEE Trans. Electron Dev. (2009). doi:10.1109/TED.2009.2030831
11. Luisier, M., Klimeck, G.: Performance comparisons of tunneling field-effect transistors made of InSb, Carbon, and GaSb-InAs broken gap heterostructures. IEDM Tech. Dig. (2009). doi:10.1109/IEDM.2009.5424280
12. Koswatta, S.O., et al.: 1D broken-gap tunnel transistor with MOSFET-like on-currents and sub-60 mV/dec subthreshold swing. IEDM Tech. Dig. (2009). doi:10.1109/IEDM.2009.5424279
13. Afzalian, A., Colinge, J.-P., Flandre, D.: Variable barrier resonant tunneling transistor: a new path towards steep slope and high on-current? In: Proceedings EUROSOI Conference (2010)
14. Afzalian, A., Colinge, J.-P., Flandre, D.: Physics of gate modulated resonant tunneling (RT)-FETs: multi-barrier MOSFET for steep slope and high on-current. Solid State Electron. (to be published)
15. Afzalian, A., et al.: A new F(ast)-CMS NEGF Algorithm for efficient 3D simulations of Switching Characteristics enhancement in constricted Tunnel Barrier Silicon Nanowire MuGFETs. J. Comput. Electron. (2009). doi:10.1007/s10825-009-0283-1

16. Colinge, J.-P., Morel, H., Chante, J.-P.: Field effect in large grain polycrystalline silicon. IEEE Trans. Electron. Dev. (1983). http://ieeexplore.ieee.org/
17. Afzalian, A., Lee, C.W., Dehdashti-Akhavan, N., Yan, R., Ferain, I., Colinge, J.-P.: Quantum confinement effects in capacitance behavior of multigate silicon nanowire MOSFETs. IEEE Trans. Nanotechnol. (2010). doi:10.1109/TNANO.2009.2039800

# Ohmic and Schottky Contact CNTFET: Transport Properties and Device Performance Using Semi-classical and Quantum Particle Simulation

**Huu-Nha Nguyen, Damien Querlioz, Arnaud Bournel, Sylvie Retailleau and Philippe Dollfus**

**Abstract** In this chapter, we investigate the device operation and performance of carbon nanotube-based field-effect transistors (CNTFETs) by means of particle Monte Carlo simulation including electron–phonon scattering. Within semi-classical approach of transport, both Ohmic- and Schottky-type source and drain contacts are considered and the effect of ambipolar transport inherent in Schottky barriers is analysed. This effect degrades the on/off current ratio and impedes the current saturation at high $V_{DS}$, which makes the transition frequency strongly dependent on the drain voltage. However, by lowering the Schottky barrier height for electrons, the ambipolar behavior is reduced, which makes the device characteristics closer to that of ohmic-contact transistors. In both cases the intrinsic current-gain cutoff frequency is shown to reach about 800 GHz for a gate length of 100 nm. The detailed analysis of scattering events in the channel shows that the fraction of ballistic electrons in the Ohmic contacts devices increases from 80 to 95% when reducing the gate length from 100 to 10 nm. Hence, the transport in the channel is likely to be strongly coherent, which is analyzed by means of quantum Wigner Monte Carlo simulation, whose algorithm is fully compatible with the semi-classical one. It is shown that in spite of significant quantum transport effects such as source-drain tunneling through the gate-induced barrier and reflection by the sharp potential drop, the microscopic quantum features, e.g. the oscillations of the Wigner function, are not strongly reflected at the macroscopic level of terminal current.

H.-N. Nguyen, D. Querlioz, A. Bournel, S. Retailleau and P. Dollfus (✉)
Institut d'Electronique Fondamentale, CNRS,
Univ. Paris-Sud, UMR 8622, F-91405 Orsay, France
e-mail: philippe.dollfus@u-psud.fr

# 1 Introduction

Exceptional transport properties of carbon nanotubes (CNTs), including very high carrier mobility [1] and mean free path [2, 3], make them very attractive for designing carbon nanotube field-effect transistors (CNTFETs) likely to overcome the limitations of Si transistors. Since the first demonstrations of CNTFETs in 1998 [4, 5], significant progress has been made in understanding the device physics and in improving the transistor performance ([6] and references therein). In parallel, efforts have been made to demonstrate CNTFET circuit operation [7–10] and integration with silicon MOS technology. Beside near-ballistic transport in CNTs, CNTFETs was rapidly expected to offer a good potential for high frequency applications [11]. The challenge was then in the enhancement of high-frequency performance [12, 13]. Indeed, the inherent mismatch between the high impedance of nanodevices and the 50 $\omega$ impedance of standard HF equipments, together with the relatively low drive current provided by a single CNT, impedes the accurate evaluation of their dynamic behaviour. Recently, the intrinsic current gain cutoff frequency of devices made of a large number of SW-CNTs rapidly increased from 8 GHz [14] to 80 GHz [15]. A transconductance of 20 µS and a transition frequency of 50 GHz were also reached at room temperature for a single-CNT transistor of 300 nm gate length [16].

It is generally observed that the electrical performance of CNTFETs is not limited by the intrinsic nanotube properties but rather by the contact properties. The first realisations of CNT-based field effect transistors offered Schottky-like source and drain contacts which strongly influence the operation and characteristics of CNTFETs. In particular, the ambipolar behaviour of Schottky transistors [17] may lead to unacceptable leakage currents [18]. Strong efforts have been made to improve the behaviour of contacts obtained from different types of metal and process [7, 19–21]. First Ohmic contact p-type CNTFETs were fabricated by depositing Pd-contact pads onto CNTs and annealing in Ar while near-Ohmic Al contacts were obtained for n-type FETs [22]. Thanks to chemically-doped contacts, they rapidly gave high performance in terms of subthreshold slope and ON/OFF ratio [23].

The semi-classical Boltzmann approach to transport in semiconductors has been very successful for many years in interpreting the physics of electron devices. It has been developed by many groups to study a wide variety of transport problems in many kinds of devices, to such a point that it is impossible to summarize here the most significant examples of its applications. Extensive overviews of this method may be found in [24–26]. In spite of disadvantages due to large computational requirements and some limitations inherent in the finite number of simulated particles, this technique of transport simulation turns out to be robust and versatile. It has been used by some groups to investigate the semi-classical operation of CNTs and CNTFETs [2, 27, 28]. Quantum simulation becomes recommended when device dimension reduces to the deca-nanometer scale, i.e. when device performance may rely on coherent quantum effects. In spite of difficulties in including scattering, the most widely used quantum simulation method is the non-equilibrium Green's function (NEGF) one [29]. Several recent works based on the NEGF simulation of CNTFETs may be found [30, 31].

Another option is to use the Wigner's formalism of quantum transport which is based on a function defined in the phase space as a Fourier transform of the density operator [32]. In the classical limit this function reduces to the classical Boltzmann distribution function and under some approximations on the treatment of scattering (weak and fast scattering limit) the collision operator of the Wigner transport equation is similar to that of the Boltzmann equation [33, 34]. Wigner Monte Carlo simulation can thus reconcile semi-classical and quantum simulations [35] and has been recently implemented for CNTFETs [36].

In this article we use extensively the Monte Carlo (MC) simulation technique to investigate the operation and performance of CNTFETs where source and drain contacts may be either Schottky- or Ohmic-like. Both semi-classical (Boltzmann) and quantum (Wigner) MC algorithms including phonon scattering are used. They are self-consistently coupled to the 2D cylindrical Poisson's equation to simulate gate-all-around CNTFETs which have been shown to be optimal in ultimately scaled devices [37]. The possibly high ballisticity of such transistors can make the transport essentially coherent and one may wonder whether the quantum transport effects have a strong impact or not on the device performance. Both quantum and semi-classical approaches are thus compared.

The article begins with a brief summary of electron and phonon dispersion in carbon nanotubes (Sect. 2). Then the MC model of Boltzmann and Wigner simulation is detailed in Sect. 3, while the numerical treatment of Ohmic and Schottky contacts is described in Sect. 4. The performance of Ohmic and Schottky CNTFETs are investigated using semi-classical simulation in Sect. 5. Finally, we analyze the ballistic and quantum transport in the channel of Ohmic-CNTFETs in Sect. 6.

## 2 Electron and Phonon Dispersion of Carbon Nanotube

Since their discovery by Ijima [38], CNTs have been considered as the ultimate carbon-based 1D system. A nanotube may be seen as a graphene sheet rolled up into a hollow cylinder and its structure depends on the circumferential chiral vector $\mathbf{C} = n\,\mathbf{a}_1 + m\,\mathbf{a}_2$ which is specified as a combination of the two unit vectors $\mathbf{a}_1$ and $\mathbf{a}_2$ of the hexagonal honeycomb lattice. The indices $(n, m)$ fully determine the electronic structure and properties of a CNT. Because the microscopic structure of CNTs is closely related to that of graphene, we can obtain the electronic band structure of a specified nanotube from that of graphene by applying appropriate periodic boundary conditions to the component of wave vector along the circumferential direction of the tube. This is the idea of the zone-folding approximation [39].

If the difference between $n$ and $m$ is divisible by 3, the nanotube is metallic, otherwise it is semiconducting. In particular, all "armchair-type" nanotubes ($n$, $n$) are metallic, one-third of "zigzag-type" nanotubes ($n,0$) are metallic and two-thirds are semiconducting [39]. In this article, we are interested in the semiconducting zigzag type.

By using the tight binding method in the nearest neighbor approximation, we can get a simple analytical expression for the $\pi$ electronic states of graphene. In a zone-folding approximation, the component of wave vector along the nanotube circumference is quantized, which is characterized by a quantum number $\eta$. In contrast, the component $k_z$ along the nanotube axis remains continuous (for infinite tubes). For a zigzag CNT of diameter $d = 2a_0 \, (n^2 + m^2 + nm)^{1/2}/2\pi$, with $a_0 = |\mathbf{a}_1| = |\mathbf{a}_2| = 0.249$ nm, the electron wave vector is

$$\mathbf{k} = k_z \mathbf{u}_z + \frac{2\eta}{d} \mathbf{u}_\theta \left( \text{with } \eta = -n+1,\ldots,n \text{ and } -\frac{\pi}{T} < k_z \leq \frac{\pi}{T} \right) \quad (1)$$

where $\mathbf{u}_z$ and $\mathbf{u}_\theta$ are the unit vectors along the tube axis and circumference, respectively, and $T = 0.43$ nm is the length of the zigzag CNT unit cell. The zone-folding method thus breaks up each band of graphene into $2n$ subbands among which $n-1$ subbands are twofold degenerate. This degeneracy leads to two equivalent valleys in the subband structure, each centered near a graphene $K$ point. The energy dispersion for a zigzag tube is finally given by

$$E(k_z, \eta) = \pm \gamma \sqrt{1 \pm 4\cos\left(\frac{Tk_z}{2}\right)\cos\left(\frac{\pi\eta}{n}\right) + 4\cos^2\left(\frac{\pi\eta}{n}\right)} \quad (2)$$

where $\gamma$ is the $\pi$-hopping integral between nearest-neighbour atoms in graphene. This approach neglects the curvature effects inherent in CNTs. It works well for a diameter larger than 0.8 nm but for very small-diameter CNTs the calculation should properly consider that the C–C bonds perpendicular and parallel to the axis may be different and that $\pi$ and $\sigma$ orbitals can mix [40].

By analogy with standard parabolic description of bands in conventional semiconductors, a convenient analytical approximation of (2) can be made to describe the energy of the first three subbands of each valley for semiconducting zigzag CNTs. The general expression of the energy dispersion is finally [27]

$$\frac{\hbar^2 k_z^2}{2m_b(n)} = \left[ E_b - E_b^0(n) \right] \left[ 1 + \alpha_b(n) \left( E_b - E_b^0(n) \right) \right] \quad (3)$$

where $E_b^0(n)$, $m_b(n)$ and $\alpha_b(n)$ are the minimum of energy, the effective mass and the non-parabolicity coefficient of subband $b$, respectively. This approximation (3) will be used in what follows for the Monte Carlo simulation, including scattering rates and transport calculation. In this article, we focus on the semiconducting zigzag CNT (19,0) with diameter $d = 1.5$ nm. Given hopping energy $\gamma = 3$ eV [39], one obtains subband energies of 0.28, 0.59, 1.09 eV, effective masses of $0.048 m_0$, $0.129 m_0$, $0.133 m_0$ and non-parabolicity coefficients of 1.572, 0.765,

0.395 eV$^{-1}$ for the first three subbands, respectively. In practice, for drain bias voltage lower than 0.5 V, it is sufficient to consider the first two subbands only.

Similarly, the phonon spectrum of CNTs can be well described by zone folding the phonon dispersion curves of 2D graphene sheet [39], in good agreement with ab initio calculation [41]. The phonon wave vector is determined in the same way as the electron wave vector

$$\mathbf{q} = q_z \mathbf{u}_z + \frac{2\mu_q}{d} \mathbf{u}_\theta \qquad (4)$$

where the quantum number $\mu_q$ specifies the confinement along the tube circumference. Each of the six phonon branches of graphene is broken in $2n$ subbands.

These approximations are commonly considered for the phonon dispersion used to calculate the electron–phonon scattering in transport simulation [2, 27, 30, 42, 43]. Longitudinal acoustic and optical modes are considered to be dominant for electron scattering within the first three subbands of both valleys [44]. For the acoustic modes, we use the same phonon energies as in [27]. A linear dispersion is used for intra-subband acoustic phonons ($\mu_q = 0$). For phonon modes involved in inter-subband/intra-valley transition ($\mu_q > 1$) the dispersion is nearly flat at $\Gamma$ point and these branches are described via a constant energy equal to the minimum value of each mode. A constant energy is also considered for acoustic phonons involved in inter-valley transitions. The longitudinal optical phonons of graphene are to be included for transport simulation, with energy values of 180 meV ($K$ point) and 200 meV ($\Gamma$ point) [45]. They induce inter-valley/inter-subband transitions. Finally, radial breathing modes (RBM) are included in the phonon description with a constant energy of 19 meV for a (19,0) tube [46, 47]. They give rise to an additional intra-subband transition.

## 3 Transport Model in Boltzmann and Wigner Formalisms

In semi-classical device simulation, the Boltzmann transport equation (BTE) is solved to obtain the distribution function $f_b$ from which we can calculate various quantities of interest such as the current and the carrier density, velocity and energy. The 1D BTE is

$$\frac{\partial f_b}{\partial t} + \frac{\hbar}{m} k \cdot \nabla_x f_b + \frac{1}{\hbar} F^{\text{ext}} \cdot \nabla_k f_b = C f_b \qquad (5)$$

where $x$ and $k$ are the position and wave vector coordinates, $m$ is the carrier effective mass, $F^{\text{ext}}$ is the external force, and $C$ is the collision operator describing the scattering processes in the system. The particle Monte Carlo method to solving the BTE consists in simulating the trajectories of individual carriers through the device under the influence of the electric field and scattering processes. The distribution function is reconstructed from the ensemble of carriers in the form

$$f_b(x,k,t) = \sum_i \delta(x-x_i(t))\delta(k-k_i(t)) \tag{6}$$

where $x_i$, $k_i$ are the position and wave vector of the $i$th particle, respectively. We have developed a model within the software MONACO to simulate the semi-classical transport in carbon nanotubes [2, 48, 49]. The analytical dispersion relation (3) is used to describe the nanotube subbands and the electron–phonon scattering is included through acoustic, optical, and radial breathing phonon modes, as described above.

The most natural approach to modeling the statistics of a quantum system is the density matrix formalism. The off-diagonal elements of the density matrix, called coherences, characterize the real-space delocalization of electrons. By Fourier transform of the density matrix, one can define a function which depends on both position and momentum [32, 50]. It is the Wigner function $f_w$. The Wigner function can be seen as a generalization of the Boltzmann distribution function in quantum mechanics formulation. However, it is not a distribution function because it can take negative value. The 1D Wigner transport equation may be written as

$$\frac{\partial f_w}{\partial t} + \frac{\hbar k}{m}\frac{\partial f_w}{\partial x} = Qf_w + Cf_w \tag{7}$$

where the non-local effect of the potential is contained in the quantum evolution term $Qf_w$

$$Qf_w(x,k) = \int dk' V_w(x,k')f_w(x,k+k') \tag{8}$$

and the Wigner potential is defined for a particle in a potential $V(x)$ as

$$V_w(x,k) = \frac{1}{2\pi\hbar}\int dx' \exp(-ikx')\left[V\left(x+\frac{x'}{2}\right) - V\left(x-\frac{x'}{2}\right)\right] \tag{9}$$

The effect of electron–phonon coupling is described by the collision term $Cf_w$. If we assume the coupling to the phonon modes to be weak and the phonon scattering phenomena to be rapid, the collision term takes the same form as the one in the Boltzmann transport equation [35]. This approximation neglects advanced effects of scattering as collisional broadening and retardation or intra-collisional field-effects. The resulting equation is often called the Wigner–Boltzmann equation [51].

The strong similarity with Boltzmann's equation suggests using similar statistical particle technique to solve the Wigner–Boltzmann transport equation. Here we choose the Monte Carlo technique that has been used in the simulation of resonant tunneling diodes and silicon nanoscaled transistors [52–54]. In this technique the Wigner function is seen as a sum of pseudo-particle contributions according to

$$f_{\rm w}(x,k,t) = \sum_i A_i \delta(x - x_i(t))\delta(k - k_i(t)) \tag{10}$$

where $x_i$, $k_i$, $A_i$ are the position, wave vector, and affinity of the $i$th particle, respectively. The affinity concept was initially proposed by Shifren et al. [55]. This weighting quantity contains all the quantum information of the electron system. It is not constant and may get negative values. In this technique, we treat statistically only the collision term. During a free flight, the coordinates of pseudo-particles evolve as for classical particles. It should be noted that, in contrast to the semi-classical trajectories, the wave vector is not changed by the potential during free flights, but only by scattering events. The potential affects only the particle affinity through the quantum evolution term. After each time step between two solutions of Poisson's equation, the affinities in each mesh $M(x, k)$ of the phase space are updated according to

$$\sum_{i \in M(x,k)} \frac{{\rm d}A_i}{{\rm d}t} = Qf_{\rm w}(x, k). \tag{11}$$

Concerning the particle injection conditions, particles with an affinity of 1 are injected at the source-drain contacts to ensure the charge neutrality. Moreover, we inject a particle of zero affinity in each mesh of the phase space where there is no particle and the quantum evolution term is non zero [34, 52]. The latter condition is mandatory to ensure the conservation of total affinity and charge.

In this article, both Boltzmann and Wigner Monte Carlo algorithms are used for the simulation of coaxially-gated CNTFET. The simulator couples the 1D transport equation, in either quantum or semi-classical formulation, with the 2D Poisson's equation for the cylindrical device symmetry.

The Pauli exclusion principle is included in the simulation by injecting particles at contacts according to the Fermi–Dirac equilibrium distribution and by using a rejection technique in the treatment of selected scattering events [48, 56], which corresponds to the introduction of a self-scattering of probability proportional to $(1 - f(x, k'))$ where $f(x, k')$ is either the Boltzmann or the Wigner function at final state. The functions $f(x, k)$ are updated after each time step. It should be noted that the band-to-band tunneling is not included in the model, which limits the range of bias operation to moderate values of gate and drain voltage with respect to the bandgap of the CNT.

## 4 Model of Ohmic and Schottky Contacts: Simulated Devices

The modelling of Ohmic and Schottky contacts is a crucial problem of computational electronics, even for ideal cases. In semi-classical Monte Carlo simulation, Ohmic contacts on highly doped regions are usually treated by assuming the vicinity of the contact to be in thermal equilibrium. The electrical neutrality of the real-space cells adjacent to the contact is thus the only condition of particle injection through the contact. After each time step, if particles are missing in some "Ohmic" cells with respect to the charge neutrality, the appropriate number of carriers (of affinity equal to 1) is injected in these cells to recover the neutrality. An equilibrium distribution is used to select randomly their wave vector components.

In Schottky barrier (SB) CNTFETs, the space charge region near the Schottky interface and the tunneling of carriers through the nanotube-metal junction must be carefully considered. In particular, specific conditions of carrier injection have to be applied.

Let's consider a contact at position $x = 0$ with a Schottky barrier $\Phi_B$. After solving the Poisson equation at the end of a time step $\Delta t$, which provides the conduction band profile $E_C(x)$, $I_{inj}\Delta t/e$ particles have to be injected from metal to CNT, where $e$ is the elementary charge. The appropriate current $I_{inj}$ is evaluated from the Landauer formula [57]

$$I_{inj} = \frac{2e}{h} \int T(E) M(E) f(E) dE \qquad (12)$$

where $h$ is the Planck's constant, $M(E)$ is the number of available subbands in the CNT, and $f(E)$ is the distribution function in the metal. The energy-dependent transmission coefficient $T(E)$ at the metal/CNT interface is calculated within the WKB approximation. The energy of injected particles is chosen randomly using a probability deduced from Landauer formula. If $E$ is higher than the Shottky barrier $\Phi_B$, electrons are injected at $x = 0$, otherwise they are injected at the position $x$ where $E_C(x) = E$. The initial kinetic energy $\varepsilon$ of tunneling electrons is 0. For thermionic particles, $E = \varepsilon + \Phi_B$ is chosen according to the distribution function in the metal.

For electrons moving in the space charge region, the possibility of tunneling from the CNT to the Schottky contact is considered when they "hit" the conduction band $E_C$, i.e. when $\varepsilon$ vanishes during the particle movement. The electron is collected by the metal only if a random number $r$ is less than $T(E_C)$. Otherwise, the particle is reflected inside the CNT as in the classical case. This model of Schottky contact was initially developed for unipolar structures and recently extended to study ambipolar transport in CNTFETs [58].

The device parameters used for the simulation are as follows (see also schematic cross-section of Fig. 1). The coaxial gate length varies from 6 to 100 nm with a 5.3 nm equivalent gate oxide thickness, unless otherwise stated. In Ohmic-contact transistors the source and drain access regions are either 20- or 30 nm-long with an N-type doping of 0.34 nm$^{-1}$. This doping is assumed to be induced electrostatically by an additional gate which is not included explicitly in the simulation. This doping level corresponds to about one electron provided for 517 carbon atoms and puts the Fermi level 59 meV above the first subband. The Schottky barriers are characterized by the barrier height $\Phi_B$. There is no access zone in Schottky-barrier (SB) transistors (see Fig. 1). A semiconducting zigzag nanotube (19,0) is considered with a bandgap of 0.55 eV. The first two subbands are taken into account with effective masses of $0.048 m_0$ and $0.129 m_0$,

**Fig. 1** Schematic cross-section of simulated devices with **a** Ohmic and **b** Schottky source and drain contacts

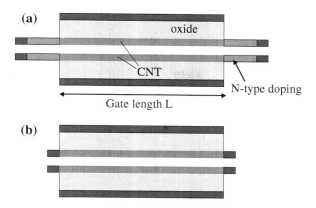

**Fig. 2** $I_D$–$V_{GS}$ characteristics of CNFETs for $L = 100$ nm and EOT $= 5.3$ nm. The S-D Schottky barrier height is $\Phi_B = 0.275$ eV

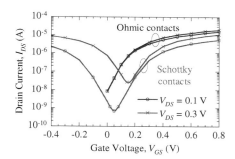

respectively, where $m_0$ is the free electron mass. Electron–phonon scattering is included with the parameters described in Sect. 2 [2] with a full set of electron–phonon interaction processes by acoustic, optical and radial-breathing phonon modes. All simulations are performed at room temperature $T = 300$ K.

## 5 Comparison of Ohmic and Schottky Contacts

All results presented in this Section were obtained using semi-classical Boltzmann simulation.

For a gate length of 100 nm, Fig. 2 compares the transfer characteristics of a Schottky-barrier (SB) CNFET with that of an Ohmic contact device. The SB height is mid-gap, i.e. $\Phi_B = 0.275$ eV, which leads to symmetric conditions of electron and hole injection at positive and negative $V_{DS}$. At high positive $V_{GS}$, the current is mostly due to the electrons going from source to drain (Fig. 3a). In contrast, at negative gate voltage, the holes injected at the drain barrier contribute mainly to the total current (Fig. 3b). We can see that the drain voltage modulates the Schottky barrier at drain-end while the potential in the channel is fixed by the gate voltage. Hence, when the drain voltage increases, the hole injection at drain contact is

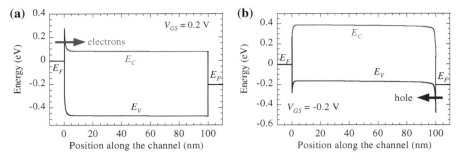

**Fig. 3** Profiles of conduction band and valence band at $V_{DS} = 0.2$ V, **a** at negative gate voltage $V_{GS} = -0.2$ V, **b** at positive gate voltage $V_{GS} = 0.2$ V

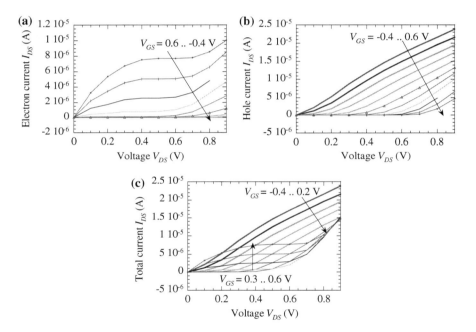

**Fig. 4** $I_D$–$V_{DS}$ characteristics of SB-FET for different values of $V_{GS}$ ($L = 100$ nm, EOT $= 5.3$ nm) **a** electron current, **b** hole current, **c** total current

enhanced, which gives a higher hole current. As a consequence, the minimum current increases.

The ambipolar transport makes SB-CNFET characteristics unconventional [59]. In particular, it degrades the current saturation behaviour (Fig. 4) and the subthreshold characteristics. For $V_{DS} = 0.3$ V the subthreshold slope is about 95 mV/dec instead of 70 mV/dec in the Ohmic contact device. The Schottky contacts are also detrimental to the on-state current and transconductance. The transconductance is limited to 20.5 μS in the SB device instead of 32 μS in the Ohmic one (not shown).

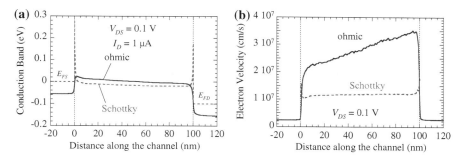

**Fig. 5** Conduction band and velocity profiles for SB and ohmic CNTFET at $V_{DS} = 0.1$ V and $I_D = 1$ μA

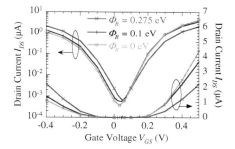

**Fig. 6** $I_D$–$V_{GS}$ characteristics for three values of Schottky barrier height at source and drain contacts ($V_{DS} = 0.1$ V)

Figure 5 shows the conduction band and electron velocity profiles in Ohmic and Schottky CNTFETs at $V_{DS} = 0.1$ V for the same value $I_D \approx 1$ μA, i.e. for $V_{GS} = 0.4$ V and $V_{GS} = 0.18$ V in SB and Ohmic devices, respectively. In the Ohmic device, the electric field, though rather weak (mean value of 2.5 kV/cm), can take significant values reaching 6 kV/cm. In such conditions of small (less than 10 kV/cm) but finite electric field, the electron mean free path is only limited by intra-subband acoustic phonon scattering and is higher than 100 nm [2]. The particles are thus accelerated in the channel to reach a relatively high velocity of 3.5 kV/cm. In the SB-device, the effective electrostatic gate control leads to a very small driving electric field in the channel, which results in a velocity nearly constant and much smaller than in the Ohmic-contact transistor.

Actually, the height and form of the Schottky barrier at contacts can be modified by interaction of metal or CNT with molecules of the environment or by the interface geometry [59]. To study the influence of the barrier height, we now consider two other barrier height values $\Phi_B = 0.1$ and 0 eV for electrons, which means that it is $E_G - \Phi_B = 0.45$ and 0.55 eV for holes, respectively. In Fig. 6 two $I_D$-$V_{GS}$ curves (at $V_{DS} = 0.1$ V) are compared to the previous case of mid-gap Schottky barrier. The electron injection at the source contact is enhanced to the detriment of hole injection at the drain, which typically increases the transconductance and the on-current which become rather close to values obtained for Ohmic-contact transistor (if compared with results of Fig. 2). However, the off-current is only slightly reduced. These results are consistent with experimental ones [60]. The ambipolar transport can thus be reduced by adjusting parameters like the Schottky barrier height.

By looking at Fig. 7a which plots the cut-off frequency as a function of gate voltage, we observed that decreasing the Schottky barrier height provides better dynamic performance. In the case of mid-gap SB, the contribution of the hole current at high $V_{DS}$ degrades the saturation of the drain current (Fig. 4) and makes the cut-off frequency strongly dependent on the drain voltage. When the SB height is reduced, the better saturation of the drain current gives a weaker dependence of $f_T$ on $V_{DS}$ and the behavior of $f_T$ becomes similar to that of Ohmic-contact CNTFET (see Fig. 7b).

However, thanks to smaller gate capacitance in Schottky devices (see Fig. 8a) the maximum transition frequency $f_T$ is similar in Ohmic and SB CNTFET and reaches typically 800 GHz for a gate length of 100 nm, as shown in Fig. 8b. In spite

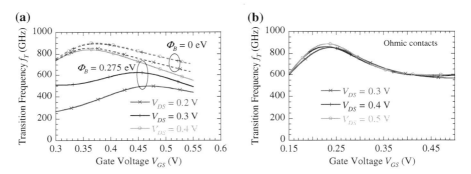

**Fig. 7** a Transition frequency $f_T$ as a function of the gate voltage for two Schottky barrier heights $\Phi_B$ of 0.27 and 0 eV for $L = 100$ nm and EOT $= 5.3$ nm, b Transition frequency $f_T$ for an Ohmic contact transistor with $L = 100$ nm

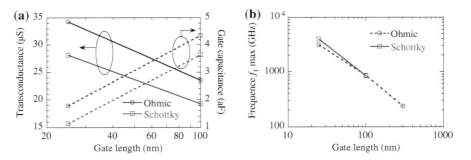

**Fig. 8** a Evolution of the maximal transition frequency $f_{Tmax}$ as a function of gate length for Ohmic and SB transistors ($\Phi_B = 0.275$ eV for SB CNTFET), EOT $= 5.3$ nm, $V_{DS} = 0.4$ V. b Evolution of transconductance and gate capacitance corresponding to $f_{Tmax}$ as a function of gate length for the same transistors at same drain bias

of drawbacks related to the ambipolar behavior (poor saturation of the $I_{DS}$–$V_{DS}$ characteristics and strong $V_{DS}$-dependence of $f_T$), SB-CNTFET may provide performance similar to that of Ohmic contact devices, which makes it possible to envision the design of new multi-function logic circuits taking advantage of ambipolar transport [61]. Additionally, as explained above, the ambipolar behavior may be reduced by lowering the Schottky barrier height $\Phi_B$ (see Fig. 6).

## 6 Ballistic and Quantum Effects in Ohmic Contacts Devices

Prior to the investigation of quantum effects, we first analyze the ballistic transport on the basis of semi-classical simulation of CNTFETs with different channel lengths. Indeed, the quantum coherence of electrons in the channel is directly linked to the ballistic nature of their trajectories. In this section, the equivalent oxide thickness of simulated devices is EOT = 0.4 nm.

### 6.1 Analysis of Semi-Classical Ballistic Transport in the Channel of CNTFET

First, we simulate Ohmic CNTFET in semi-classical formalism to study the ballistic transport. In Fig. 9, the transfer characteristics of the Ohmic-contact CNTFET are plotted for different channel lengths ranging from 10 to 100 nm. It appears that the drain current is quite weakly dependent on the gate length, which is a first indication of a strong contribution of ballistic transport in the gated part of the channel. As previously observed, due to the low driving field in this region the electron mean free path, limited only by weakly effective intra-subband scattering with acoustic phonons, may reached values higher than 100 nm [2]. At high electric field, typically greater than 10 kV/cm, the mean-free path becomes limited by intervalley scattering due to interaction with high energy phonons. It is then

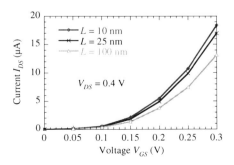

**Fig. 9** $I_{DS}$–$V_{GS}$ characteristics at $V_{DS} = 0.4$ V for the Ohmic CNTFET with different channel lengths

strongly dependent on the electric field and decreases sharply to 20 nm under a field of 100 kV/cm [2].

The conduction band profile of the CNTFET is plotted in Fig. 10a for three gate lengths at $V_{DS} = 0.4$ V. The intervalley phonon interactions happen only beyond the gated part of the channel, i.e. beyond the sharp potential drop, where the energy spectrum exhibits a second low-energy peak, distance of which from the first high-energy peak is exactly equal to the optical (LO-K) phonon energy of 180 meV (Fig. 10b). Thanks to the excellent electrostatic gate control of the channel which generates a weak electric field in the active region this optical phonon scattering effect thus plays a minor role in the performance of the transistor and the transport is strongly ballistic. It is a consequence of the quantum capacitance regime typical of low-dimensional structures with low density of states, as CNTs [62].

The ballistic character of transport in the active region of the channel is confirmed by Fig. 11 where we plot the fraction of electrons as a function of the number of scattering events experienced between the source-end and the drain-end of the intrinsic channel. The fraction of ballistic electrons is 80% for $L = 100$ nm, 89% for $L = 25$ nm and reaches 95% for $L = 10$ nm. For comparison, it is only

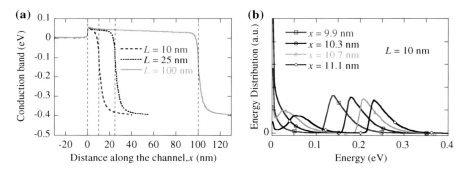

**Fig. 10** **a** Profiles the first subband energy for different gate lengths ($L = 10, 25, 100$ nm), **b** Particle energy spectrum at different position in the the 10 nm-long device for $V_{GS} = 0.2$ V and $V_{DS} = 0.4$ V, EOT $= 0.4$ nm, $x$ is the distance from the beginning of the gated part of the channel

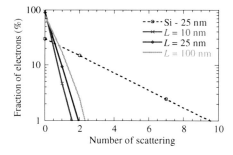

**Fig. 11** Fraction of electrons crossing the channel as a function of the number of scattering events in CNTFET ($V_{DS} = 0.4$ V) of different gate lengths and in DG-MOSFET ($V_{DS} = 0.7$ V) of gate length $L = 25$ nm

31% in an Si double-gate MOSFET with 25 nm channel length (dashed black line in Fig. 11).

This highly ballistic transport raises a question regarding the validity of the semi-classical transport approximation used here. Indeed, the phase coherence of the electron wave functions is likely to be preserved over long distance which may give rise to quantum effects such as tunneling and quantum reflection which are associated with the wave nature of electrons.

To study the effect of quantum and coherent transport on the behavior of CNTFETs, we now use the Wigner MC simulation to model Ohmic-contact CNTFETs. In the next sub-section, the quantum transport analysis is performed with careful comparison with semi-classical results, which is made possible thanks to the full compatibility between Wigner and Boltzmann MC algorithms.

## 6.2 Effect of Quantum Transport

We first consider the on-state of the transistor, i.e. at $V_{GS} = 0.2$ V, $V_{DS} = 0.4$ V. In Fig. 12 we plot the cartography in the phase space of the Boltzmann function $f_b$ (from semiclassical calculation) and the Wigner function $f_w$ (from quantum calculation) for the gate length of 25 nm. The two functions look very different, much more different than in the case of 6 nm-long DG-MOSFET [53]. In the semi-classical case, the stream of hot electrons shows that carriers are abruptly accelerated by the electric field at the drain-end of the channel. In contrast, in the quantum case, the acceleration seems much slower as if the carriers feel the potential fall in advance, which is consistent with the idea of delocalized electrons with finite extension of their wave function. It also reflects the consequence of the non local effect of potential in the Wigner formalism. Moreover, the positive–negative oscillations of the Wigner function at the drain-end of the gate are the

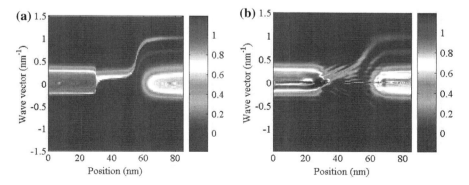

**Fig. 12** Cartography of Boltzmann **a** and Wigner functions **b** from semi-classical and Wigner MC simulation of CNTFET, respectively, for $V_{GS} = 0.2$ V, $V_{DS} = 0.4$ V. The 25 nm-long gate is located from $x = 30$ nm to $x = 55$ nm

signature of a strongly coherent transport, with typical quantum effects such as tunneling through the gate-induced barrier and reflection on the sharp fall of potential at the drain-end of the channel. The quantum coherence between incident and reflected electrons also appears in the rapid oscillations about $k = 0$ [34].

For $L = 25$ nm, the direct source-drain tunneling is always negligible and the main possible quantum effect is the quantum reflection occurring at the drain-end of the channel, as already observed in nano-MOSFETs [53]. In Fig. 13a the conduction band profile is plotted for $L = 25$ nm at $V_{GS} = 0.2$ V, i.e. for the same bias as in Fig. 12. The main difference between Wigner and Boltzmann simulations is in the small spike occurring at the source-end of the channel in the Wigner case. In spite of possible tunneling through this spike, the main effect is a weaker injection of electrons in the channel than in the Boltzmann case, as illustrated by the corresponding density profiles plotted in Fig. 13b.

In spite of these significant differences observed at microscopic level, it is noticeable to see that the two types of simulation give quite close terminal currents, as shown in Fig. 14 for the two gate lengths $L = 25$ nm and $L = 6$ nm. The current is weakly dependent on the gate lengths. This is the result of strongly ballistic transport in the channel. It is clearly and quantitatively shown in Fig. 11 where we plot the fraction of electrons as a function of the number of scattering events experienced between the source-end and the drain-end of the channel.

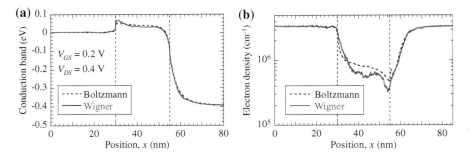

**Fig. 13** Conduction band **a** and density **b** profiles obtained from Wigner and Boltzmann methods for $L = 25$ nm ($V_{GS} = 0.2$ V, $V_{DS} = 0.4$ V)

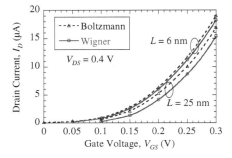

**Fig. 14** $I_D$-$V_{GS}$ characteristics obtained from Boltzmann and Wigner simulations for the gate length of 25 and 6 nm at $V_{DS} = 0.4$ V

Whatever the gate voltage the Wigner current is systematically smaller than the Boltzmann current. The relative difference of drain current between Wigner and Boltzmann simulation is quite small in on-state, i.e. less than 10%, and reaches only 60% in nearly off-state, as shown in Fig. 15. Surprisingly enough, this difference is smaller for $L = 6$ nm than for $L = 25$ nm, which deserves to be explained.

For smaller channel length, because of the thinner channel barrier, the source-drain tunnelling is now possible. It tends to slightly enhance the drain current in comparison to the semi-classical case, which leads to smaller difference between Wigner and Boltzmann currents for $L = 6$ nm than for $L = 25$ nm, as observed in Fig. 14. This overall behaviour of CNTFETs demonstrates the importance of the self-consistence in this type of device simulation, which tends to reduce the differences between quantum and semi-classical results, in spite of strong apparent differences at the microscopic level.

The occurring of source-to-drain tunnelling in 6 nm-long channel is clearly illustrated in Fig. 16 where we plot the Wigner function cartography of the device biased in off-state ($V_{GS} = 0.05$ V) in comparison to the case $L = 25$ nm. While the Wigner function takes only very small values in the channel for $L = 25$ nm

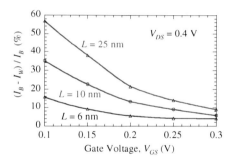

**Fig. 15** Relative difference between the currents obtained from Wigner and Boltzmann simulation for three different gate lengths ($V_{DS} = 0.4$ V)

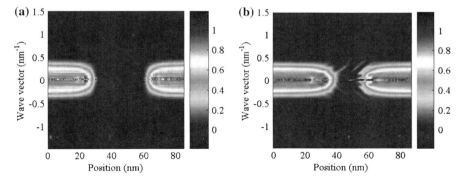

**Fig. 16** Cartography of the Wigner function for the gate length of 25 nm **a** and 6 nm **b** at $V_{GS} = 0.05$ V, $V_{DS} = 0.4$ V

(Fig. 16a), significant oscillations are visible for $L = 6$ nm at positive and negative values of wave vector, which is the manifestation of the persistence of both tunnelling and quantum reflections.

## 7 Conclusion

We have taken advantage of the strength of the Wigner/Boltzmann Monte Carlo approach including electron–phonon scattering to investigate and to bridge the main features of semi-classical and quantum transport in CNTFETs with either Ohmic or Schottky source and drain contacts.

The characteristics and performance of Ohmic and Schottky CNTFETs have been carefully compared. While Ohmic-contact transistors exhibit standard FET characteristics with high on/off current ratio, the ambipolar transport behavior inherent in Schottky-contact transistors strongly influences the device electrical characteristics. In particular, the ambipolar character of carrier transport impedes the current saturation at high $V_{DS}$ and degrades the on/off current ratio. Additionally, the poor saturation of current makes the transition frequency strongly dependent on the drain voltage, which may be critical for circuit application. However, it has been shown that by lowering the Schottky barrier height for electrons, the ambipolar behavior reduces, which makes the device characteristics closer to that of Ohmic-contact transistors. Finally, in both cases the intrinsic transition frequency is shown to reach about 800 GHz for a gate length of 100 nm and enters the THz range for sub-100 nm devices.

The highly ballistic character of carrier transport has been clearly evidenced with a fraction of ballistic electrons reaching 89% in the active region for a gate length of 25 nm, to be compared to 31% in Si DG-MOSFET of the same gate length. This high ballisticity manifests in particular in the weak gate length-dependence of drain current.

The analysis of quantum and semi-classical Monte Carlo simulation of short channel CNTFETs shows that, in spite of significant quantum transport effects such as tunneling through potential barrier and reflection by sharp potential drop, the microscopic quantum features, e.g. the oscillations of the Wigner function, are not strongly reflected at the macroscopic level of terminal current. Quantum and semi-classical simulations for 25 and 6 nm long devices even provide closer results for the shorter of these channel lengths. This study highlights the importance of including the self-consistence in device simulation which tends here to weaken the consequences of quantum effects. This kind of consideration seems to make acceptable the semi-classical approach of transport much beyond its theoretical domain of validity.

**Acknowledgments** This work was partially supported by the European Community, through Network of Excellence NANOSIL (ICT-216171), and by the French ANR, through project ACCENT (ANR-06-NANO-069).

# References

1. Durkop, T., et al.: Extraordinary mobility in semiconducting carbon nanotubes. Nano Lett. **4**, 35–39 (2004)
2. Cazin d'Honincthun, H., et al.: Electron–phonon scattering and ballistic behavior in semiconducting carbon nanotubes. Appl. Phys. Lett. **87**, 172112 (2005)
3. Javey, A., et al.: Ballistic carbon nanotube field-effect transistors. Nature **424**, 654–657 (2003)
4. Tans, S.J., Verschueren, A.R.M., Dekker, C.: Room-temperature transistor based on a single carbon nanotube. Nature **393**, 49–52 (1998)
5. Martel, R., et al.: Single- and multi-wall carbon nanotube field-effect transistors. Appl. Phys. Lett. **73**, 2447–2449 (1998)
6. Rutherglen, C., Jain, D., Burke, P.: Nanotube electronics for radiofrequency applications. Nat. Nanotechnol. **4**, 811–819 (2009)
7. Bachtold, A., et al.: Logic circuits with carbon nanotube transistors. Science **294**, 1317–1320 (2001)
8. Javey, A., et al.: Carbon nanotube transistor arrays for multistage complementary logic and ring oscillators. Nano Lett. **2**, 929–932 (2002)
9. Tseng, Y.C., et al.: Monolithic integration of carbon nanotube devices with silicon MOS technology. Nano Lett. **4**, 123–127 (2003)
10. Chen, Z., et al.: An integrated logic circuit assembled on a single carbon nanotube. Science **311**, 1735 (2006)
11. Jing, G., et al.: Assessment of high-frequency performance potential of carbon nanotube transistors. IEEE Trans. Nanotechnol. **4**, 715–721 (2005)
12. Frank, D.J., Appenzeller, J.: High-frequency response in carbon nanotube field-effect transistors. IEEE Electron Device Lett. **25**, 34–36 (2004)
13. Rosenblatt, S., et al.: Mixing at 50 GHz using a single-walled carbon nanotube transistor. Appl. Phys. Lett. **87**, 153111 (2005)
14. Bethoux, J.M., et al.: An 8-GHz $f_t$ carbon nanotube field-effect transistor for gigahertz range applications. IEEE Electron Device Lett. **27**, 681–683 (2006)
15. Nougaret, L., et al.: 80 GHz field-effect transistors produced using high purity semiconducting single-walled carbon nanotubes. Appl. Phys. Lett. **94**, 243505 (2009)
16. Chaste, J., et al.: Single carbon nanotube transistor at GHz frequency. Nano Lett. **8**, 525–528 (2008)
17. Martel, R., et al.: Ambipolar electrical transport in semiconducting single-wall carbon nanotubes. Phys. Rev. Lett. **87**, 256805 (2001)
18. Radosavljevic, M., et al.: Drain voltage scaling in carbon nanotube transistors. Appl. Phys. Lett. **83**, 2435–2437 (2003)
19. Freitag, M., et al.: Controlled creation of a carbon nanotube diode by a scanned gate. Appl. Phys. Lett. **79**, 3326–3328 (2001)
20. Derycke, V., et al.: Controlling doping and carrier injection in carbon nanotube transistors. Appl. Phys. Lett. **80**, 2773–2775 (2002)
21. Appenzeller, J., et al.: Field-modulated carrier transport in carbon nanotube transistors. Phys. Rev. Lett. **89**, 126801 (2002)
22. Javey, A., et al.: Advancements in complementary carbon nanotube field-effect transistors. In IEDM '03 Technical Digest, pp. 741–744 (2003)
23. Javey, A., et al.: High performance n-type carbon nanotube field-effect transistors with chemically doped contacts. Nano Lett. **5**, 345–348 (2005)
24. Jacoboni, C., Lugli, P.: The Monte Carlo method for semiconductor device simulation. Springer, Wien/New York (1989)
25. Moglestue, C.: Monte Carlo simulation of semiconductor devices. Chapman & Hall, London (1993)

26. Jungemann, C., Meinerzhagen, B.: Hierarchical device simulation: the Monte Carlo perspective. Springer, Wien/New York (2003)
27. Pennington, G., Goldsman, N.: Semiclassical transport and phonon scattering of electrons in semiconducting carbon nanotubes. Phys. Rev. B **68**, 045426 (2003)
28. Hasan, S., et al.: Monte Carlo simulation of carbon nanotube devices. J. Comput. Electron. **3**, 333–336 (2004)
29. Datta, S.: Quantum transport: atom to transistor. Cambridge University Press, Cambridge (2005)
30. Koswatta, S.O., et al.: Nonequilibrium Green's function treatment of phonon scattering in carbon-nanotube transistors. IEEE Trans. Electron Devices **54**, 2339–2351 (2007)
31. Pourfath, M., Kosina, H.: The effect of phonon scattering on the switching response of carbon nanotube field-effect transistors. Nanotechnology **18**, 424036 (2007)
32. Jacoboni, C., et al.: Quantum transport and its simulation with the Wigner-function approach. Int. J. High Speed Electron. Syst. **11**, 387–423 (2001)
33. Nedjalkov, M.: Wigner transport in presence of phonons: particle models of the electron kinetics. In: From Nanostructures to Nanosensing Applications. Societa Italiana Di Fisica, IOP Press, Amsterdam. pp. 55–103 (2005)
34. Querlioz, D., Dollfus, P.: The Wigner Monte Carlo method for nanoelectronic devices. ISTE-Wiley, London (2010)
35. Querlioz, D., et al.: Wigner Monte Carlo simulation of phonon-induced electron decoherence in semiconductor nanodevices. Phys. Rev. B **78**, 165306 (2008)
36. Querlioz, D., et al.: Wigner–Boltzmann Monte Carlo approach to nanodevice simulation: from quantum to semiclassical transport. J. Comput. Electron. **8**, 324–335 (2009)
37. Guo, J., et al.: Metal–insulator–semiconductor electrostatics of carbon nanotubes. Appl. Phys. Lett. **81**, 1486–1488 (2002)
38. Iijima, S.: Helical microtubules of graphitic carbon. Nature **354**, 56–58 (1991)
39. Saito, R., Dresselhaus, G., Dresselhaus, M.S.: Physical properties of carbon nanotubes. Imperical College Press, London (1998)
40. Charlier, J.C., Blase, X., Roche, S.: Electronic and transport properties of nanotubes. Rev. Mod. Phys. **79**, 677 (2007)
41. Dubay, O., Kresse, G.: Accurate density functional calculations for the phonon dispersion relations of graphite layer and carbon nanotubes. Phys. Rev. B **67**, 035401 (2003)
42. Verma, A., Kauser, M.Z., Ruden, P.P.: Ensemble Monte Carlo transport simulations for semiconducting carbon nanotubes. J. Appl. Phys. **97**, 114319 (2005)
43. Hasan, S., Alam, M.A., Lundstrom, M.S.: Simulation of carbon nanotube FETs including hot-phonon and self-heating effects. IEEE Trans. Electron Devices **54**, 2352–2361 (2007)
44. Hertel, T., Moos, G.: Electron–phonon interaction in single-wall carbon nanotubes: a time-domain study. Phys. Rev. Lett. **84**, 5002–5005 (2000)
45. Koswatta, S.O., et al.: Computational study of exciton generation in suspended carbon nanotube transistors. Nano Lett. **8**, 1596–1601 (2008)
46. Machon, M., et al.: Strength of radial breathing mode in single-walled carbon nanotubes. Phys. Rev. B **71**, 035416 (2005)
47. Verma, A., Kauser, M.Z., Ruden, P.P.: Effects of radial breathing mode phonons on charge transport in semiconducting zigzag carbon nanotubes. Appl. Phys. Lett. **87**, 123101 (2005)
48. Cazin d'Honincthun, H., et al.: Monte Carlo study of coaxially gated CNTFETs: capacitive effects and dynamic performance. C R Phys. **9**, 67–77 (2008)
49. Fregonese, S., et al.: Computationally efficient physics-based compact CNTFET model for circuit design. IEEE Trans. Electron Devices **55**, 1317–1327 (2008)
50. Feynman, R.: Statistical mechanics : a set of lectures. Westview Press, Boulder (1998)
51. Nedjalkov, M., et al.: Unified particle approach to Wigner–Boltzmann transport in small semiconductor devices. Phys. Rev. B **70**, 115319 (2004)
52. Querlioz, D., et al.: An improved Wigner Monte-Carlo technique for the self-consistent simulation of RTDs. J. Comput. Electron. **5**, 443–446 (2006)

53. Querlioz, D., et al.: A study of quantum transport in end-of-roadmap DG-MOSFETs using a fully self-consistent Wigner Monte Carlo approach. IEEE Trans. Nanotechnol. **5**, 737–744 (2006)
54. Querlioz, D., et al.: On the ability of the particle Monte Carlo technique to include quantum effects in nano-MOSFET simulation. IEEE Trans. Electron Devices **54**, 2232–2242 (2007)
55. Shifren, L., Ringhofer, C., Ferry, D.K.: A Wigner function-based quantum ensemble Monte Carlo study of a resonant tunneling diode. IEEE Trans. Electron Devices **50**, 769–773 (2003)
56. Lugli, P., Ferry, D.K.: Degeneracy in the ensemble Monte Carlo method for high-field transport in semiconductors. IEEE Trans. Electron Devices **32**, 2431–2437 (1985)
57. Datta, S.: Electronic transport in mesoscopic system. Cambridge University Press, Cambridge (1995)
58. Nguyen, H.N., et al.: Monte Carlo study of ambipolar transport and quantum effects in carbon nanotube transistors. In: SISPAD 2009 Proceedings, IEEE, pp. 277–280 (2009)
59. Heinze, S., et al.: Carbon nanotubes as Schottky barrier transistors. Phys. Rev. Lett. **89**, 106801 (2002)
60. Chen, Y.F., Fuhrer, M.S.: Electric-field-dependent charge-carrier velocity in semiconducting carbon nanotubes. Phys. Rev. Lett. **95**, 236803 (2005)
61. Yu, W.J., et al.: Adaptive logic circuits with doping-free ambipolar carbon nanotube transistors. Nano Lett. **9**, 1401–1405 (2009)
62. John, D.L., Castro, L.C., Pulfrey, D.L.: Quantum capacitance in nanoscale device modeling. J. Appl. Phys. **96**, 5180–5184 (2004)

# Quantum Simulation of Silicon-Nanowire FETs

Marco Pala

**Abstract** We present numerical calculations of transport properties of semiconductor nanowires based on a three-dimensional (3D) self-consistent Keldysh Green's function approach, which is able to treat quantum confinement, quasi-ballistic transport, out-of-equilibrium effects as well as the influence of elastic and inelastic scattering. We investigate the role of main scattering mechanisms responsible for mobility degradation at room temperature in ultrashort electron devices like Silicon-nanowire FETs. We consider spatial fluctuations as surface-roughness (SR) or remote-charge scattering (RCS) as main sources of elastic scattering, whereas electron–phonon (PH) interaction is assumed responsible of inelastic scattering processes. We apply these techniques to evaluate the effects of SR and RCS on the transfer characteristics and electron mobility of short-channel Silicon nanowires at room temperature and then focus on scattering-limited mobilities. Our results show that SR and RCS are mainly responsible for threshold voltage shift and sub-threshold voltage slope degradation, whereas PH scattering remains the main scattering mechanism limiting the mobility at room temperature.

## 1 Physical Models and Numerical Methods

Numerical calculations are carried out through the self-consistent solution of the three-dimensional Schrödinger and Poisson equations with open boundary

M. Pala (✉)
IMEP-LAHC (UMR 5130), Grenoble INP, Minatec, 3 Parvis Louis Néel,
BP 25738016, Grenoble, France
e-mail: pala@minatec.inpg.fr

conditions along the transport direction. We adopt a coupled mode-space approach [4, 8] within the Keldysh Green's function formalism [2, 5] to calculate the electron charge and transport properties as low-field mobility. In such a formalism we compute the retarded Green's function $G^R$, describing carrier dynamics inside the conductor and the lesser-(greater-) than Green's function $G^<(G^>)$, describing occupied (non-occupied) states. In steady-state conditions the two kinetic equations describing the transport properties of the system read out:

$$G^R = \left[E - H_d - \Sigma^R\right]^{-1}$$
$$G^{<(>)} = G^R \Sigma^{<(>)} G^A,$$

where a matrix notation has been adopted in the real space or in the mode-space domain. Here, $H_d$ is the device Hamiltonian, $\Sigma^R$ the retarded self-energy, $G^A$ is the advanced Green's function and $\Sigma^{<(>)}$ is the lesser(greater) self-energy describing the in(out)-scattering rates including the tunnel coupling with both contacts and the phonon scattering in the whole device. In case of dissipative transport such equations are non-linearly coupled and have to be self-consistently solved [10, 13]. The Hamiltonian is described in the material-dependent effective mass approximation in order to reduce the computational burden, which is still a good approximation with respect to tight-binding methods for a SiNW width larger than 1.7 nm [19].

Further, we focus on three main scattering mechanisms limiting the mobility in decananometric nanowire FETs: (1) surface-roughness scattering due to the spatial fluctuations at the interface between the channel and the oxide layer, (2) remote-Coulomb scattering due to fixed charges present in the gate stack, especially at the interface between Silicon oxide and high-k dielectrics like $HfO_2$, and (3) phonon scattering due to different modes of lattice vibrations including acoustic and optical phonons. In particular, scattering from acoustic phonons in the elastic approximation and dispersionless optical phonon scattering accounting for both f- and g-type processes are considered.

The effect of surface roughness is included via a geometric method as an abrupt randomly varying interface between Si and $SiO_2$. A quasi-continuous function $\Delta(r)$ representing the two-dimensional Si–$SiO_2$ interface displacement is assumed to be characterized by an exponential autocorrelation function

$$C(r) = \Delta_m^2 e^{-\sqrt{2}r/L_m},$$

where $\Delta_m$ is the root mean square (RMS) of the fluctuations, and $L_m$ is the correlation length [3]. Each particular realization of SR is generated by starting from the power density spectrum obtained by Fourier transforming the correlation function and then by adding a random odd phase to the square root of the power spectrum. The presence of surface roughness is considered only on the channel region in order to properly evaluate the SR-limited mobility in the channel [1].

A similar geometrical treatment is adopted to evaluate the effects of random charged defects present in the oxide region and responsible of remote-charge scattering (RCS) [9]. An interfacial layer of silicon oxide with thickness $t_{IL}$ is placed between the silicon channel and the high-$\kappa$ gate dielectric. Positive and negative impurities are supposed to be randomly localized at the interface with a uniform probability distribution over the two-dimensional interfacial plane with a density $N_{Fix}$. Similarly to the case of SR the presence of fixed charges is considered only in the region covered by the gate, thus allowing us to focus on channel transport properties. In [14], the use of fixed charges of opposite sign is at the basis of the observed mobility degradation connected to RCS. A uniform charge distribution located at the gate/oxide interface is indeed supposed to generate a simple shift of the flatband voltage with small impact on the channel mobility due to the distant location. On the other side, dipoles located at the interfacial layer are able to affect the carrier transport through potential oscillations, whereas no significant shift of the flatband voltage is supposed due to a quantitative balance between charges of the opposite sign. The magnitude and the impact of the potential oscillations will depend on the choice of specific values of $N_{Fix}$ and $t_{IL}$, the only two parameters introduced in the model.

The electron–phonon interaction is accounted for via the self-consistent Born approximation and is included with local self-energies. We assume bulk band structures for all phonons and neglect effects due to the confinement on the transverse plane which start to be important for cross-section smaller than $5 \times 5$ nm$^2$ [12]. The lesser(greater)-than self-energy of acoustic phonons for n-th mode at the $i$th discrete space site along the transport direction reads: $\Sigma_{v;n,n}^{<(>)}(i,i;E) = |M_q|^2 \sum_m G_{v;m,m}^{<(>)}(i,i;E) I_{m,v}^{n,v}(i,i)$, where $v$ is the valley index, $|M_q|^2$ the electron–phonon matrix element and $I_{m,v}^{n,v'}(i,i) = \int dy dz \left|\chi_{i,v'}^n(y,z)\right|^2 \left|\chi_{i,v}^m(y,z)\right|^2$ represents the usual form factor, where $\chi_{i,v}^n$ are the eigenfunctions at the transverse plane. Similarly, for the j-th optical mode the lesser(greater)-than self-energy reads out: $\Sigma_{j,v;n,n}^{<(>)}(i,i;E) = |M_q|^2 (N_j + \frac{1}{2} \pm \frac{1}{2}) \sum_{m,v'} g_j^{v,v'} G_{v';m,m}^{<(>)}(i,i;E \pm \hbar\omega_j) I_{m,v}^{n,v}(i,i)$ where $N_j = (\exp(\hbar\omega_j/KT) - 1)^{-1}$ is the average phonon density at the energy $\hbar\omega_j$ and the degeneracy factor is $g_j^{v,v'} = \delta_{v,v'}$ for g-type and $g_j^{v,v'} = 2(1 - \delta_{v,v'})$ for f-type optical phonons. The two electron–phonon matrix elements are $|M_q|^2 = \frac{D_{AC}KT}{\rho u_s^2}$ and $|M_q|^2 = \frac{\hbar D_{OP}^2}{2\rho\omega_j}$ for the acoustic and optical phonon case, respectively, where $D_{AC}$ and $D_{OP}$ represent the acoustic and optical deformation potentials, $\rho$ the mass density, $u_s$ the acoustic velocity and $\omega_j$ the optical phonon angular frequency. The phonon retarded self-energies are assumed as purely imaginary for both optical and acoustic scatterings.

After the solution of the Schrödinger problem for any energy E the 3D electron density at the discretized space site $(i, j, k)$ is evaluated as

$$n(i,j,k) = \frac{g_v g_s}{\Delta x \Delta y \Delta z} \text{Im} \sum_{n,m,v} \int \frac{dE}{2\pi} G_{v;n,m}^<(i,i,E) \chi_{i,v}^n(j,k) \chi_{i,v}^m(j,k)$$

where $g_v$ and $g_s$ are the valley and spin degeneration coefficients, respectively. In order to obtain a self-consistent solution the electron charge is then inserted in the charge density $\rho$ in the 3D Poisson equation $\vec{\nabla}\cdot\varepsilon(\vec{r})\vec{\nabla}\phi(\vec{r}) = -\rho(\vec{r})$, with $\varepsilon$ the material permittivity and $\varphi$ the electrostatic potential.

Hence, the current flowing from the coordinate $x_i$ to $x_{i+1}$ is evaluated according to the Meier–Wingreen formalism [6] as: $J_{i,i+1}(j,k) = -\frac{q g_v g_s}{\hbar \Delta y \Delta z} 2\mathrm{Re} \sum_{n,m,\nu} \int \frac{dE}{2\pi} V_{i,i+1} G^<_{\nu;n,m}(i+1,i,E) \chi^n_{i+1,\nu}(j,k)\, \chi^m_{i,\nu}(j,k)$ where $V_{i,i+1}$ is the coupling strength between slice i and i + 1.

The extraction of the effective mobility $\mu_{\mathrm{eff}}$ is performed in the linear transport following the method presented in [8]. This is achieved after separately computing the conductance G and the linear electron charge density $N_{1D}$ according to

$$\mu_{\mathit{eff}} = \frac{G L_{ch}}{q N_{1D}},$$

where $L_{ch}$ is the channel length.

In order to avoid spurious effects in the evaluation of the linear charge density due to the charge penetration in the channel region from doped S/D regions we evaluate $N_{1D}$ in a restricted channel region. Moreover, the evaluation of the effective mobility for such short-channel devices accounts for the nonlocal effect of the apparent or ballistic mobility component arising from scatterings occurring at the contact interfaces and linearly depending on the channel length. In order to isolate a purely scattering-limited mobility $\mu_{\mathrm{sc}}$, we adopt the Shur's model [15]

$$\mu_{sc} = \left(\mu_{\mathit{eff}}^{-1} - \mu_{bal}^{-1}\right)^{-1}$$

where $\mu_{\mathrm{bal}}$ is the ballistic mobility evaluated on a clean device, and $\mu_{\mathrm{sc}}$ represents any kind of scattering among PH, SR and RCS. This decomposition is physically justified by the fact that all the scattering mechanisms are each other independent [10].

## 2 Surface Roughness

Surface roughness at the Si/SiO$_2$ interfaces is a major scattering mechanism in ultra short devices like SiNWs due to the small surface/volume ratio. It first influences the electrostatics properties and hence the transport characteristics of such devices [7]. As it is shown in Fig. 1 SR affects differently the various slices along the longitudinal direction, generating random potential wells and hence spatial fluctuations of the charge density. A direct consequence is the change of the eigenfunction solutions of the 2D Schrödinger equation on the transverse plane. This is connected to the rise of the mixing between transverse modes of adjacent slides [18].

# Quantum Simulation of Silicon-Nanowire FETs 241

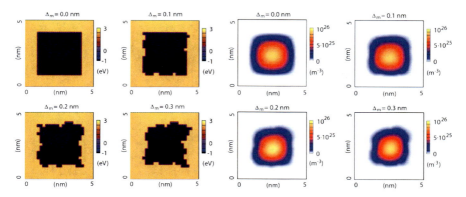

**Fig. 1** Example of cross-section of the conduction band profile (*left*) and of the charge density (*right*) for different SR parameters. The average wire section is $3 \times 3$ nm$^2$. SR parameters $\Delta_m = 0.0, 0.1, 0.2$ and $0.3$ nm, $L_m = 1.0$ nm

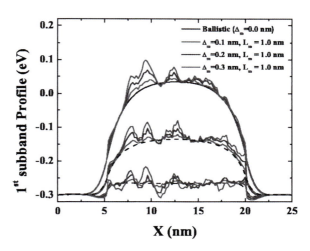

**Fig. 2** First subband profile with ideal (*black*) and rough (*color*) interfaces for three different gate voltages ($V_{GS} = 0.1, 0.3$ and $0.6$ V). From the bottom to the top SR parameters are $\Delta_m = 0$, $0.1, 0.2, 0.3$ nm and $L_m = 1.0$ nm. The wire section is $3 \times 3$ nm$^2$, $V_{DS} = 1$ mV

More precisely, we can distinguish between two main effects due to SR in nanowires: one is to create large fluctuations of the effective-potential profile seen by electrons, while the second is to enhance the mode mixing between transverse eigenfunctions at adjacent sections [1, 7, 8]. Usually the former is more important at low electron density whereas the second is more relevant at large overdrive when electrons are pushed towards the surfaces by the gate action. Subband fluctuations along the transport direction can be observed in Fig. 2, where the potential profile of the first subband of an unprimed valley is reported at low and high gate biases, both for the (solid lines) SR and the (dashed lines) ballistic case for a narrow nanowire with average cross-section of the $3 \times 3$ nm$^2$. Several potential wells in the transport direction can give rise to electron localization, which can be of some importance in a quasi-ballistic transport regime.

**Fig. 3** Transfer characteristics of SiNWFETs as a function of $V_{GS} - V_T$. From the bottom to the top SR parameters are $\Delta_m = 0$, 0.1, 0.2, 0.3 nm and $L_m = 1.0$ nm. The wire section is $3 \times 3$ nm$^2$, $V_{DS} = 1$ mV

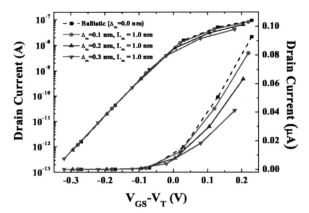

The interplay between subband fluctuations and mode-mixing mechanism is responsible of the different dependence on the inversion charge of the effective mobility in nanowires of different lateral sizes [8].

The effect of SR on the drain current of SiNWs is addressed in Fig. 3 showing the transfer characteristics at low-drain bias $V_{DS} = 1$ mV and different values of roughness RMS $\Delta_m$. In particular, we can study electrical performances in terms of the threshold voltage $V_T$ and of the sub-threshold voltage slope $S = (\partial \log I_D / \partial V_G)^{-1}$. Threshold voltage is evaluated via the comparison of the output characteristics in the exponential current regime. Dashed line represents the characteristics of the ballistic device, and it is adopted as reference in order to quantify the impact of the SR on the transport properties, whereas solid lines represent the characteristics of the SiNWT with rough interfaces.

First we find that SR determines a monotonic threshold voltage shift $\Delta V_T$ with respect to the smooth case with $\Delta V_T \approx 20$, 40, and 70 mV for $\Delta_m = 0.1, 0.2$, and 0.3 nm, respectively. This can be explained by considering that in the low-drain bias regime the current is limited by the smallest channel constriction generated by SR. Threshold voltage shift attains about 70 mV for $\Delta_m = 0.3$ nm and is compatible with the maximum variation of the subband profile in Fig. 2. Therefore, a monotonic reduction of the output current is also reported for a fixed gate overdrive as roughness RMS is increased.

Second, we find that all lines in Fig. 4 show the same $S \approx 70$ mV/dec, which is slightly higher than the ideal value of 60 mV/dec at room temperature due to a combination of finite source–drain tunneling current and short-channel effects occurring in such a scaled device. Therefore, SR is not effective in altering the slope which is independent of $\Delta_m$. This can be explained by noticing that the amplitude of potential fluctuations due to SR are negligible with respect to the source/drain barrier in the subthreshold regime and almost independent of $V_{GS}$.

Further, we focus on the SR-limited mobility $\mu_{SR}$ which is computed after the evaluation of the effective mobility $\mu_{\text{eff}}$ and of the so-called ballistic mobility $\mu_{\text{bal}}$. The former refers to clean devices, where electrons are not backscattered and it has not to be considered as a standard mobility. It has been studied in several

**Fig. 4** Surface-roughness-limited mobility of SiNWFETs as a function of the 1D electron charge. From the bottom to the top SR parameters are $\Delta_m = 0.1, 0.2, 0.3$ nm and $L_m = 1.0$ nm. Average wire section is $3 \times 3$ nm$^2$, $V_{DS} = 1$ mV

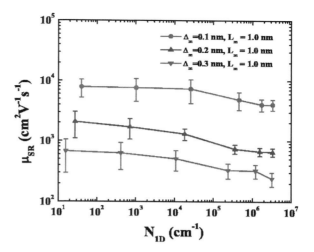

theoretical works [20] and since it arises from contact resistance, it scales linearly with the device length. Ballistic mobility decreases with the inversion charge due to the increasing importance of degeneracy effects [8], described by the analytical flux theory for a single occupied 1D subband [11], which shows a decrease of the effective mobility at large gate voltage

$$\mu_{bal} \propto \frac{F_{-1}(\eta_F)}{F_{-1/2}(\eta_F)},$$

where $F_n(\eta_F)$ is the nth order Fermi integral and $\eta_F = (E_F - E_1)/(kT)$, $E_F$ is the Fermi level, and $E_1$ is the first subband energy.

The presence of roughness at the Si/SiO$_2$ interfaces adds another source of scattering in the device thus reducing the absolute value of the effective mobility when increasing $\Delta_m$.

Hence, SR-limited mobility can be interpreted as the mobility of electrons scattered by SR and $\mu_{bal}$ as the mobility of ballistic electrons that are limited only by contact resistance. This decomposition is physically reasonable since the two scattering mechanisms are independent. In Fig. 4, we show the SR-limited mobility as a function of the 1D electron charge of SiNWs with a $3 \times 3$ nm$^2$ cross-section, as well as the errorbars arising from the uncertainty in calculating the linear charge. We point out a relevant and monotonic mobility reduction as a function of $\Delta_m$, which is consequence of the subband fluctuation mechanisms, whereas the mode-mixing mechanism starts to be important only for significant $\Delta_m$ and large gate overdrives.

## 3 Remote Coulomb Scattering

The impact of RCS on SiNWs is evaluated by considering rectangular wires with channel lateral dimensions of $5 \times 5$ nm$^2$ and gate length of 40 nm. The channel orientation is supposed along the [100] direction. Bulk silicon effective masses ($m_l = 0.916\ m_0$, $m_t = 0.191\ m_0$), where $m_0$ is the free electron mass, are used. The silicon oxide effective mass is fixed at one half of $m_0$. The high-$\kappa$ oxide has a thickness of $t_{ox} = 2$ nm, whereas $t_{IL}$ varies from 0.6 to 1.4 nm. The dielectric constant of the external layer has been set to twice the value of the silicon oxide.

In analogy to the case of SR at the interface the presence of fixed-charge centers in the gate stack generates spatial fluctuations the potential profile and, hence, of the 1-D subband experienced by electrons. Moreover, the variation of the confinement condition along the transport direction determines an additional enhancement of the coupling effect between transverse modes. The two competing mechanisms affect the carrier transport with different strength at different gate biases. Importantly, the inclusion of the impurities directly in the Poisson's equation as additional charge-density contribution determines a solution that self-consistently accounts for the screening effect due to the channel carriers. This is shown in Fig. 5, where the first subband profile of one of the unprimed valleys is reported for different gate-bias conditions. As the gate voltage is increased and a higher carrier density is accumulated in the channel, the Coulomb potential generated by the fixed charges is screened by the electrons at the surface, thus reducing the perturbation effect. A progressive smoothing of the potential energy fluctuations is shown, while the potential profile approaches the ballistic curve. At low gate voltages, the presence of marked and sharp oscillations of the potential is the limiting scattering mechanism. At higher gate overdrives, the squeezing of the carriers toward the $Si/SiO_2$ interfaces induces an increased mode mixing and determines a current reduction.

The effect of RCS on the transfer characteristics for different values of the interface layer thickness and impurity densities is carried out with a qualitative analysis of the leakage current and of the threshold voltage. In Fig. 6, the output characteristics in both linear and logarithmic scale are reported for devices with interfacial layer thickness of $t_{IL} = 1.4, 1.0, 0.6$ nm and fixed $N_{Fix} = 10^{13}$ cm$^{-2}$,

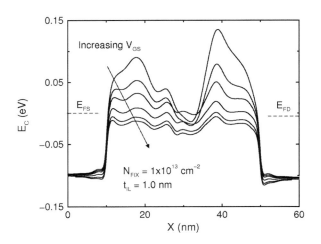

**Fig. 5** First subband profile along the transport direction for a SiNW with high-k gate stack for various overdrive values. RCS parameters are $N_{FIX} = 10^{13}$ cm$^{-2}$ and $T_{IL} = 1$ nm

considering a single realization of an impurity distribution for each parameter value. The fixed-charge density is chosen similar to typical values presented in literature [14]. The curves are plotted as a function of $V_{GS} - V_T$, where $V_T$ is extracted from the linearization of the output characteristic at high overdrive. First, we observe a progressive increase in $V_T$ with decreasing interface layer thickness. The random generation of discrete fixed charges deviates from the ideal condition of a uniform distribution of dipoles causing a non-compensation of the impact on the potential. In particular, especially at large $N_{Fix}$ and thin $t_{IL}$, we observe the formation clusters of charges of the same sign, strongly influencing the transfer characteristics at low inversion density. An example of such a phenomenon is reported in Fig. 5, where the subband profiles at different gate voltages are shown. Due to the increasing importance of screening effects electrons in the channel subband fluctuations are progressively smoothed when $V_{GS}$ increases.

A clustering of positive charges induces a large threshold voltage shift of several mV with respect to the ballistic case. Moreover, the presence of charge clusters has a direct effect on the current characteristics. In fact, depending on the specific realization of fixed charges, favorable source-to-drain tunneling paths can be possible. The direct-tunneling component of the current has a detrimental impact on the performances of the device with a sharp increase in the current levels in the OFF condition and with a reduction of the gate-control capability, reflected by a progressive increase in the subthreshold voltage slope S. An increasing importance of the tunneling current can be observed for the device with the smaller value of $t_{IL}$ and the larger $N_{Fix}$, reflected by a spreading of the current values over almost one order of magnitude in the subthreshold regime. In Fig. 6 we report a progressive increase of the inverse subthreshold slope from 63 to 78 mV/dec, compared with the ideal condition of $S = 60$ mV/dec of the ballistic device.

RCS also influences the electrical performance at large inversion density, where we notice a current reduction for thinner $t_{IL}$. This is easily explained by the

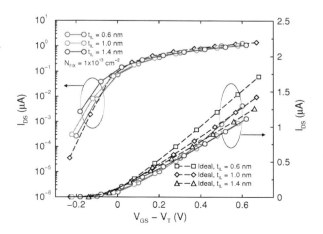

**Fig. 6** Turn-on characteristics for devices with different $t_{IL}$ values and a fixed value of $N_{Fix} = 10^{13}$ cm$^{-2}$ (*Solid lines with symbols*). Reference ballistic case is reported for comparison (*Black dashed line with symbols*). $V_{DS} = 5$ mV

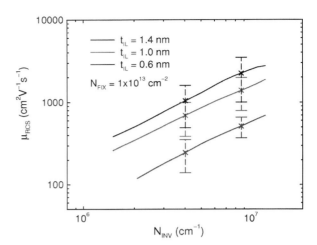

**Fig. 7** Mean value of the RCS-limited mobility as a function of the channel electron density evaluated over ten device samples for different $t_{IL}$ values and a fixed value of $N_{FIX} = 10^{13}$ cm$^{-2}$. From the top to the bottom $t_{IL} = 1.4, 1.0, 0.6$ nm

increasing importance of the coupling of transverse modes for large electronic densities: although electronic screening determines a drastic reduction of the spatial fluctuations and carriers experience a thermionic transport, the increased number of conducting modes implies a non-negligible RCS inversely proportional to $t_{IL}$ [9].

In order to analyze the impact of the mere RCS on the mobility, we have subtracted the apparent mobility component following the Mathiessen decomposition $\mu_{RCS} = \left(\mu_{eff}^{-1} - \mu_{bal}^{-1}\right)^{-1}$ where $\mu_{RCS}$ is the RCS-limited mobility. Again, this factorization is physically justified since the two scattering mechanisms are independent.

In Fig. 7, we show the average RCS-limited mobility as a function of the linear electron density in the channel. To compare the devices in the same condition of charge density, an interpolation of the simulated data is performed before averaging. The evaluated standard deviation is reported for the two limit conditions of high and low electron densities. We find clear power-law dependence of $\mu_{RCS} \sim (N_{Inv})^\alpha$, where $\alpha$ is close to unity, with a slow deviation only at high-density values. Such a result is consistent with the power law found in experiments on double-gate and ultrathin-body MOSFETs [17], where the exponent $\alpha$ was observed to increase by reducing the body thickness. Physically, the power-law behavior is originated by the screening effect at room temperature and is characteristic of the increase in the inversion screening length with the linear charge density [16].

## 4 Channel-Length Dependence of Low-Field Mobility

In order to study the influence of inelastic scattering arising from electron–phonon interaction we consider rectangular gate-all-around SiNWs with a cross section of

$5 \times 5$ nm² and a high-$\kappa$ gate stack of 2 nm with a SiO$_2$ interfacial layer of 1 nm. We adopt ideal ohmic source and drain contacts with a uniform donor doping of $2 \times 10^{20}$ cm$^{-3}$. The channel length L$_{ch}$ can vary from 10 to 40 nm, whereas the source/drain extensions are 10 nm.

In Fig. 8 we show the output characteristics at a low-drain bias in the presence of only PH scattering and of both PH and SR scattering for the different channel lengths in logarithmic scale. A ballistic reference curve is added for a device with L$_{ch}$ = 20 nm. PH scattering is responsible of a uniform current degradation at all inversion densities, whereas SR is responsible for a threshold voltage shift (few mV depending on the channel length), but leaves the sub-threshold slope unchanged, accordingly to [1]. Moreover, besides the performance deterioration due to short-channel effects, we notice the presence of direct tunneling for the shortest device with L$_{ch}$ = 10 nm. Hence, PH scattering is not effective in reducing source-to-drain tunneling in short devices.

The channel-length dependence of low-field mobility mobility is analyzed in Fig. 9, where the effective and scattering-limited mobilities for various channel lengths are shown as a function of the channel electron density N$_{Inv}$. While the dependence of $\mu_{eff}$ on L$_{ch}$ can be easily explained as the effect of the ballistic component linearly depending on the gate length the non trivial dependence of $\mu_{sc}$ on L$_{ch}$ is due to the importance of non homogeneous scattering in the channel and the source/drain extensions. While $\mu_{eff}$ always increases with L$_{ch}$, at low values of N$_{Inv}$ the scattering-limited mobility increases with decreasing channel length, whereas a faster mobility reduction is present at large channel densities for shorter devices [10].

Such a result is further investigated in Fig. 10 which shows the PH-limited ($\mu_{PH}$) and SR-limited ($\mu_{SR}$) mobilities obtained by means of the Matthiessen rule. Since phonon scattering is proportional to the charge density the phonon-limited mobility presents a rapid convergence to the same mobility curve as the channel length is increased to more than 20 nm due to the reduced impact of scattering occurring in the doped regions with respect to scattering in the channel.

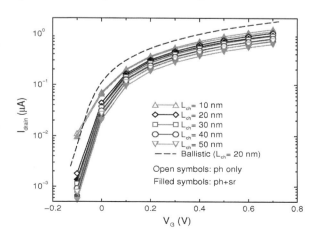

**Fig. 8** Turn-on characteristics as a function of V$_G$ in logarithmic scale for different channel lengths in the presence of (opens symbols) only PH and (filled symbols) both PH and SR scattering (Black dashed line). Ballistic reference for a device with L$_{ch}$ = 20 nm. V$_{DS}$ = 5 meV

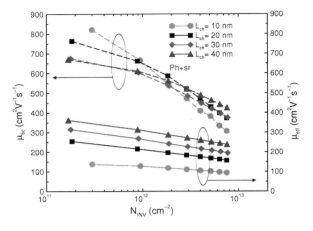

**Fig. 9** Effective and scattering limited mobility as a function of $N_{Inv}$ for different channel lengths. SR parameters: $\Delta_m = 0.2$ nm, $L_m = 1$ nm

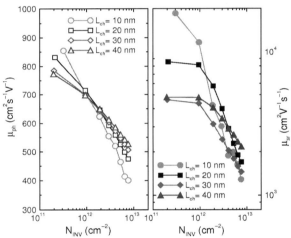

**Fig. 10** *Left* PH–limited mobility for different channel lengths as a function of $N_{Inv}$. *Right* SR-limited mobility for different channel lengths as a function of $N_{Inv}$. SR parameters: $\Delta_m = 0.2$ nm, $L_m = 1$ nm

Significantly, $\mu_{PH}$ is always much smaller than $\mu_{SR}$ thus showing that electron–phonon interaction is the dominant scattering mechanism at room temperature for the chosen parameters.

A different behavior is found out for $\mu_{SR}$, for which long devices present smaller mobility degradation as a function of $N_{Inv}$, whereas small devices show a much stronger mobility reduction for increasing densities. This can be interpreted as due to the interplay between the two main mechanisms generated by SR: subband fluctuations and mode mixing. The former is effective at small gate overdrives and responsible for electron backscattering and localization in the channel region, whereas the latter is important at high electrons densities when carriers are pushed toward the rough interfaces. Since in shorter devices electrons encounter a smaller number of potential barriers, they have a smaller probability to be backscattered or localized, being more sensitive to mode mixing rather than to subband spatial fluctuations. At low $N_{Inv}$, the larger localization effects present in long devices imply a reduced mobility with respect to short device values, while, as the charge density increases, the mode-mixing mechanism gives rise to a similar slope of mobility degradation for all devices [10].

## References

1. Buran, C., Pala, M.G., Bescond, M., et al.: Three-dimensional real-space simulation of surface roughness in silicon nanowire FETs. IEEE-Trans. Elec. Dev. **56**, 2186–2192 (2009)
2. Ferry, K., Goodnick, S.M.: Transport in Nanostructures. Cambridge University Press, Cambridge (1997)
3. Goodnick, S.M., Ferry, D.K., Wilmsen, C.W., et al.: Surface roughness at the Si(100)-SiO2 interface. Phys. Rev. B **32**, 8171–8186 (1985)
4. Luisier, M., Schenk, A., Fichtner, W.: Quantum transport in two and three-dimensional nanoscale transistors: coupled mode effects in the nonequilibrium Green's function formalism. J. Appl. Phys. **100**, 043713 (2006)
5. Mahan, G.D.: Many-Particle Physics. Kluwer, New York (1981)
6. Meier, Y., Wingreen, N.S.: Landauer formula for the current through an interacting electron region. Phys. Rev. Lett. **68**, 2512–2515 (1992)
7. Pala, M.G., Buran, C., Poli, S., Mouis, M.: Full quantum treatment of surface roughness effects in Silicon nanowire and double gate FETs. J. Comput. Electron. **8**, 374–381 (2009)
8. Poli, S., Pala, M.G., Poiroux, T., Deleonibus, S., Baccarani, G.: Size dependence of surface-roughness-limited mobility in Silicon-nanowire FETs. IEEE Trans. Elecron. Dev. **55**, 2968–2976 (2008)
9. Poli, S., Pala, M.G., Poiroux, T.: Full quantum treatment of remote Coulomb scattering in Silicon-nanowire FETs. IEEE Trans. Electron. Dev. **56**, 1191–1198 (2009)
10. Poli, S., Pala, M.G.: Channel-length dependence of low-field mobility in Silicon-nanowire FETs. IEEE Electron. Dev. Lett. **30**, 1212–1214 (2009)
11. Rahman, A., Lundstrom, M.S.: A compact scattering model for the nanoscale double-gate MOSFET. IEEE Trans. Electron. Dev. **49**, 481–489 (2002)
12. Ramayya, E.B., Vasileska, D., Goodnick, S.M., Knezevic, I.: Electron transport in silicon nanowires: the role of acoustic phonon confinement and surface roughness scattering. J. Appl. Phys. **104**, 063711 (2008)
13. Rogdakis, K., Poli, S., Bano, E., Zekentes, K., Pala, M.G.: Phonon- and surface-roughness-limited mobility of gate-all-around 3C-SiC and Si nanowire FETs. Nanotechnology **20**, 295202 (2009)
14. Saito, S., Torii, K., Shimamoto, Y., et al.: Remote charge scattering limited mobility in field-effect transistors with $SiO_2$ and $Al_2O_3/SiO_2$ gate stacks. J. Appl. Phys. **98**, 113706 (2005)
15. Shur, M.S.: Low ballistic mobility in submicron HEMTs. IEEE Electron. Dev. Lett. **23**, 511–513 (2002)
16. Stern, F., Howard, W.E.: Properties of semiconductor surface inversion layers in the electric quantum limit. Phys. Rev. **163**, 816–835 (1967)
17. Uchida, K, Koga, J, Takagi, S: Experimental study of carrier transport mechanisms double- and single-gate ultrathin-body MOSFETs—Coulomb scattering, volume inversion, and $\delta T$ SOI-induced scattering. IEDM Tech. Dig., pp. 805–808 (2003)
18. Wang, J., Polizzi, E., Gosh, A., Datta, S., Lundstrum, M.: Theoretical investigation of surface roughness scattering in silicon nanowire transistors. Appl. Phys. Lett. **87**, 043101 (2005)
19. Zheng, Y., Rivas, C., Lake, R., Alam, K., Boykin, T.B., Klimeck, G.: Electronic properties of silicon nanowires. IEEE Trans. Electron. Dev. **52**, 1097–1103 (2005)
20. Zilli, M., Esseni, D., Palestri, P., Selmi, L.: On the apparent mobility in nanometric n-MOSFETs. IEEE Electron. Dev. Lett. **28**, 1036–1039 (2007)

# Single Dopant and Single Electron Effects in CMOS Devices

M. Sanquer, X. Jehl, M. Pierre, B. Roche, M. Vinet and R. Wacquez

**Abstract** For the first time a state-of-the-art CMOS foundry (CEA-LETI-MINATEC) has been used to design silicon nanostructures with single or multiple gates dedicated to the study of single electron effects. The nanofabrication uses two e-beam lithography steps to define an active region formed by silicon-on-insulator (SOI) nanowires of cross section down to 20 nm × 10 nm and polysilicon gates of lengths down to 20 nm. The pitch at the gate level (distance between centers of the successive gates) is as small as 70 nm. Several technological splits (SOI thickness, channel doping, LDD doping, nitride spacer's length, trimming of active layer) have been made to compare devices differing only by one crucial parameter. Several dozen of designs have been introduced in the e-beam data base to analyse the impact of key geometrical parameters. As a whole more than 40,000 samples have been fabricated and several hundreds of them have been studied electrically, mostly at low temperature. Some of them are described in this contribution.

## 1 Introduction

This contribution focuses on quasi-1D 10/20 nm thin SOI etched nanowire field effect transistors (FETs). The advantage of this geometry is the good electrostatic integrity of the device even at small gate length. As channel dimensions shrink, engineering junctions to source and drain (nitride spacers and LDD/HDD doping) plays a key role in the final transistor performances (current drive, short channel effect, threshold voltage value and sub threshold swing).

M. Sanquer (✉), X. Jehl, M. Pierre, B. Roche and R. Wacquez
INAC-SPSMS-LaTEQS, CEA-Grenoble, 17 rue des Martyrs, 38054, Grenoble France
e-mail: marc.sanquer@cea.fr

M. Vinet
CEA-LETI-MINATEC, CEA-Grenoble, 17 rue des Martyrs, 38054, Grenoble France

We have extensively studied tri-gate transistors at low temperature by varying cross section of the nanowire, doping, spacer thickness, gate length, oxide thickness. Experiments performed at low temperature capture with an unprecedented precision the presence of single dopant and single electron effects. We have shown how to engineer the access regions and to finely tune the access resistance to make accessible the control one-by-one of electrons in the channel thanks to the Coulomb blockade phenomena, i.e. to change the MOSFET into a MOS-SET (Metal–Oxide–Semiconductor Single Electron Transistor).

The paper is organized as follows: first we discuss the nature and the role of access resistance in the Coulomb blockade phenomena. Then we discuss the potentialities of the MOS-SET in terms of scalability and performance. We show how to combine two (or more) MOS-SETs to envision new functionalities. Finally we discuss single dopant effects.

## 2 The Access Resistance and the One-by-one Control of Electrons in the Channel of a MOS-FET

To demonstrate the impact of single electron effects in silicon devices it is valuable to investigate a simple geometry directly inspired from the MOS-FET. To this goal single gate etched nanowire MOSFET is considered—usually named trigate MOSFET because the gate wraps three sides of the nanowire, the fourth one being an interface with a good quality buried oxide (BOX). The crystalline silicon nanowire is etched from a ultra thin SOI by e-beam lithography. Very small sizes are required to observe single electron effects because this electrostatic effect is inversely proportional to the diameter of the considered electron island. The studied trigate MOSFETs can be split into two categories depending on the source–drain junctions architecture: either overlap or underlap geometry. We will consider an overlapped source-gate geometry in the last section (single dopant effects) and concentrate first on the under lapped source-gate case. This underlap geometry has some virtues: it permits to limit the Miller Cgs and Cgd capacitances and to improve the channel electrostatic integrity and to limit the effect of lateral dopant diffusion from the HDD. Its drawback is the increase of the access resistance especially for very thin SOI MOS-FETs. This drawback can be turned into an advantage if we are interested to promote single electron effects in the channel [1]. In fact single electron effects require two conditions: first the charging energy should be larger than the temperature (and than the energy provided by the source–drain voltage source when one electron in transferred). Second the conductance associated to tunnel barriers between the dot and the electrodes should be smaller than the quantum of conductance $e^2/h$. The microscopic nature of these tunnel barriers is not crucial for observing Coulomb blockade: the tunnel barriers can be pure thin oxide barriers or thick disordered barriers or even pinch off constrictions... only their (elastic) conductance matters. By elastic we mean that the energy of carriers is conserved during tunnelling. For our case the silicon nanowires below the spacers are two excellent tunnel barriers at low temperature provided that they are clean and short enough. At high temperature energy relaxation time is short and short nanowires—that is short spacers length—are needed to avoid inelasticity. For increasing SET operating temperature small nanowire cross section are desirable. Therefore scaling down the MOS-SET— short spacers—small cross section—makes it works better at higher temperature.

The scaling rules are slightly different for the MOS-FET and for the MOS-SET as far as the gate oxide thickness and the spacer length are concerned. For the MOS-FET decreasing the gate length implies to decrease the gate oxide thickness by the same factor, to preserve good channel electrostatic control by the gate. If Cg = Cox is the gate-channel capacitance and Cs = Cd is the channel—source (drain) capacitance good electrostatic control means that Cg ≫Cs + Cd. For a MOS-SET this rule does not apply: the MOS-SET works perfectly if the so-called lever arm parameter Cg/(Cg + Cs + Cd) is small. Only the voltage gain of the MOS-SET which is Cg/Cd is small in that case.

Therefore in our devices where the SOI thickness is 20 nm we used a $SiO_2$ gate oxide thickness of 5 nm or more (up to 20 nm without limitation). For an optimized MOS-FET this parameter's choice will induce strong short channel effects.

The spacer length is the second parameter we choose to optimize the MOS-SET. For the MOS-SET a small Cs, Cd are necessary to have a large charging energy $e^2$/(Cg + Cs + Cd). The main component to Cd, Cs is the silicon nanowire below the spacer which is a dielectric between the HDD source–drain and the accumulated layer of electrons below the gate. To a first approximation this capacitive component is proportional to the nanowire cross section and to the inverse of the spacer length. It increases also with the polarizability of the silicon nanowire which depends on gate voltage and dopant concentration [2]. Long spacers (underlap) are a priori a good option to reduce Cs, Cd. Nevertheless there is a trade-off between the capacitance Cs, d and the conductance Gs, d of the nanowire below the spacers. The conductance of the nanowire should be kept below $e^2/h$ but a too small conductance will be very detrimental for any application (already the MOS-SET has an intrinsic impedance of 100 kOhms when optimized that limits its application to low power–low frequency applications or to memory applications) .

In fact the maximum conductance—neglecting resonant tunnelling effects which appear for very small dots—is GsGd/2(Gs + Gd) when sequential tunnelling through the channel in the Coulomb blockade regime is permitted [at CgVg = e(N + 1/2)]. The current through the MOS-SET will vary periodically between zero and GsVds/4 (if Gs = Gd) for Vds ≪ e/(Cg + Cs + Cd). Therefore Gs, d should be kept as close as possible to $e^2/h$ if a large transconductance is targeted.

Unfortunately Gs, d is very sensitive to the geometry of the nanowire: Gs, d varies exponentially (not linearly!) with the spacer length if tunnelling through the nanowire is considered. This is valid for direct tunnelling but also in average over energies for barriers including localized states [3]. Therefore the spacer

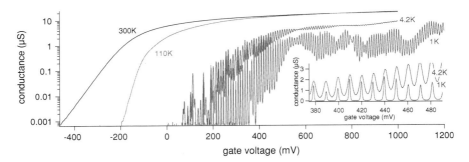

**Fig. 1** Source–drain conductance versus gate voltage at various temperature in a nanowire MOS-SET. $L = 30$ nm, $W = 40$ nm, Tsi = 17 nm, Spacers length = 40 nm, Tox = 10 nm. The quantum of conductance is $4 \times 10^{-5}$ S, from [2]. *Inset* detail in linear scale

length should be tuned to optimize the conductance as a priority, the optimization of capacitance being adjusted also with the nanowire cross section.

In this Gs, $d$ optimization we are helped by a remarkable observation that in our nanowire devices, at room temperature and at large gate voltage—i.e. when Coulomb blockade does not exist anymore ($kT > e^2/(Cg + Cs + Cd)$) and when source–drain conductance is GsGd/(Gs + Gd) (source–drain resistance dominated by the access regions: Gs, d $<$ W/L $\mu$ Cg(Vg − Vth))—the better MOS-SET are obtained for a source–drain conductance close to the quantum $e^2/h$.

This is expected if Gd = Gs = $e^2/h$ from the previous considerations because this corresponds precisely to the largest possible value for the conductance compatible with the Coulomb blockade phenomena. What is remarkable and phenomenological at this stage is that a room temperature criteria is very efficient to deduce what is happening at low temperature: we observe that when Gd = Gs = $e^2/h$ at room temperature, despite a non monotonous behaviour for the temperature dependence of the source–drain conductance, the low temperature conductance at large gate voltage is also close to $e^2/h$. By contrast if at room temperature and large gate voltage the source–drain conductance is much smaller than the quantum the source–drain conductance decreases to almost zero at low temperature, and if on the contrary the source–drain conductance is larger than the quantum, no MOS-SET effect is observed at low temperature—only eventually some stochastic Coulomb blockade phenomena [3].

Examples of MOS-SETs-among hundred other tested samples from the same batch are shown on Figs. 1 and 2. As discussed the room temperature source–drain conductance near Vg = 1 V (Vth $<$ 0 V) is about one half of the quantum $e^2/h \approx 4 \times 10^{-5}$ Siemens. The variation between 300 and 4.2 K at this gate voltage is about a factor of four of reduction.

Coulomb blockade oscillations (CBO) appear at low temperature for gate voltage (well) above 0 V that is when the accumulation electron channel exists: the threshold voltage (at 300 K) is approximately −0.150 V for our non intentionally doped channel nanowires (n++ polysilicon gate, 5 nm gate oxide thickness and 20 nm for nanowire thickness) [4].

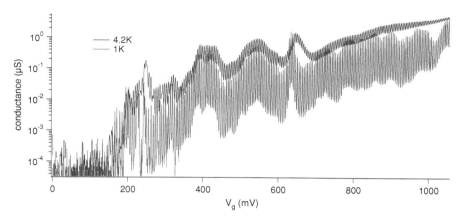

**Fig. 2** Source–drain conductance of a MOS-SET versus gate voltage. 300 periodic Coulomb blockade oscillations are recorded both at $T = 4.2$ K and $T = 1$ K. The period of oscillation ($\Delta V_g = 4.46$ mV) is in agreement with the calculated geometrical gate-channel capacitance (4.38 mV). The SOI cross section is 10 nm × 45 nm the gate length is 80 nm and the spacer width is 40 nm (undoped channel). $T_{ox}$ is 5 nm

$$V_{\text{Threshold}} = \Phi_{\text{MS}} + k_B T/q \ln \left( 3C_{ox} / q^2 N_{it} Si \right)$$

where $N_{it}$ is the intrinsic density of carrier's. The first electrons in the channel are poorly coupled to the source and drain at low temperature, therefore the amplitude of the drain source conductance is small. This amplitude is a strong—exponential-like—function of the gate voltage as it can be seen on Figs. 1 and 2. Sample specific features appear in addition to this mean behaviour. The exponential behaviour of the drain source current versus gate voltage for the envelop of the Coulomb blockade fluctuations reflects the exponential increase of the tunnelling rates between the source–drain and the channel with carrier's energy. If the silicon nanowire below the nitride spacers is disorder-free they can be viewed as tunnel barriers with an effective barrier height which is linearly decreasing with the gate voltage, therefore the exponential dependence of the tunnelling rate. Because there is disorder in the silicon nanowire below the spacers the tunnelling rate is not a monotonic function of the gate voltage. Indeed some localized states inside the barrier promote resonant transmission at certain energies. The dependence of the tunnelling rate with energy in a disordered tunnel barrier is very complex, both because of quantum interference and charging effects. Nevertheless thanks to the small volume of the nanowire below the spacers it is possible to obtain very precise information for the localized states in the nanowire using the MOS-SET as a detector [5]. In particular some of these localized states have been identified without doubt as single arsenic donors implanted in the nanowire (or diffused from source and drain HDD). Their location, ionization energy and spin have been measured with precision [5].

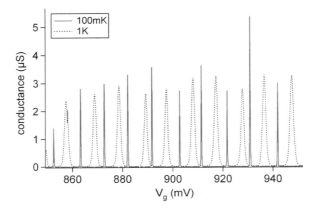

**Fig. 3** Source–drain conductance in the linear regime (small Vds = 100 microV) at $T = 0.1$ and 1 K versus gate voltage in a $W = 60$ nm Lg $= 50$ nm nanowire MOS-SET. The conductance is plotted in linear scale (instead of log scale in Figs. 1 and 2 to emphasize the sharpness of the CBO at low temperature). The transfer curve is a series of resonance with a periodic spacing given by $e/Cg$. The half width at half height is proportional to the temperature of the carriers; The value of the conductance (therefore of the drain current) is not periodic in gate voltage reflecting the non-monotonic gate voltage, energy and temperature dependence of the tunnelling rate between the Coulomb island and the source and drain

If the envelop of Coulomb blockade oscillations is sample specific and depends on microscopic details the period of the oscillation is fully controlled by the geometry. For a MOS-SET containing a large number of electrons, that is when the mean level spacing due to the discretization of the excitation spectrum in the electron island is negligible, peaks of source–drain current appear for $CgVg = e(N + 1/2)$ Fig. 3. Here we have neglected offset charges as well. The period of the Coulomb blockade oscillations is therefore approximately $e/Cg$. In several MOS-SETs the measured period is fully in agreement with the geometrical capacitance estimated between the gate and the nanowire segment below it [6]. In particular this capacitance does not change with the gate voltage that means that the surface of the MOS-SET is not increasing with the number of electrons in it. The MOS-SET can accommodate a large number of electrons without changing it shape, therefore the density of carriers in it can be varied in large proportion as illustrated on Fig. 2. Also the gate oxide can be varied in large proportion without affecting the MOS-SET function [6], only the gate capacitance changes.

## 3 Scalability and Performance of the MOS Single Electron Transistor

The MOS-SET is very well characterized and controllable at low temperature ($T = 4.2$ K and below). Compared to other SETs—in particular to other silicon

SETs it presents several advantages: compactness, simplicity (only one gate), yield and reproducibility, and straightforward co-integration with traditional CMOS circuits.

To obtain the best results however a state-of-the-art CMOS foundry is required.

Compared to silicon SET made of intrinsic silicon (bulk or SOI) using two levels of gates (top and upper levels) [7], it minimizes the number of gates, their cross talk and the parasitic capacitances.

The compactness and scalability are well illustrated by the fabrication of coupled MOS-SETs (see next section) [8].

As other silicon SETs [9] and in contrast to metallic SETs, whose applications are severely limited due to large low frequency noise coming from the low dynamics of offset charges, the MOS-SET exhibits very high stability in time. The pattern of CBO is identical between two records separated by 5 years [10]. The charge noise itself is very low of order of $10^{-4}$ e/Hz$^{1/2}$ @1 Hz and $T = 1$ K [11]. There are offset charges in the immediate environment of the MOS-SET-particularly in the silicon nanowire under the spacers—but they present no dynamics except at peculiar gate voltage values [5].

The main question is certainly to know if the MOS-SET can be scaled down to be used at room temperature. As already noted, scaling down the MOS-SET makes room temperature operation possible in principle. Size in the range of few nanometers are necessary to reach Coulomb energy (much) larger than 300 K, of order 0.2–1 eV. We have recently obtained MOS-SET of size 20 nm whose charging energy is as large as 10 meV (not shown). Recently Shin et al. [12] have observed room temperature CBO in a device which, in our opinion, is an ultimate MOS-SET: their sample is a few nanometers silicon nanowire obtained by top-down approach and covered by a central local gate with spacers. Our conclusion is that there is plenty of space to improve the charging energy of the MOS-SET and technological progresses towards the next node of the ITRS roadmap will benefit for the MOS-SET itself.

## 4 Coupled MOS-SETs

Figures 4, 5 show compact coupled MOS-SETs micrographs. Two MOS-SETs are put in series by covering a silicon nanowire (in blue) of length 200 nm by two independent gates (in red) with nitride spacers (in green). By playing on the nitride spacer thickness (40 nm in Fig. 4 and 15 nm in Fig. 5) one can obtain two MOS-SETs in series with tunable inter-coupling (Fig. 4), or a very compact triple dot system, where in addition to the two MOS-SET a central implanted quantum dot is inserted [8].

Electrical characterization of the tuneable coupled silicon quantum dot is shown on Fig. 6 [8]. The so-called stability diagram is represented: the source–drain conductance is plotted (in blue) as function of the gate voltage applied to both gates. The source–drain current vanishes due to Coulomb blockade when the

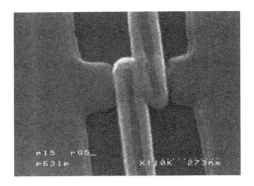

**Fig. 4** False colours SEM micrograph of SOI nanowires covered by two polysilicon gates. The *blue part* is the SOI, the *red parts* are the PolySi gates and the *green parts* are the nitride spacers; The nitride spacers are 40 nm thick; the gate length is 60 nm and the spacing is 60 nm

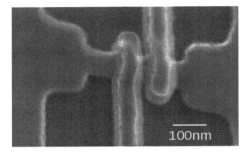

**Fig. 5** False colours SEM micrograph of SOI nanowires covered by two polysilicon gates. The *blue part* is the SOI, the *red parts* are the PolySi gates and the *green parts* are the nitride spacers; The nitride spacers are 15 nm thick; the gate length is 60 nm and the spacing is 60 nm. The only difference with sample presented on Fig. 4 is the thickness of the spacers. Because the spacer thickness is less than half the distance between gate edges the central region of the SOI nanowire is exposed to source–drain implantation (and eventually to silicidation step)

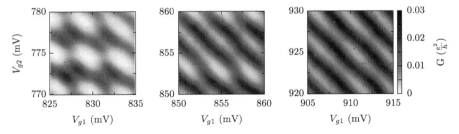

**Fig. 6** The source–drain conductance in quantum units as function of the two gate voltages. The data are recorded at $T = 1$ K. For this sample the gate length is 60 nm, the spacing between gates is 40 nm and the nitride spacers are 40 nm thick. The SOI cross section is 20 nm × 60 nm. At low gate voltage (*left panel*) the two MOS-SET below the gates are capacitively coupled. The pattern exhibits a typical honeycomb structure. At large gate voltage (*right panel*) the two MOS-SETs are tunnel coupled by the accumulation channel created in the central region: the two MOS-SETs merge as a single island and an anti diagonal pattern appears. The *central panel* describes an intermediate situation (from Ref. 8)

electron number on both MOS-SETs is fixed (in principle it is enough to have one MOS-SET blocking because the MOS-SETs are put in series, but the co tunnelling effect is enough to produce a measurable current when one MOS-SET is on). For two MOS-SETs with only a capacitive inter coupling this produces a typical honeycomb pattern as in Fig. 6, left panel. But when the tunneling becomes too strong between the two MOS-SETs—that happens at larger gate voltage—the number of electrons is no more a good quantum number on each MOS-SET. Only the sum of the number of electrons is well defined. This produces a series of anti-diagonal patterns as in Fig. 6, right panel.

This illustrates that two MOS-SETs can be coupled in a very compact design and that the number of electrons on each of these MOS-SETs can be measured and controlled. Moreover, the correlation between electrons can be used for new functionalities [13], for instance electron pumps or coherent single electron shuttling [14].

## 5 Single Dopant Effects

In this section we will consider the overlap source-gate geometry instead of the underlap geometry used for MOS-SETs in previous parts. Single dopant effects in the access region have been reported in MOS-SETs too (see Sect. 13.2 and Ref. [5]) and are indeed responsible for the variability of the source–drain current that is modulated by the Coulomb Blockade oscillations. In this section we are interested to measure a direct source–drain current which is due to a single donor centered in the channel. With an underlap geometry this is not possible because the distance separating the centred donors and the source and drain is several ten's of nanometers. Only strongly hybridized state between a dopant and a surface accumulation layer of electrons will extend to 10 nm and above in order to have sufficient overlap between the centred state wave function and the source–drain reservoirs [15]. This is also the reason why the first electrons in the MOS-SET are not detected easily.

To have a sufficient overlap between the wave function of an electron localized around a centred shallow donor and the electronic orbitals in the source–drain, the channel length should be comparable to the Bohr radius, i.e. few nanometers up to ten nanometers. For a single dopant, one can estimate the tunneling rates to source (1) and drain (2) by: $h\Gamma_{1,2} \sim \Delta E_1 \exp(-2r_{1,2}/a_B)$, where $a_B$ is the Bohr radius, $r_{1,2}$ the distance from the dopant to the source and drain and $\Delta E_1$ the typical energy mean level spacing of the excitation spectrum for the donor. Due to the exponential dependence on the distance, a small shift from the centre of the channel for the dopant position induces a large effect on the tunneling rate asymmetry. Fortunately the resonant tunnelling helps to increase the source–drain current when the electronic state of the donor is aligned with the Fermi energy in the source–drain. The resonant tunnelling enhancement is particularly important for well centred donor because the resonant tunnelling is exponentially sensitive to the ratio between the in and out tunnelling rates: The resonant conductance could be as large as Gmax = $e^2/h$ for equal tunneling rates from the resonant state to source and drain: $\Gamma_1 = \Gamma_2$, but is (exponentially) reduced for unequal rates.

Inelastic events can also reduce the resonant conductance. Suppose a centred dopant in a long channel sample $\Gamma_1 = \Gamma_2$ are very small and the electron should stay for a very long time $\Gamma_1^{-1}$ without inelastic event to fully build the constructive interference responsible for the peaked perfect transmission. For a 30 nm long channel and $a_B = 3$ nm, and $\Delta E_1 = 10$ meV this gives $\Gamma_1^{-1} = 10^{-8}$ s much larger than any inelastic relaxation time!

Therefore, even taking into account resonant tunneling, a centred single donor cannot produce a substantial drain–source current if the electrical channel length exceed several tens of nanometers.

We thus consider samples with an overlap geometry where the electrical channel length (10 nm) is much smaller than the gate length itself (30 nm). We define the channel length as the distance separating the regions with continuous arsenic doping, i.e. doping level leading to overlapping orbitals. This concentration is given by the Mott transition, at $8 \times 10^{18}$ at cm$^{-3}$. Following our process simulation the channel length is 10 nm [14]. Of course the channel length is fluctuating because positions of donors are random. In particular for our channel cross section of Tsi = 20 nm times channel width = 50 nm the process simulation gives in average five donors in the whole channel volume. These donors come

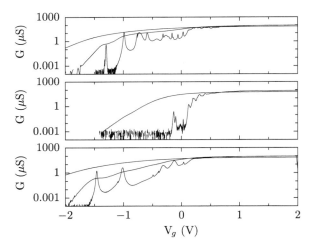

**Fig. 7** Source–drain conductance versus gate voltage for three samples over ten nominally identical samples. The temperature is 300, 90 K (*top* and *bottom panels*) and 4.2 K. The *top* and *bottom panels* shows samples with a small sub-threshold swing at 300 K associated with a low threshold voltage. At $T = 4.2$ K these samples exhibit very large sub-threshold resonances. On the contrary the middle panel shows a sample with a better sub threshold swing associated with a less negative threshold voltage. In this sample there is no sub-threshold resonance at $T = 4.2$ K. Resonances appears in all the samples above (or in the vicinity of) the threshold voltage. These resonances cannot be attributed directly to single isolated donors, but are likely occurring in the accumulation channel

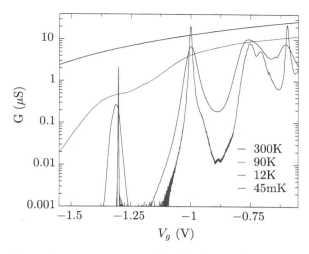

**Fig. 8** Source–drain conductance versus gate voltage for the sample presented on the preceding figure (*top panel*) at different temperature. The first sharp resonance around Vg = −1.27 V is due to resonant tunnelling through the As + → As0 transition when the electronic ground state of the donor is aligned with the Fermi level in the source (and drain). The associated ionization energy is 108 meV. The resonating donor is centred in the channel but close to the BOX The second resonance near VG = −1.0 V is due to a second distinct As donor. The associated ionization energy is 50 meV. The donor is centred in the channel but closer to the gate. The third resonance is due to the double occupation of the second donor, as explained in Ref. [16]. Resonance at larger gate voltage are close or above to the threshold voltage which is −0.5 V in sample without centred donors (see Fig. 7 *middle panel*)

from the lateral diffusion of Arsenic highly doped source and drain because the channel itself is undoped. Note that the volume of the silicon channel is $10^{-17}$ cm$^{-3}$ such that the average concentration of arsenic in the channel is $0.5 \times 10^{18}$ at cm$^{-3}$. The average distance between these donors is a little larger than 10 nm, and more importantly the probability to find a cluster of two donors separated by less than 5 nm is 5% only [16]. Limiting the statistical occurrence of cluster is very important because such a cluster has very large ionization energy and will spoil the analysis of our data in terms of single donor's physics only.

Because it is nowadays impossible to control the exact position of a single donor in silicon we should rely on a statistics over several samples to extract some samples where the presence of centred arsenic is evidenced. The robust criteria we have chosen is the observation of a small sub threshold swing (several 100 mv/decade) associated with a large negative threshold voltage. It is indeed expected that a sample with a centred donor will be less easily switched off than a sample without a centred donor.

We studied 10 nominally identical samples, some of them presented on Fig. 7.

Low temperature characteristics ($T = 4.2$ K) of the samples provide a decisive answer to the origin of the small sub threshold swing. In fact systematically samples with small sub threshold swing exhibit a few very sharp resonances at low temperature (Figs. 7, 8). Moreover one can explain the current observed at room temperature just by convoluting the low temperature resonant features by the thermally broadened Fermi distribution function [16]. Thanks to the resolution of the transport spectroscopy at low temperature the resonances can be attributed to single donor.

The most interesting output is probably that the ionization of these centred donors is very large: the energy ground state of these arsenic donors is $\approx 100$ meV below the conduction band, fixed by the appearance of the accumulation channel. This value is twice as large as the ionization energy for arsenic donors in bulk silicon. This large ionization energy means of course that these donors are far from ionized at room temperature. The origin of this large increase in the ionization energy lies in the phenomena called dielectric confinement [17]. When a donor is sitting close to the buried oxide the effective dielectric constant is a mix of the dielectric constant for silicon and silica. In other word there is an image charge of the As+ donor inside the BOX, which is also a positive charge. The As+ and the positive image charge are attracting more strongly the bounded electron that produces an increase of the ionization energy.

This study is the first experimental evidence that a single donor centred in the channel is responsible for the sub threshold characteristic degradation of a short MOS-FET at room temperature [18].

## 6 Conclusions

In very small silicon nanowire MOS-FETs single electron can be controlled one-by-one by adjusting the access resistance close to the quantum of resistance (25,813 Ohms). Reducing the cross section, the gate and spacers length around few nanometers make the single electron effect important even at room temperature. The variability due to microscopic details will become very large. For instance we have shown that a single centred shallow donor is responsible for the sub threshold swing at 300 K in a 10 nm long SOI MOSFET.

**Acknowledgments** Work done in collaboration with the AFSiD Partners. The research leading to these results has received funding from the European Community's seventh Framework (FP7 2007/2013) under the Grant Agreement Nr:214989. The samples subject of this work have been designed and made by the AFSID Project Partners http://www.afsid.eu.

## References

1. Hofheinz, M., Jehl, X., et al.: A simple and controlled single electron transistor based on doping modulation in silicon nanowires. Appl. Phys. Lett. **89**, 143504–143506 (2006)
2. Hofheinz, M., Jehl, X., et al.: Capacitance enhancement in Coulomb blockade tunnel barriers. Phys. Rev. B **75**, 235301–235304 (2007)
3. Jehl, X., Sanquer, M., et al.: Nanoelectronics with CMOS transistors: electrostatic and quantum effects. In: Lévy L. P. (ed.) Special Issue: on Nanotechnology in France II: C'NANO Rhône-Alpes Guest. Lévy Int. J. Nanotechnol. **7**, 4–8 (2010)

4. Deleonibus, S., et al.: Physical and technological limitations of NanoCMOS devices to the end of the roadmap and beyond. In: Deleonibus, S. (ed.) Electronic Device Architectures for the Nano-CMOS Era. Pan Stanford Publishing, Singapore (2009)
5. Hofheinz, M., Jehl, X., et al.: Individual charge traps in silicon nanowires: Measurements of location, spin and occupation number by Coulomb blockade spectroscopy. Eur. Phys. J. B **54**, 299–307 (2006)
6. Boehm, M., Hofheinz, M., et al.: Size scaling of the addition spectra in silicon quantum dots. Phys. Rev. B **71**, 033305–033308 (2005)
7. Ono, Y., Fujiwara, A., et al.: Manipulation and detection of single electrons for future information processing. J. Appl. Phys. **97**, 031101–031103 (2005)
8. Pierre, M., Wacquez, R., et al.: Compact silicon double and triple dots realized with only two gates. Appl. Phys. Lett. **95**, 242107–242109 (2009)
9. Zimmerman, N.M., Simonds, B.J., et al.: Charge offset stability in tunable-barrier Si single-electron tunneling devices. Appl. Phys. Lett. **90**, 033507–033509 (2007)
10. Pierre, M., Jehl, X., et al.: Sample variability and time stability in scaled silicon nanowires. ULIS2009 proceedings. (2009). doi:10.1109/ULIS.2009.4897583
11. Jehl, X., Sanquer, M., et al.: Random telegraph noise in ultimate MOSFETS at very low temperature in the subthreshold regime. J. Phys. **4**(12), Pr3–Pr107 (2002)
12. Shin, S.J., Jeong, C.S., et al.: Enhanced quantum effects in an ultra-small coulomb blockaded. device operating at room-temperature. arXiv:1003.2112 (2010)
13. Gautier, J., Jehl, X., Sanquer, M.: Single electron devices and applications. In: Stanford Pan and World Scientific Publishing Corporation (ed.) Electronic Devices Architectures for the NANO-CMOS Era (2008)
14. Buehler, T.M., Chan, V., et al.: Controlled single electron transfer between Si:P dots. Appl. Phys. Lett. **88**, 192101–192103 (2006)
15. Lansbergen, G.P., Rahman, R., et al.: Gate-induced quantum-confinement transition of a single dopant atom in a silicon. FinFET. Nat. Phys. **4**, 656–661 (2008)
16. Pierre, M., Wacquez, R., et al.: Single donor ionization energies in a nanoscale CMOS channel. Nat. Nanotechnol. **5**, 133–135 (2009)
17. Diarra, M., Niquet, Y.M., et al.: Ionization energy of donor and acceptor impurities in semiconductor nanowires: importance of dielectric confinement. Phys. Rev. B **75**, 045301–045304 (2007)
18. Wacquez, R., Vinet, M., et al.: Single dopant impact on electrical characteristics of SOI NMOSFETs with effective lengths down to 10 nm. Invited contribution VLSI Symposium 2010. (2010)

# Part III
# Diagnostics of the SOI Devices

# SOI MOSFET Transconductance Behavior from Micro to Nano Era

J. A. Martino, P. G. D. Agopian, E. Simoen and C. Claeys

**Abstract** The transconductance is one of the main device parameters used to analyze the electrical characteristics of the MOSFET. From the transconductance versus gate voltage characteristic it is possible to extract many electrical and technological parameters like threshold voltage, carrier mobility, electric field mobility degradation and others. However, partially and fully depleted SOI (planar and multi-gate) devices present second order effects that have to be well understood in order to avoid any mistake of the parameter extraction. This chapter is devoted to show the main second order effects that modify the transconductance behavior from micro to nano era of SOI devices like: partially-depleted, fully depleted, planar and multi-gate, standard and strained, DTMOS and GC SOI MOSFETs. The impact of the gate stack composition such as cap layer and metal gate thickness is also outlined. For example multiple gm peaks are sometimes observed and can be related with different origins like gate induced floating body effects, multiple threshold voltages, quantum effects and others.

---

J. A. Martino (✉) and P. G. D. Agopian
LSI/PSI/USP, University of Sao Paulo, Av. Prof. Luciano Gualberto, trav. 3, no 158, 05508-010, São Paulo, Brazil
e-mail: martino@lsi.usp.br

P. G. D. Agopian
Centro Universitario da FEI, Av. Humberto de Alencar Castelo Branco, no 3972, 09850-901, Sao Bernardo do Campo, Brazil

E. Simoen and C. Claeys
IMEC, Kapeldreef 75, B-3001, Leuven, Belgium

## 1 Introduction

The MOSFET transconductance (gm) is defined mathematically as the derivative of the drain current ($I_{DS}$) to the gate voltage ($V_{GF}$) as shown by (1.1). Physically gm is a measure of the effectiveness of the control of the drain current by the gate voltage. The gm versus gate voltage (gm × $V_{GF}$) is one of the more important curves used for extraction of the main device and technology parameters.

$$\mathrm{gm} = \frac{dI_{DS}}{dV_{GF}} \quad (1.1)$$

For example, the first order approximation model for the drain current and transconductance for a bulk MOSFET in the saturation region is shown in (1.2) and (1.3), respectively.

$$I_{Dsat} = \frac{1}{2n} \cdot \mu_{eff} \cdot C_{ox} \cdot \frac{W}{L} (V_{GF} - V_T)^2 \quad (1.2)$$

$$\mathrm{gm} = \frac{\mu_{eff} \cdot C_{ox} \cdot W}{n \cdot L} \cdot (V_{GF} - V_T) \quad (1.3)$$

where $C_{ox}$ is the gate oxide capacitance ($\varepsilon_{ox}/t_{ox}$), $t_{ox}$ is the oxide thickness, $\mu_{eff}$ is the effective carrier mobility, $V_{GF}$ is the gate voltage, $V_T$ is the threshold voltage, $n = n_{bulk} = (1 + \varepsilon_{Si}/x_{dmax}C_{ox})$, $x_{dmax}$ is the maximum depletion region, and $W$ and $L$ is the width and length of the transistor.

However, some special phenomena in gm behavior have been observed for SOI (planar and MuGFET) devices compared to the bulk ones and require a carefully study. The evolution from the micro to nano era has modified the gm × $V_{GF}$ curve as will be discussed in this chapter. The correct understanding of the second order effects on gm is important in order to enable a correct parameter extraction from this curve.

## 2 SOI MOSFET Transconductance

Figure 1 shows a cross section of a Silicon-On-Insulator (SOI) nMOSFET. The presence of the buried oxide under the active area of the transistor changes the electrical behavior of the transistor significantly.

The physics of a SOI MOSFET strongly depends on the channel doping concentration ($N_a$) and the silicon film thickness ($t_{Si}$). Devices where the channel silicon film never can be completely depleted are called partially depleted (PD), while devices that can be totally depleted under certain conditions (when the threshold voltage is applied to the front gate, for example) are called fully depleted (FD) ones. The transconductance behavior is very different in each case as described below.

**Fig. 1** Cross section of a SOI MOSFET

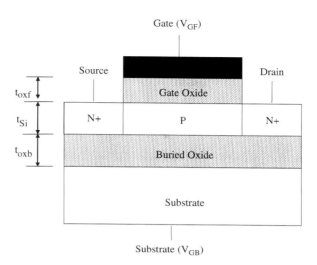

## 2.1 Partially Depleted SOI

In terms of transconductance, the PD-SOI MOSFET works like a bulk MOSFET if the transistor channel (body) is grounded. If this is not the case "floating body effects" can strongly modify the shape of gm × $V_{GF}$ curve. A special effect will be described in Sect. 5, for the case that the gate oxide is very thin. For thicker gate oxides (where no gate tunneling is observed) the gm equation is identical to that showed for a bulk MOSFET (1.3).

## 2.2 Fully Depleted SOI

The gm of FD-SOI MOSFETs is very dependent on the back gate voltage ($V_{GB}$) once that it influences the front surface potential which is not only dependent on $V_{GF}$, as is the case of PD-SOI MOSFETs. The modulation of the front surface potential by the $V_{GB}$ changes the $I_{DS} \times V_{GF}$ and consequently the gm × $V_{GF}$ due to the $V_T$ variation (which causes a shift on the $V_{GF}$ axis) and the electrostatic coupling between front and back gate (which changes the magnitude) as shown in Fig. 2.

A simple model for gm in the saturation region can still be described by (1.3) but now the body factor ($n$) has to be replaced by:

$n = n_{acc} = 1 + C_{Si}/C_{oxf}$ for devices with back interface accumulated

$n = n_{depl} = 1 + C_{Si}C_{oxb}/C_{oxf}.(C_{Si} + C_{oxb})$ for devices with back interface depleted.

where $C_{Si}$ is the silicon capacitance ($\varepsilon_{si}/t_{si}$), $C_{oxf}$ is the front gate oxide capacitance ($\varepsilon_{ox}/t_{oxf}$) and $C_{oxb}$ is the buried oxide capacitance ($\varepsilon_{ox}/t_{oxb}$).

**Fig. 2** Transconductance behavior of a FD SOI MOSFET for different back gate voltages

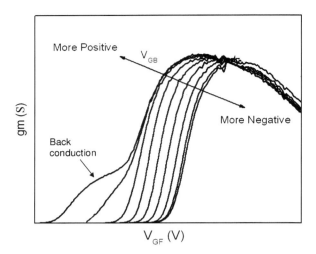

Comparing the magnitude of the body factor for different cases it is possible to notice that typically:

$$n_{depl} < n_{bulk} < n_{acc}$$

From this simple gm model it is easy to see that for fully depleted devices the gm is larger ($n_{depl}$ is lower) than in bulk ($n_{bulk}$) and also than in fully depleted devices with back accumulation ($n_{acc}$). Also it is possible to derive by Eq. 1.3 and Fig. 2 that the gm curve shifts to lower values of $V_{GF}$ due to the well known decrement of $V_T$ caused by the $V_{GB}$ increment when the back interface is depleted [1].

Figure 2 also shows a particular "plateau" (back conduction region) that can be observed under certain conditions in the gm curves of FD SOI devices and which is not found for PD SOI or bulk devices. This "plateau" is caused by a current that flows at the back interface due to the back threshold decrease when $V_{GF}$ is increased [2, 3].

## 3 GC SOI MOSFET

The Graded-Channel (GC) SOI MOSFET consists of an asymmetric channel transistor (Fig. 3) proposed in 2000 by Ref. [4] where a $V_T$ implantation is performed only at the source side while at the drain side ($L_{LD}$) the substrate doping concentration ($1 \times 10^{15}$ cm$^{-3}$) is kept. As a result, the surface of the lightly doped region becomes inverted at a $V_{GF}$ well below the $V_T$ resulting in an extension of the drain. This structure has a shorter effective channel length ($L_{eff} = L - L_{LD}$), where $L$ is the mask gate length.

Figure 4 shows gm × $V_{GF}$ for different $L_{LD}/L$ ratios. If $L_{LD}/L$ increases gm also increases so that $L_{eff}$ decreases for the same transistor channel length ($L$).

**Fig. 3** Cross section of a graded channel (GC) SOI MOSFET

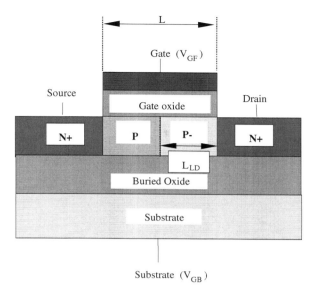

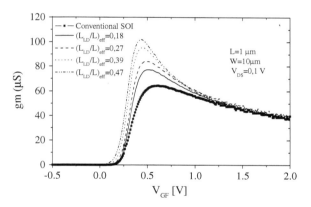

**Fig. 4** Experimental transconductance as a function of gate voltage for different $L_{LD}/L$ ratios

The improved GC performance is clearly observed in the $I_{DS} \times V_{DS}$ curves (Fig. 5), where it is easy to see the better output conductance (and consequently better Early Voltage) of the GC compared to the conventional SOI devices.

## 4 DTMOS

The Dynamic Threshold MOS (DTMOS) or voltage-controlled bipolar-MOS devices were first reported in 1987 by Ref. [5]. These devices are basically a PD SOI MOSFET in which the gate electrode is connected to the neutral part of the channel (body). Then, when $V_{GF} = V_{body}$ increases, the $V_T$ decreases

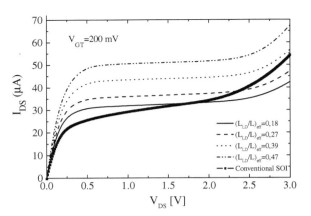

**Fig. 5** Experimental $I_{DS} \times V_{DS}$ for conventional SOI and GC SOI with different $L_{LD}/L$ ratios

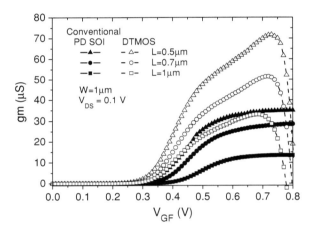

**Fig. 6** Transconductance of DTMOS and conventional SOI devices

(due to positive body bias) increasing $I_{DS}$. Thanks to that this devices has an almost ideal subthreshold slope, reduced body effect ($n$) and improved drain current characteristics. However, these devices cannot be used for $V_{GF}$ higher than 0.6 V due to forward bias of the channel/source junction. They are mainly used for very low voltage applications.

The gm × $V_{GF}$ curves for DTMOS and conventional PD SOI devices are shown in Fig. 6. As expected, the DTMOS structure has always a higher gm compared to the conventional SOI MOSFET for the same transistor channel length.

## 5 Thin Gate Oxide SOI Devices

With the technology improvement and consequently the gate oxide scaling, the MOS technology becomes susceptible to high electric fields, resulting in carrier

**Fig. 7** The gate current components in a MOS structure

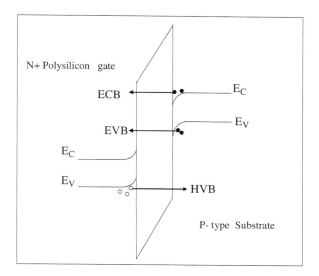

transport across the gate oxide. For gate oxides thinner than 5 nm the predominant carrier transport phenomenon is the direct tunneling current [6]. This current is for a gate stack formed by N+ Polysilicon/gate oxide/silicon film composed by the ECB (electron conduction band), EVB (electron valence band) and HVB (hole valence band) components as can be seen in Fig. 7.

Although the ECB is the predominant component of the gate current, EVB and HVB are components that have influence on the transconductance behavior for thin gate oxide devices. When nMOSFET devices with thin gate oxides are submitted to a high enough vertical electric field, the substrate valence band reaches the polysilicon gate conduction band and EVB starts to flow through the gate oxide. HVB current flows even for lower gate bias.

In case of partially depleted SOI devices with floating body, the EVB and HVB currents cause a body potential increase that in turn, causes the threshold voltage reduction resulting in a drain current increase. This increase in the drain current can be observed as the transconductance second peak. This phenomenon is called the gate induced floating body effect (GIFBE) [7] or linear kink effect (LKE) [8]. Figure 8 presents the simulated transconductance and body potential behavior when considering or not the gate tunneling current [9].

Figure 8 shows that only the transconductance for which the gate tunneling current was considered in the simulation, presents the second gm peak. Focusing on the body potential for both situations (with and without tunneling current) it is possible to see that when the gate current is neglected the body potential shows a slight increase and when the tunneling current is considered, the body potential increases exponentially and the GIFBE onset is observed.

Although GIFBE occurs predominantly in partially depleted devices, it can also be observed in Fully Depleted ones if the back gate voltage ($V_{GB}$) is negative enough to accumulate the back interface as can be seen in Fig. 9.

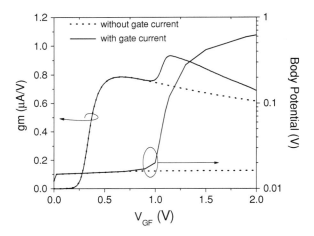

**Fig. 8** Simulation results of the transconductance and body potential behavior with and without the influence of the gate tunneling current for a PD SOI MOSFET with 2.5 nm gate oxide [9]

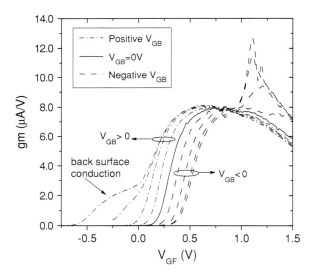

**Fig. 9** Experimental transconductance behavior for various $V_{GB}$ of a FD SOI nMOSFET with 2.5 nm gate oxide

Firstly, evaluating the curves for which the applied back gate voltages are positive, when the $V_{GB}$ becomes more positive, a plateau in the gm curve appears due to the back interface conduction as described in Sect. 2.2. The transconductance curve for $V_{GB}$ larger or equal to zero does not present GIFBE because the body is totally depleted and the tunneling current does not cause a body potential modulation in this condition. However, when the focus is the negative $V_{GB}$, the more negative the polarization the higher the transconductance second peak, because the back interface becomes accumulated and the tunneling current promotes an early body potential change.

Making use of the numerical simulator, it is possible to keep all the physical and geometrical parameters constant, while changing only the gate oxide thickness

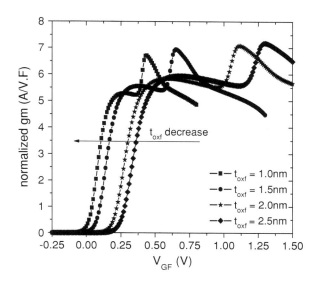

Fig. 10 Normalized gm for different gate oxide thickness

from 2.5 nm down to 1 nm. The transconductance is normalized by the gate oxide capacitance. After the normalization, one can observe a shift of the threshold voltage ($V_T$) as a consequence of the gate oxide thickness reduction ($t_{oxf}$) and a more pronounced shift is found for the transconductance second peak than for the $V_T$. As a result of these shifts, both peaks come closer to each other, as shown in Fig. 10.

## 5.1 Temperature Operation

It is known that the maximum transconductance depends on the operating temperature due to the temperature influence on carrier mobility. Besides increasing the maximum transconductance value, the temperature has also influence on GIFBE. A "C" shape of the GIFBE onset ($Vt_2$) can be observed [10] in Fig. 11 when the temperature is varied from 100 to 450 K.

Assuming that the transconductance curve at 300 K is the reference, $Vt_2$ occurs for higher gate voltage values as the temperature is increased beyond 300 K. For temperatures lower than 300 K the behavior is the opposite, with $Vt_2$ shifting towards higher gate voltage values with decreasing temperature.

The reason of the "C" shape behavior occurrence is a competition between the recombination process and the effective mobility degradation factor ($\theta$) [10], as shown in Fig. 12.

Evaluating the curve where the mobility effect was not considered in the simulations, $Vt_2$ shifts toward a lower gate voltage as the temperature decreases. This shift is due to the reduction of the recombination process with a temperature

**Fig. 11** Experimental gm behavior for temperatures varying from 100 to 450 K

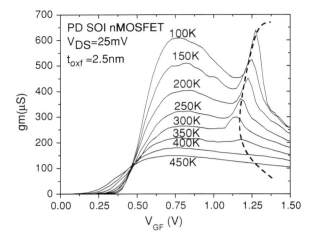

**Fig. 12** Experimental and simulated relationship between temperature, mobility degradation factor ($\theta$) and the GIFBE onset ($Vt_2$)

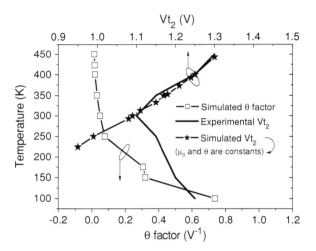

decrease. The lower recombination process allows a higher accumulation of charges and the body potential becomes exponential earlier, which results in a smaller $Vt_2$. Focusing on the effective mobility degradation ($\theta$), it is possible to see that it reduces when the temperature increases.

Looking at the experimental $Vt_2$ values, it can be stated that as the $\theta$ factor is negligible at high temperatures, the mechanism responsible for the $Vt_2$ shift is the recombination process. In this way, the simulated $Vt_2$ behavior, neglecting the mobility behavior, is the same as obtained experimentally. On the other side, when the temperature is reduced, the $\theta$ factor can no longer be neglected and its influence becomes higher than the recombination process. As a result, the $Vt_2$ value grows once again, due to the higher mobility degradation.

# 6 MuGFET

Multiple gate field effect transistors (MuGFETs) are considered as promising structures for replacing the planar single gate devices in future technologies. This section will look at the gm behavior for devices with lowly doped channel (around $1 \times 10^{15}$ cm$^{-3}$) and highly doped channel.

## 6.1 Lowly Doped Devices

The measured MuGFETs used for the transconductance analysis in this section are triple gate devices with a high-k dielectric for gate oxide with an effective oxide thickness (EOT) of 1.5 nm. Although the gate tunneling current is smaller due to the high thickness of the high-k material, the influence of the gate tunneling current components on the gm shape still cannot be neglected.

From Fig. 13a it is possible to observe the gm lateral positive shift as the back gate bias $V_{GB}$ is reduced (more negative), owing to the threshold voltage increment [11]. When the back surface reaches accumulation, the gm second peak appears and the gate voltage that corresponds to the gm second peak ($V_{gmax2}$) increases. It occurs because when the device is near to the accumulation region, the EVB and HVB tunneling components allow the body potential modulation. Looking at the gm curve with $V_{GB}$ equal to $-10$ V, it is noticed that the first and second gm peaks are so close that they overlap and only one peak can be observed.

However, when the applied $V_{GB}$ becomes more negative (lower than $-10$ V), both the strong coupling between the gates and the accumulation region expansion cause a higher recombination process, that in turn results in an later GIFBE as can be seen in Fig. 13b.

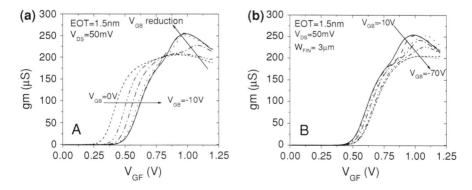

**Fig. 13** Experimental gm versus $V_{GF}$ for a triple gate device with negative $V_{GB}$ from 0 to $-10$ V (**a**) and between $-10$ and $-70$ V (**b**)

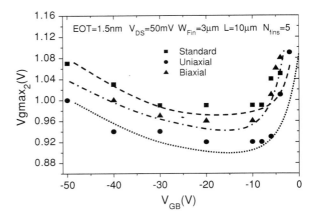

**Fig. 14** Experimental gm second peak as a function of back gate bias for different MuGFETs types (standard, uniaxial and biaxial strained)

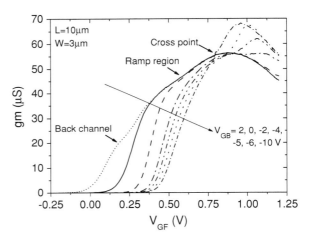

**Fig. 15** Experimental gm versus $V_{GF}$ for different $V_{GB}$ of a triple gate device with high-k gate oxide and metal gate (TiN)

This results in a "U" shaped curve when $V_{gmax2}$ is plotted as a function of $V_{GB}$ (Fig. 14). If devices without and with strained silicon (uniaxial and biaxial) are compared, the tensile strained devices present an earlier GIFBE. This effect occurs for tensile strained devices because the polysilicon gate is compressive so that HVB increases and results in an earlier GIFBE onset (low $V_{gmax2}$).

Biaxially strained triple gate devices have also been measured (Fig. 15) and the gm curve shows the presence of a back channel, a ramp region and a cross point when $V_{GB}$ changes from +2 V down to −10 V. For $V_{GB} = 2$ V the back channel conduction can be observed at the beginning of the gm curve. The ramp region is related to the current variation in the back channel with the front gate voltage increment, i.e., due to the current composition between the back and front interface conduction [12]. For $V_{GB} = -10$ V (back interface accumulated), only one gm peak is observed. Also in this case it is possible to try to extract some parameter from the $gm_{max}$. However the $gm_{max}$ observed in this curve is enhanced

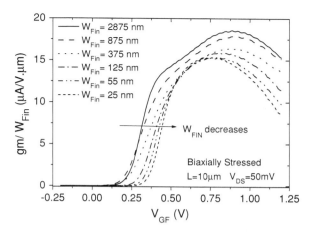

Fig. 16 Normalized experimental gm as function of the effective channel width for a triple gate device with a high-k gate oxide and metal gate (TiN)

by the decrease of the threshold voltage, due to the increase of body potential caused by the gate tunneling current.

If this enhanced $gm_{max}$ is used to extract the maximum mobility, an overestimation will be obtained. The easiest (simplest) way to determine whether or not the maximum gm peak is enhanced due to the GIFBE is through the presence of the gm cross point (where for a certain $V_{GF}$ the gm is almost constant as a function of $V_{GB}$), that happens just before the $gm_{max}$ point. If GIFBE does not occur the gm cross point is not observed.

Since the triple gate devices usually have a narrow fin width ($W_{Fin}$), the influence of a fin width reduction on the transconductance behavior was evaluated. Figure 16 presents the transconductance behavior for different fin widths for devices with a high-k gate dielectric and metal gate (TiN). The transconductance was normalized by the channel width. From Fig. 16, it can be concluded that as the fin width is reduced, the lateral gates are closer to each other and the coupling between them becomes stronger. Therefore the back gate voltage has a smaller influence on the back interface potential and the GIFBE diminishes. For narrow fin widths the lateral gates are so close that the triple gate FET becomes immune to the substrate polarization and, consequently, the GIFBE does not occur. It is also worth to notice that when the $W_{Fin}$ decreases (Fig. 16), the ramp in the gm curve disappears due to the better control of the back channel by the sidewall gates (suppression of the back interface conduction).

The degradation of the electron mobility for the sidewall (110) crystal orientation, with respect to the (100) at the top surface plane [13], leads to a decrement of $gm_{max}$.

It is known that with the use of a high-k dielectric the front interface mobility in MuGFETs becomes smaller than the backside one. Three-dimensional numerical simulations were performed with different mobility ratios between front interface mobility ($\mu_F$) and back interface mobility ($\mu_B$), as shown in Fig. 17.

When no high-k mobility degradation is considered $\mu_F = \mu_B$ the ramp is not observed. However, when $\mu_F$ is reduced and consequently $\mu_F/\mu_B$ decreases, the conduction from the front gate reduces and the back gate conduction starts to have more influence on the drain current. As a result, the $gm_{max}$ also reduces and the gm ramp becomes more evident. For $\mu_F/\mu_B = 0.6$ a clear gm ramp can be seen and the ramp slope decreases with $\mu_F/\mu_B$ [12].

## 6.2 Highly Doped Devices

MuGFETs with a highly doped channel may present multiple conduction paths, which can be easily seem in the gm derivative behavior. Figure 18 shows the cross section of a triple gate SOI MOSFET, indicating the top and bottom corners.

The gm and $dgm/dV_{GF}$ of MuGFETs with a channel doping concentration of $5 \times 10^{19}$ cm$^{-3}$ are given in Fig. 19 [14].

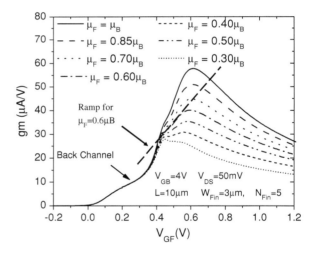

**Fig. 17** 3D-Simulated transconductance curves for different front ($\mu_F$) and back mobility ($\mu_B$) ratios

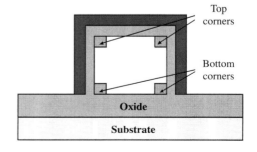

**Fig. 18** Cross section of a triple gate MuGFET

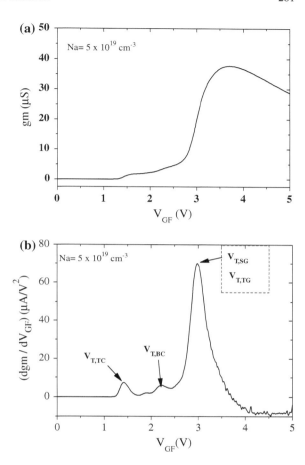

**Fig. 19** Transconductance (**a**) and its derivative (**b**) of a MuGFET with a high channel doping concentration [14]

When $V_{GF}$ increases, the top corner is the first region that reaches inversion ($V_{T,TC}$) due to the sum of the electric field from the top and sidewall gates. If $V_{GF}$ is more positive the bottom corner is the next region that gets inversion ($V_{T,BC}$). Finally for larger $V_{GF}$ the top and sidewall surfaces are inverted for almost the same gate voltage ($V_{T,TG} = V_{T,SG}$). For each threshold voltage $I_{DS}$ shows a kink that is reflected as a peak on the gm derivative. For lower channel doping it is not possible to see any difference among these threshold voltages.

## 7 Strain Technology

The use of mechanical stress has been shown as a key to enhance the device performance by increasing the mobility and consequently the drive capability [15]. Several techniques have been used in order to obtain mechanical strain in the devices. One may use either a global wafer-level (biaxial strain) technique or the

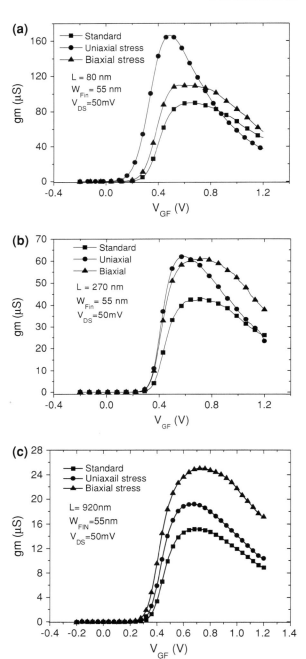

**Fig. 20** The transconductance behavior of standard and strained (uniaxial and biaxial) triple gate MuGFETs as a function of $V_{GF}$, for three different channel length: 80 nm (**a**), 270 nm (**b**) and 920 nm (**c**)

implementation of stressors resulting in locally stressed regions (uniaxial strain) [16]. In the latter case a strained contact etch-stop layer (CESL) is often used.

Besides the mobility increase, the strained devices show a threshold voltage reduction. In the uniaxially strained devices the threshold voltage change results from both the band gap reduction and the generated density of states. The threshold voltage shift for the biaxially strained devices is due to the band gap change and the electron affinity. The higher variation in the band gap is responsible for the higher threshold voltage reduction [17], [18].

Focusing on the channel length, it is possible to notice from Fig. 20 that for short channel devices both a higher transconductance and a larger threshold voltage shift occur for the uniaxially strained devices. This is a result of the higher effectiveness of the stress in uniaxial devices as the channel length becomes shorter. Knowing that for biaxial devices the stress effectiveness gets higher as the device becomes larger, in this case the better transconductance and higher gm variation occur for the device with a channel length of 920 nm (Fig. 20c). Therefore, from Fig. 20b it is observed that for an intermediate channel length both uniaxial and biaxial devices are equivalent. In all cases both types of strained devices have better gm values than standard (non strained) devices.

## 8 Gate Stack Influence on gm

The transistor gate stack with a different cap layer also modifies the gm behavior as can be seen in Fig. 21. In summary, based on electrical characterizations it is shown that the presence of a cap layer like $Dy_2O_3$ increases EOT and reduces the gate effective work-function which decrease $V_{FB}$ and $V_T$. As a consequence the transconductance and the gate current (due to the EVB) also decrease and the onset of GIFBE ($Vt_2$) increases [19].

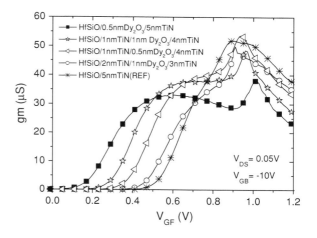

Fig. 21 Experimental gm for different gate stacks [19]

## 9 TiN Thickness Influence on gm

The gm derivative behavior for MuGFET devices with different gate electrode (TiN) thicknesses can be seen in Fig. 22. For a thicker metal gate electrode the threshold voltage increases due to the increase of the gate work-function [20]. The GIFBE can also be observed.

## 10 Quantum-Effect on gm

If a MOS transistor is made in a thin silicon film electron transport can become two-dimensional. Figure 23 shows the transconductance as a function of gate voltage, of a double-gate SOI MOSFET measured at very low temperature (0.3 K) [21] with a silicon film thickness of 40 nm.

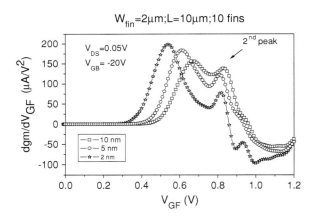

**Fig. 22** Experimental gm derivative for different TiN thicknesses

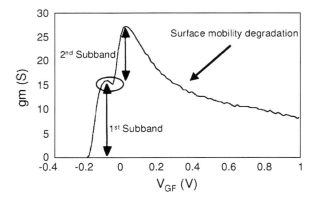

**Fig. 23** Transconductance behavior of a double-gate device, two dimensional SOI MOSFET [21]

When for this device $V_{GF}$ is below $-0.20$ V, there are no electrons in the conduction subbands and the current is equal to zero. When $V_{GF}$ is larger than $-0.2$ V (but lower than zero) the gm increases once that the lowest energy subband is populated by electrons. At higher gate voltages ($V_{GF} > 0$) electrons populate also the second subband and the gm presents a second gm peak. The transconductance decreases for $V_{GF} > 0$ due to the classical surface mobility degradation.

# References

1. Colinge, J.P.: Silicon-On-Insulator Technology: Materials to VLSI, 3rd edn. Kluwer Academic Publishers, London (2004)
2. Colinge, J.P.: Transconductance of Silicon-on-insulator (SOI) MOSFET's. IEEE Electron Dev. Lett. **6**, 573–574 (1985)
3. Ouisse, T., Cristoloveanu, S., Borel, G.: Influence of series resistances and interface coupling on the transconductance of fully-depleted silicon-on-insulator MOSFETs. Solid State Electron. **35**, 141–149 (1992)
4. Pavanello, M.A., Martino, J.A., Flandre, D.: Graded-channel fully depleted silicon-on-insulator nMOSFET for reducing the parasitic bipolar effects. Solid-State Electron. **44**, 917–922 (2000)
5. Colinge, J.P.: An SOI voltage-controlled bipolar-MOS device. IEEE Trans. Electron Dev. **34**, 845–849 (1987)
6. Schuegraf, K.F., Hu, C.: Hole injection $SiO_2$ breakdown model for very low voltage lifetime extrapolation. IEEE Trans. Electron Dev. **41**, 761–767 (1994)
7. Pretet, J., Matsumoto, T., Poiroux, T., Cristoloveanu, S., Gwoziecki, R., Raynaud, C., et al.: New mechanism of body charging in partially depleted SOI-MOSFETs with ultra-thin gate oxides. In: Proceedings of ESSDERC, pp. 515–518 (2002)
8. Mercha, A., Rafi, J.M., Simoen, E., Augendre, E., Claeys, C.: Linear kink effect induced by electron valence band tunneling in ultrathin gate oxide bulk and SOI MOSFETS. IEEE Trans. Electron Dev. **50**, 1675–1682 (2003)
9. Agopian, P.G.D., Martino, J.A., Simoen, E., Claeys, C.: Study of the linear kink effect in PD SOI nMOSFETs. Microelectron. J. **38**, 114–119 (2007)
10. Agopian, P.G.D., Martino, J.A., Simoen, E., Claeys, C.: Temperature influence on the gate-induced floating body effect parameters in fully depleted SOI nMOSFETs. Solid-State Electron. **52**, 1751–1754 (2008)
11. Poiroux, T., Vinet, M., Faynot, O., Widiez, J., Lolivier, J., Ernst, T., Previtali, B., Deleonibus, S.: Multiple gate devices: advantages and challenges. Microelectron. Eng. **80**, 378–385 (2005)
12. Agopian, P.G.D., Martino, J.A., Simoen, E., Claeys, C.: Transconductance ramp effect in high-k triple gate sSOI nFinFETs. In: IEEE International SOI Conference 2009 (2009). doi:10.1109/SOI.2009.5318766
13. Rudenko, T., Collaert, N., De Gendt, S., Kilchytska, V., Jurczak, M., Flandre, D.: Effective mobility in FinFET structures with HfO2 and SiON gate dielectrics and TaN gate electrode. Microelectron. Eng. **80**, 386–389 (2005)
14. Andrade, M.G.C., Martino, J.A.: Threshold voltages of SOI MuGFETs. Solid-State Electron. **52**, 1877–1883 (2008)
15. Giusi, G., Crupi, F., Simoen, E., Eneman, G., Jurczak, M.: Performance and Reliability of Strained-Silicon nMOSFETs With SiN Cap Layer. IEEE Trans. Electron. Dev. **54**, 78–82 (2007)

16. Claeys, C., Simoen, E., Put, S., Giusi, G., Crupi, F.: Impact strain engineering on gate stack quality and reliability. Solid-States Electron. **52**, 1115–1126 (2008)
17. Goo, J.-S., Xiang, Q., Takamura, Y., Arasnia, F., Paton, E.N., Besser, P., et al.: Band offset induced threshold variation in strained-Si nMOSFETs. IEEE Electron. Dev. Lett. **24**, 568–570 (2003)
18. Lim, J.-S., Thompson, S.E., Fossum, J.G.: Comparison of threshold-voltage shifts for uniaxial and biaxial tensile-stressed n-MOSFETs. IEEE Electron. Dev. Lett. **25**, 731–733 (2004)
19. Martino, J.A., Rodrigues, M., Mercha, A., Simoen, E., Veloso, A., Collaert, N., Jurczak, M., Claeys, C.: Gate Stack Influence on GIFBE in nFinFETs. ECS Trans. **19**(4), 133–138 (2009)
20. Rodrigues, M., Martino, J.A., Collaert, N., Mercha, A., Simoen, E., Claeys, C.: Impact of the TiN metal gate thickness on gate induced floating body effect. In: Conference Proceeding of conference EuroSOI 2009, pp. 131–132 (2009)
21. Baie, X., Colinge, J.P.: Two-dimensional confinement effects in gate-all around (GAA) MOSFETs. Solid-State Electron. **42**, 499–504 (1998)

# Investigation of Tri-Gate FinFETs by Noise Methods

N. Lukyanchikova, N. Garbar, V. Kudina, A. Smolanka, E. Simoen and C. Claeys

**Abstract** New noise methods for investigation of SOI MOSFETs are developed. The methods are based on the analysis of the BGI (Back-Gate-Induced) and LKE (Linear Kink Effect) Lorentzian fluctuations of the drain current. The results of application of those methods as well as the methods based on measuring the 1/f noise for studying strained and non-strained fully depleted tri-gate FinFETs with HfSiON/SiO$_2$ or HfO$_2$/SiO$_2$ gate dielectrics are presented. The following effects were observed for the first time: the electron valence-band tunneling currents $I_{EVB}$ flowing through the gate dielectric and the dependences of $I_{EVB}$ on SOI or sSOI substrates are different for HfSiON/SiO$_2$- and HfO$_2$/SiO$_2$-devices; the value of $[m'\beta^2/C_{eq}]$ where $\beta$ is the body factor, $C_{eq}$ is the body-source capacitance and $m' \approx 1$ increases with increasing $|V^*|$ under strong inversion conditions for the HfSiON/SiO$_2$-FinFETs; the value of $(C_{eq}/\beta^2)$ is independent of the fin width $W_{eff}$ at $W_{eff} \leq 0.37$ μm; the value of $\beta$ for FinFETs investigated is higher than for their planar counterparts; the bulk oxide trap density $N_{ot}$ decreases with the distance $x$ from the Si/SiO$_2$ interface, and the distributions $N_{ot}(x)$ are different for different gate dielectrics; in very narrow ($W_{eff} = 0.02$ μm) SOI devices with a HfO$_2$/SiO$_2$ dielectric the values of $N_{ot}$ are relatively low and homogeneously distributed over $x$.

---

N. Lukyanchikova (✉), N. Garbar, V. Kudina and A. Smolanka
V. Lashkaryov Institute of Semiconductor Physics, Prospect Nauki 45,
03028 Kiev, Ukraine
e-mail: natali@isp.kiev.ua

E. Simoen and C. Claeys
Imec, Kapeldreef 75, B-3001 Leuven, Belgium

C. Claeys
KU Leuven, Kasteelpark Arenberg 10, B-3001 Leuven, Belgium

# 1 Introduction

It is well known that there are a lot of noise methods for determination of different physical values. For example, by measuring the spectral density of voltage fluctuations, $S_V(f)$, on the contacts of some open-circuit resistor (the Nyquist noise [1]) one can find the value of the resistor, $R$, and the parasitic capacitance, $C$, between the contacts. In fact, the spectrum of such fluctuations is of a Lorentzian type, namely: $S_V(f) = [S_V(0)]/[1 + (2\pi f\tau)^2]$ where $f$ is the frequency at which the value of $S_V$ is measured, $S_V(0) = 4kTR$ is the Lorentzian plateau, $k$ and $T$ are the Boltzmann constant and the temperature, respectively, $\tau = RC$ is the Lorentzian time constant. Therefore, one finds the value of $R$ from the Lorentzian plateau $S_V(0)$ and then calculates the value of $C$ by $C = \tau/R$ where $\tau$ is determined by $\tau = (2\pi f_0)^{-1}$ from the turn-over frequency $f_0$ at which the value of $S_V(f)$ becomes equal to $[S_V(0)/2]$.

Very popular among the noise methods is the so-called noise spectroscopy of levels [2]. The method is based on the fact that the electron exchange between the levels and the conduction and valence bands modulates the conductivity of a semiconductor sample and gives rise to its fluctuations (so-called Generation-Recombination noise). If some voltage is applied to the sample, such fluctuations are transformed into current fluctuations and can be measured. The spectra of the GR noise are also of a Lorentzian type. By determining the GR Lorentzian time constants for different temperatures, the energy position of a corresponding level and the electron capture cross-section can be found. The concentration of the levels is appreciated from the GR Lorentzian plateau.

Another very popular noise method is based on the analysis of the so-called $1/f$ noise [3]. The point is that as a rule this noise is connected with the electron exchange between the semiconductor and the surface levels located in the oxide covering the surface of a semiconductor (the McWhorter noise). Frankly speaking, such a noise is a kind of GR fluctuations. The difference is that in this case the electrons exchange with the levels located at different distances from the interface by tunneling. As a result, the noise is characterized by a distribution of $\tau$ which is responsible for the $1/f^\gamma$ shape of the noise spectrum. For a homogeneous distribution of the level concentration $N_{ot}$ over the distance $x$ from the interface inside the oxide, one has $\gamma = 1$ and the value of $N_{ot}$ can be determined from the value of $[f \times S_{VG}(f)]$ which is independent of $f$, where $S_{VG}$ is the input-referred voltage $1/f$ noise. If the distribution is not homogeneous, one has $\gamma \neq 1$ where the value of $\gamma$ can be somewhat lower or higher than unity and can be different for different intervals of $f$ [3, 4]. In this case the distribution $N_{ot}(x)$ is found from the values of $[f \times S_I(f)] \propto f^{1-\gamma}$ measured at different $f$. This method is widely used for noise characterization of the Metal-Oxide-Semiconductor Field-Effect-Transistors (MOSFETs).

It should be noted that the microminiaturization of MOSFETs makes it possible to develop new noise methods for studying modern transistor structures where Silicon-On-Insulator (SOI) substrates are used. It is well known that the floating body effects are typical for the SOI MOSFETs and the new noise methods are based on the analysis of the BGI (Back-Gate-Induced) and LKE (Linear Kink Effect) Lorentzian components in the drain current noise spectra where the Lorentzians of both types can be regarded as floating body noise effects [5, 6].

It should be noted that when one wants to observe the BGI Lorentzians, it is necessary to apply an accumulation back-gate voltage $|(V_{GB})_{acc}|$ to the back contact of the SOI substrate. The source of the BGI Lorentzians is the Nyquist noise of the equilibrium conductivity of the drain and source p–n junctions, $G_0$, that increases with increasing $|(V_{GB})_{acc}|$ [6]. It should be also noted that, as a rule, application of the accumulation back-gate voltage is one of the conditions for observation of the LKE Lorentzians in the case of fully depleted SOI MOSFETs [7]. The source of LKE Lorentzians is the shot noise of the electron valence-band (EVB) tunneling current, $I_{EVB}$, flowing through the gate dielectric and of the forward current, $I_F$, flowing through the source p–n junction under conditions of EVB tunneling where $I_F = I_{EVB}$ [5, 6]. The capacitive character of the body-source impedance is responsible for the Lorentzian shape of both BGI and LKE noise.

## 2 Theory

### 2.1 The Methods Based on Measurements of the BGI and LKE Lorentzians

The following relations are valid for the parameters of the BGI and LKE Lorentzians [5, 6]:

$$[S_I(0)]_{BGI} = 4kT\tau_{BGI}\beta^2 g_m^2 / C_{eq} \tag{1}$$

$$\tau_{BGI} = C_{eq}/G_0, \tag{2}$$

$$[S_I(0)]_{LKE} = 4kTm'\tau_{LKE}\beta^2\mu^2 Z^2 C_0^2 V_{DS}^2 / L_{eff}^2 C_{eq}, \tag{3}$$

$$\tau_{LKE} = m'kTC_{eq}/qI_{EVB}, \tag{4}$$

where $\beta = (\partial V_{th}/\partial V_{BS})$ is the body factor, $V_{th}$ is the threshold voltage, $V_{BS}$ is the body-source voltage, $g_m$ is the transconductance, $C_{eq}$ is the body-source capacitance, $m' = q/[(\eta+\nu)kT]$, $\eta = (q/nkT)$, $n = (1-2)$, $\nu = (1/I_{EVB})(\partial I_{EVB}/\partial V^*)$, $V^* = (V_{GF} - V_{th})$ is the gate overdrive voltage, $V_{GF}$ is the front-gate voltage, $\mu$ is the electron (hole) mobility in the channel, $Z$ and $L_{eff}$ are the channel width and length, respectively, and $C_0$ is the front-gate dielectric capacitance per cm$^2$, $V_{DS}$ is the drain voltage. It follows from Eq. 2 that $\tau_{BGI}$ is independent of $V_{GF}$. On the contrary, $\tau_{LKE}$ decreases with increasing $|V_{GF}|$ due to the increase of the EVB current (see Eq. 4).

The BGI Lorentzians are observed at sufficiently low gate voltages corresponding to depletion and low inversion and are transformed into the LKE Lorentzians at strong inversion [6].

It follows from Eqs. 1 and 2 that by measuring the parameters of the BGI Lorentzians at different gate voltages and the values of the transconductance corresponding to those gate voltages, one can find the dependence $[S_I(0)/\tau]_{BGI} \propto (g_m)^2$ and then calculate the value of $(C_{eq}/\beta^2)$ by the following formula (see Eq. 1):

$$(C_{eq}/\beta^2) = (C_{eq}/\beta^2)_{BGI} = 4kTg_m^2/[S_I(0)/\tau]_{BGI}. \quad (5)$$

Then the values of $[I_{EVB}/m'\beta^2]$ are easily determined by

$$[I_{EVB}/m'\beta^2] = (kT/q\tau_{LKE})(C_{eq}/\beta^2). \quad (6)$$

An interesting information concerning the behaviour of the values of $\mu$ and $(\beta^2/C_{eq})$ with increasing $|V^*|$ under strong inversion conditions can be gained from the dependences of the value of $[S_I(0)/\tau]_{LKE}$ on $|V^*|$. For example, if $(\beta^2/C_{eq})$ does not change with increasing $|V^*|$—which is not always the case (see below)—one obtains from Eq. 3:

$$\mu^2 \propto [S_I(0)/\tau]_{LKE}. \quad (7)$$

Therefore, the dependence of the mobility of charge carriers in the channel on $|V^*|$ can be appreciated with the help of the dependences $[S_I(0)/\tau]_{LKE}$ vs. $|V^*|$.

It should be noted that in the case where the following "classical" relation is valid for the value of $g_m$ under strong inversion conditions [8]

$$g_m = \mu C_0 V_{DS} Z/L_{eff} \quad (8)$$

the values of $(C_{eq}/m'\beta^2)$ and $[I_{EVB}/(m'\beta)^2]$ can be also determined from the LKE data. Indeed, from Eqs. 8 and 3 one obtains:

$$[S_I(0)]_{LKE} = 4kTm'\tau_{LKE}\beta^2 g_m^2/C_{eq}. \quad (9)$$

Therefore, in this case, by measuring the parameters of the LKE Lorentzians and the values of $g_m$ corresponding to those Lorentzians, one can find the values of $(C_{eq}/m'\beta^2)$ and $[I_{EVB}/(m'\beta)^2]$ by the following formulas:

$$(C_{eq}/m'\beta^2) = (C_{eq}/m'\beta^2)_{LKE} = 4kTg_m^2/[S_I(0)/\tau]_{LKE}, \quad (10)$$

$$[I_{EVB}/(m'\beta)^2] = (2kT)^2 g_m^2/q[S_I(0)]_{LKE}. \quad (11)$$

Note that Eq. 8 corresponds to the case where the charge carrier density in the channel, $N_S$, increases linearly with increasing $|V^*|$ and, hence, only small deviation of the curve $I(|V^*|)$ from the straight line $I \propto |V^*|$ due to a decrease of $\mu$ with increasing $|V^*|$ has to take place for sufficiently long devices. However, the leveling off of the drain current at high gate voltages which cannot be attributed to the decrease of $\mu$ with increasing $|V^*|$ or to the influence of the series resistance has been observed in some cases [9]. In those cases Eq. 8 is not fulfilled (an abnormally quick decrease of $g_m$ with increasing $|V^*|$ takes place) and, hence, Eqs. 10 and 11 are not valid.

## 2.2 The Methods Based on Measurements of the 1/f Noise

The following formulas are valid for the McWhorter noise under strong inversion conditions [3, 9]:

$$(S_{VG})_{1/f} = N_{ot} \times \frac{q^2 kT \lambda}{fZL_{eff}C_0^2},\qquad(12)$$

$$[S_I(V^*)]_{1/f} = N_{ot} \times \frac{q^2 kT \lambda ZV_{DS}^2}{L_{eff}^3 f} \times \frac{[\mu(V^*)]^4}{\mu_0^2}\qquad(13)$$

where $(S_I)_{1/f}$ is the spectral density of the drain current 1/f noise, $= 10^{-8}$ cm is the tunneling parameter, $\mu_0$ corresponds to the value of $\mu$ at $V^* = 0$ V in $\mu = \mu_0/(1+\theta \cdot V^*)$, $\theta$ is the mobility degradation coefficient [10]. Equations 12 and 13 correspond to the case where the levels are distributed homogeneously over the thickness of the gate oxide.

It follows from Eqs. 12 and 13 that:

- the value of $(S_{VG})_{1/f}$ has to be independent of $V^*$;
- the voltage dependence of $(S_I)_{1/f}$ is determined by the voltage dependence of $\mu$.

Note that the experimental value of $(S_{VG})_{1/f}$ can be found by measuring the values of $(S_I)_{1/f}$ and $g_m$ and introducing them into the formula $(S_{VG})_{1/f} = [(S_I)_{1/f}]/(g_m)^2$. If the value of $(S_{VG})_{1/f}$ appears to be independent of $V^*$ then the concentration of the noisy levels $N_{ot}$ is determined by using relation (12).

In the case where the gate oxide is characterized by an inhomogeneous distribution of $N_{ot}$ over $x$, this distribution is found as follows. One determines the value of $(S_{VG})_{1/f} = [S_{VG}(f_{meas})]_{1/f}$ at a given frequency $f_{meas}$ corresponding to the distance $x = x_{meas} = -\lambda \ln(2\pi f_{meas} \tau_{min})$ where $\tau_{min} = 10^{-10}$ s [3, 4] and ensures that the value of $[S_{VG}(f_{meas})]_{1/f}$ does not depend on $V^*$. Then the concentration of the levels $N_{ot}(x_{meas})$ located at the distance $x = x_{meas}$ from the Si/SiO$_2$ interface inside the oxide is determined by the formula:

$$N_{ot}(x) = N_{ot}(x_{meas}) = f_{meas} S_{VG}(f_{meas}) ZL_{eff} C_0^2 \lambda / q^2 kT.\qquad(14)$$

By performing such measurements for different $f_{meas}$, the distribution $N_{ot}(x)$ can be found.

The paper is devoted to the results of investigations of tri-gate fully depleted FinFETs by the BGI, LKE and 1/f noise methods described above.

## 3 Devices and Experimental

The devices investigated were strained and non-strained $n$- and $p$-channel fully depleted tri-gate FinFETs with HfSiON/SiO$_2$ or HfO$_2$/SiO$_2$ gate dielectrics. In the case of the HfSiON/SiO$_2$ dielectric, the gate stack consisted of 2.3 nm HfSiON (50% Hf) on the top of 1 nm interfacial SiO$_2$. The following five types of devices with HfSiON/SiO$_2$ gate dielectric were studied: sSOI, sSOI + SEG, SOI + SEG, sSOI + CESL and SOI + CESL where sSOI indicates the devices processed in the strained Si-On-Insulator substrates prepared by using the biaxial global tensile-strain-induced technique [11], CESL means that a uniaxial local tensile-strain-induced technique where strain is induced by a tensile strained nitride Contact Etch Stop-Layer has been applied [11] and SEG means that the corresponding devices were processed with the Selective Epitaxial Growth of source and drain regions [12].

In the case of the HfO$_2$/SiO$_2$ dielectric, the gate stack consisted of 2 nm HfO$_2$ on top of 1 nm interfacial SiO$_2$. The devices studied in this case were processed on both standard (SOI) and strained (sSOI) SOI wafers.

The parameters of the devices were as follows: $h_{fin}$ = 65 and 55 nm for SOI and sSOI FinFETs, respectively, $W_{eff}$ = 0.02–9.87 μm, $L_{eff}$ = 0.15–0.9 μm, where $h_{fin}$ is the fin height, $W_{eff}$ and $L_{eff}$ are the effective fin width and length, respectively. The full device width $Z$ was calculated by the formula $Z = N_{fin} \times (W_{eff} + 2 \cdot h_{fin})$ where $N_{fin}$ is the number of fins (we used $N_{fin}$ = 1, 5 and 30). The thickness of the buried oxide was 145 nm for the SOI devices and 130 nm for the sSOI ones.

The drain current noise spectral density $S_I(f)$ in the frequency range $f$ = 1 Hz to 100 kHz, the drain current $I$ and the transconductance $g_m$ were measured on wafer at 0.4 V $\leq |V_{GF}| \leq$ 1.5 V and $|V_{DS}|$ = 25 mV where $V_{GF}$ and $V_{DS}$ are the gate and drain voltage, respectively.

## 4 Results and Discussion

It has been found that LKE and BGI Lorentzians appear in the noise spectra of the FinFETs studied if the accumulation back-gate voltage $|(V_{GB})_{acc}| \geq 7.8$ V is applied. Those Lorentzians manifest themselves as the maxima in Fig. 1 where the noise spectra are presented as the dependences of $(f \times S_I)$ vs. $f$. Since a Lorenzian component is described as $S_I = [S_I(0)]/[1 + (2\pi f \tau)^2]$, the following relations are valid for the frequency $f_0$ corresponding to the maxima $(f \times S_I)_{max}$ and for the value of those maxima: $f_0 = (2\pi\tau)^{-1}$ and $(f \times S_I)_{max} = [f_0 \times S_I(0)/2]$. Therefore, the Lorentzian parameters $S_I(0)$ and $\tau$ can be determined from the experimental values of $f_0$ and $(f \times S_I)_{max}$.

The Lorentzians observed at $|V_{GF}| \leq 0.6$ V in Fig. 1 where $f_0 \neq f_0(V_{GF})$ and, hence, $\tau \neq \tau(V_{GF})$ are the BGI Lorentzians (Fig. 1, curves 1 to 2) [6]. At the same

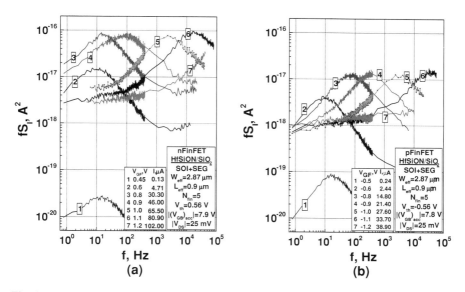

**Fig. 1** Drain current noise spectra multiplied by the frequency for the SOI + SEG $n$-type (**a**) and $p$-type (**b**) HfSiON/SiO$_2$-FinFETs measured at $|(V_{GB})_{acc}| \approx 7.8$ V and different gate voltages

time, the Lorentzians observed at $0.8\text{ V} \leq |V_{GF}| \leq 1.2\text{ V}$ where $f_0$ increases ($\tau$ decreases) with increasing $V_{GF}$ by three orders of magnitude are the LKE Lorentzians (Fig. 1, curves 3 to 7) [6].

Figure 2a and b shows the dependences of the value of $[S_I(0)/\tau]_{BGI}$ on the value of the transconductance $g_m$ measured for the $n$- and $p$-channel FinFETs with the HfSiON/SiO$_2$-gate dielectric, respectively. The gate overdrive voltage is denoted near every experimental point. It is seen that the BGI Lorentzians are observed in depletion and weak inversion conditions. It is also seen that the dependences $[S_I(0)/\tau]_{BGI} \propto (g_m)^2$ are observed experimentally. This means that the value of $(C_{eq}/\beta^2)$ is independent of $|V^*|$ in depletion and weak inversion (see Eq. 1). The observation of the relation $[S_I(0)/\tau]_{BGI} \propto (g_m)^2$ means also that one may use Eq. 5 for determination of the value of $(C_{eq}/\beta^2) = (C_{eq}/\beta^2)_{BGI}$.

The results of such a determination are presented in Fig. 3 where the dependences of the value of $(C_{eq}/\beta^2)$ on the fin width $W_{eff}$ are shown for the devices with different gate dielectrics. It is seen that those dependences coincide with each other for the HfO$_2$/SiO$_2$- and HfSiON/SiO$_2$-devices. It is also seen that $(C_{eq}/\beta^2) \propto W_{eff}$ at $W_{eff} \geq 0.87$ μm while the decrease of $(C_{eq}/\beta^2)$ with decreasing $W_{eff}$ becomes slower at $W_{eff} < 0.87$ μm and the value of $(C_{eq}/\beta^2)$ stops to be dependent on $W_{eff}$ at $W_{eff} \leq 0.37$ μm. Such a behavior means that while at $W_{eff} \geq 0.87$ μm the value of $(C_{eq}/\beta^2)$ is determined by the width of the fin top interface, at $W_{eff} < 0.87$ μm the contribution of the fin side interfaces to that value increases and at $W_{eff} \leq 0.37$ μm the value of $(C_{eq}/\beta^2)$ is fully determined by the width of the side interfaces equal to $2h_{fin}$.

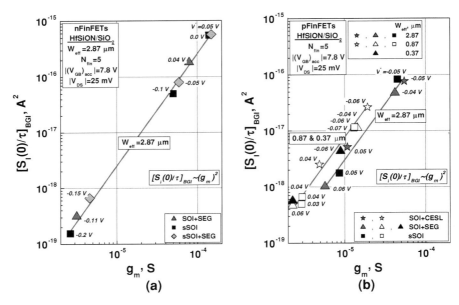

**Fig. 2** Dependences of $[S_I(0)/\tau]_{BGI}$ on $g_m$ for $n$-type (**a**) and $p$-type (**b**) SOI and sSOI HfSiON/SiO$_2$-FinFETs

**Fig. 3** Dependences of $(C_{eq}/\beta^2)$ on $W_{eff}$ for $n$- and $p$-type SOI and sSOI FinFETs with HfSiON/SiO$_2$ and HfO$_2$/SiO$_2$ gate dielectrics determined by Eq. 5

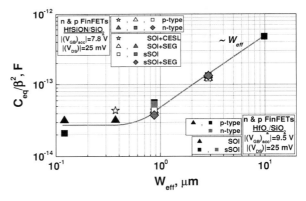

When considering the values of $(C_{eq}/\beta^2)$ at $W_{eff} \geq 0.87$ μm for the FinFETs studied where the relation $(C_{eq}/\beta^2) \propto W_{eff}$ is observed, it is useful to recollect that the following relations are valid for the planar SOI MOSFETs: $\beta = (C_D/C_0)$ and $C_{eq} = (C_{Deff} + C_{js} + C_{jd})$ where $C_D$ is the capacitance of the depletion layer per cm$^2$, $C_0 = \varepsilon_0/t_{EOT}$, $\varepsilon_0$ is the dielectric constant for SiO$_2$, $t_{EOT}$ is the equivalent oxide thickness, $C_{Deff} = C_D Z L_{eff}$ is the capacitance of the depletion layer, $C_{js}$ and $C_{jd}$ are the capacitances of the source and drain $p$–$n$ junctions [5, 6]. Note also that $(C_{js} + C_{jd}) \propto Z$ and the value of $(C_{js} + C_{jd})$ is independent of $L_{eff}$, so that

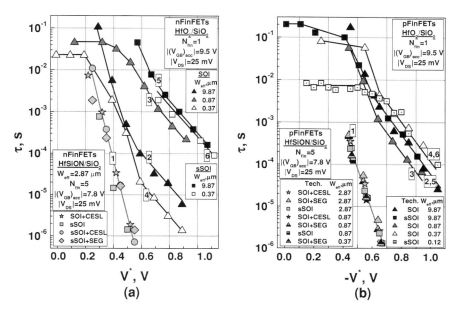

**Fig. 4** Dependences of Lorentzian time constant on the gate overdrive voltage for SOI and sSOI FinFETs of *n*-type (**a**) and *p*-type (**b**) with HfSiON/SiO$_2$ (*curves 1*) and HfO$_2$/SiO$_2$ (*curves 2–6*) gate dielectrics

$C_{eq} = C_{Deff} = C_D Z L_{eff}$ for sufficiently long devices for which $C_{Deff} > (C_{js} + C_{jd})$. For a fully depleted planar MOSFET one has: $C_D = \varepsilon_{Si}/t_{Si}$ and, hence, $\beta = (\varepsilon_{Si} t_{EOT}/\varepsilon_0 t_{Si})$ where $\varepsilon_{Si}$ and $t_{Si}$ are the dielectric constant for Si and the thickness of the silicon layer, respectively.

It has been shown in [13] that $C_D > (\varepsilon_{Si}/t_{Si})$ in the case of fully depleted FinFET and, hence, we can write for our devices: $C_{eq} > (\varepsilon_{Si}/t_{Si}) Z L_{eff}$. Then one finds for $W_{eff} = 10$ μm, $L_{eff} = 0.9$ μm and $N_{fin} = 1$ that $C_{eq} > 1.4 \times 10^{-14}$ F. Since the experimental value of $(C_{eq}/\beta^2)$ in this case is equal to $5 \times 10^{-13}$ F (see Fig. 3) we have: $\beta = [C_{eq}/(5 \times 10^{-13})]^{0.5} > [(1.4 \times 10^{-14})/(5 \times 10^{-13})]^{0.5} = 0.17$.

At the same time, by using the formula $\beta \approx (\varepsilon_{Si} t_{EOT}/\varepsilon_0 t_{Si})$ valid for the fully depleted planar MOSFETs, one obtains $\beta = 0.08$ (for $t_{EOT} = 1.5$ nm) which is lower than $\beta = 0.17$ found for the FinFETs investigated. The fact that the gates exist not only at the top but also at the sidewalls in the tri-gate FinFETs may be the reason for the higher value of $\beta$ found for such devices.

Consider now the results of application of Eq. 6 to the experimental data on the LKE Lorentzians.

Figure 4a and b shows the dependences of the Lorentzian time constants on the gate overdrive voltage $|V^*|$ for the *n*- and *p*-type devices, respectively. First of all, it should be noted that the value of $|V^*|$ at which the LKE Lorentzians begin to show themselves in the noise spectra of the devices considered are as low as 0.25 V (curves 1 in Fig. 4a, b). This contradicts the classical model of the EVB tunneling

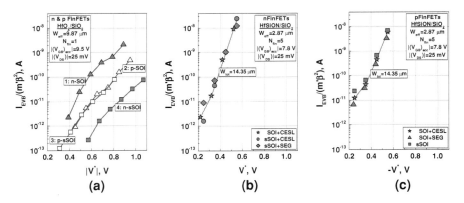

**Fig. 5** Dependences of $[I_{EVB}/m'\beta^2]$ on $|V^*|$ for $n$- and $p$-type SOI (*curves 1 and 2*) and sSOI (*curves 3 and 4*) HfO$_2$/SiO$_2$-FinFETs of $W_{tot} = N_{fin} \times W_{eff} = 9.87$ μm (**a**) and for $n$-type (**b**) and $p$-type (**c**) SOI and sSOI HfSiON/SiO$_2$-FinFETs of $W_{tot} = 14.35$ μm

according to which the condition $|V^*| \geq 1$ V has to be fulfilled for the appearance of the EVB current [14]. However, as shown in [15], in the case of devices with a metal gate and/or high-$k$ gate dielectric the onset of EVB tunneling can occur at significantly lower values of $|V^*|$ due to a shift of the work function at the gate/dielectric interface.

It is seen from Fig. 4 that in the case of HfSiON/SiO$_2$-gate dielectric the dependences $\tau_{LKE}(|V^*|)$ can be presented as $\tau_{LKE}(|V^*|) \propto \exp(-|V^*|)$ where $\alpha = \alpha_n = 31.3$ V$^{-1}$ and $\alpha = \alpha_p = 26.6$ V$^{-1}$ for the $n$FinFETs and $p$FinFETs, respectively.

As to the devices with the HfO$_2$/SiO$_2$-gate dielectric, it follows from Fig. 4 that $\alpha_n \approx 7$ V$^{-1}$ and $\alpha_p \approx 12$ V$^{-1}$ for $n$- and $p$FinFETs at $V^* > 0.6$ V, respectively. At the same time, the value of increases at lower values of $|V^*|$. For example, $\alpha_n \approx 12$ V$^{-1}$ and 15 V$^{-1}$ at 0.3 V $< |V^*| < 0.6$ V for sSOI and SOI $n$FinFETs of $W_{eff} = 9.87$ μm.

It is also seen from Fig. 4 that the values of $\tau_{LKE}$ for HfSiON/SiO$_2$-devices (curves 1) are much lower than for HfO$_2$/SiO$_2$ ones (curves 2–6) and are independent of the substrate type (SOI or sSOI). At the same time, the substrate type influences the value of $\tau_{LKE}$ for the $n$-channel HfO$_2$/SiO$_2$-FinFETs as follows: $(\tau_{LKE})_{SOI} < (\tau_{LKE})_{sSOI}$ (Fig. 4a, curves 2 and 5, 4 and 6). However, $(\tau_{LKE})_{SOI} = (\tau_{LKE})_{sSOI}$ appears to be valid for the $p$-channel HfO$_2$/SiO$_2$ ones (Fig. 4b, curves 2 and 5).

The dependences of $[I_{EVB}/m'\beta^2]$ on $|V^*|$ calculated by Eq. 6 are shown in Fig. 5. Note that $m' \approx 1$ is found for the devices studied. It is seen from Fig. 5a that those dependences for the $p$-channel HfO$_2$/SiO$_2$-FinFETs with SOI or sSOI substrate coincide (curves 2 and 3). However, for the $n$-channel HfO$_2$/SiO$_2$-devices the value of $[I_{EVB}/m'\beta^2]$ appears to be much higher in the case of the SOI substrate (curve 1) than for the sSOI one (curve 4) at one and the same overdrive voltage. This can be explained by the fact that the barrier at the

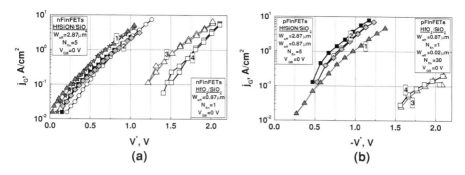

**Fig. 6** Dependences of the leakage gate current density on the gate overdrive voltage for *n*-type (**a**) and *p*-type (**b**) SOI (*curves 1 and 3*) and sSOI (*curves 2 and 4*) FinFETs with HfSiON/SiO$_2$ (*curves 1 and 2*) and HfO$_2$/SiO$_2$ (*curves 3 and 4*) gate dielectrics

Si/SiO$_2$ interface that has to be overcome by electrons tunneling through the gate dielectric is lower for the SOI substrate than for the sSOI one [16].

The increase of that barrier in the case of the sSOI *n*FinFETs is supported by the results presented in Fig. 6 where the densities of the gate leakage current $j_G = j_{ECB}$ associated with the electron conduction band (ECB) tunneling are shown for SOI (curves 1 and 3) and sSOI (curves 2 and 4) devices. It is seen that the values of $j_{ECB}$ are also lower in the case of the sSOI substrate (Fig. 6a). Note that the values of $j_{ECB}$ appear to be the same in the case of the HfO$_2$/SiO$_2$-*p*FinFETs with sSOI and SOI substrates (Fig. 6b, curves 3 and 4). This observation also correlates with the behavior of the value of $[I_{EVB}/m'\beta^2]$ in such devices (Fig. 5a, curves 2 and 3).

As to the HfSiON/SiO$_2$-devices, no influence of the substrate type on the value of $[I_{EVB}/m'\beta^2]$ at a given value of $|V^*|$ is observed either for *n*- or for *p*FinFETs (Fig. 5b, c). At the same time, like in the case of the HfO$_2$/SiO$_2$-dielectric, the values of $j_{ECB}$ appear to be lower for the *n*-channel HfSiON/SiO$_2$-FinFETs with sSOI substrates than with the SOI ones (Fig. 6a, curves 2 and 1). Moreover, the opposite influence of the substrate type on the value of $j_{ECB}$ manifests itself for the *p*FinFETs (Fig. 6b, curves 1 and 2). Therefore, the correlation between the influence of the substrate type on the values of $j_{ECB}$ and $[I_{EVB}/m'\beta^2]$ found for the HfO$_2$/SiO$_2$-FinFETs has not been observed for the HfSiON/SiO$_2$-devices.

It should be also noted that the HfSiON/SiO$_2$-devices are characterized by much higher values of $j_{EVB}$ and $j_{ECB}$ than the HfO$_2$/SiO$_2$-ones (compare Fig. 5b, c with Fig. 5a as well as curves 1 and 2 with 3 and 4 in Fig. 6).

Figure 7a and b shows the dependences of $[I_{EVB}/m'\beta^2]$ on $Z$ for the *n*- and *p*-channel FinFETs with HfSiON/SiO$_2$-dielectric. It is seen that $[I_{EVB}/m'\beta^2] \propto Z$ for *n*FinFETs (Fig. 7a) while the sub-linear dependence of $[I_{EVB}/m'\beta^2]$ on $Z$ is typical for the *p*FinFETs where $[I_{EVB}/m'\beta^2] \propto Z^{0.78}$ at $W_{eff} > 0.87$ μm and $[I_{EVB}/m'\beta^2]$ does not change with $W_{eff}$ at $W_{eff} < 0.87$ μm (Fig. 7b).

Figure 7c and d presents the dependences of $[I_{EVB}/m'\beta^2]$ on $Z$ for the *n*- and *p*-channel HfO$_2$/SiO$_2$-FinFETs. It is seen from Fig. 7d that in this case the

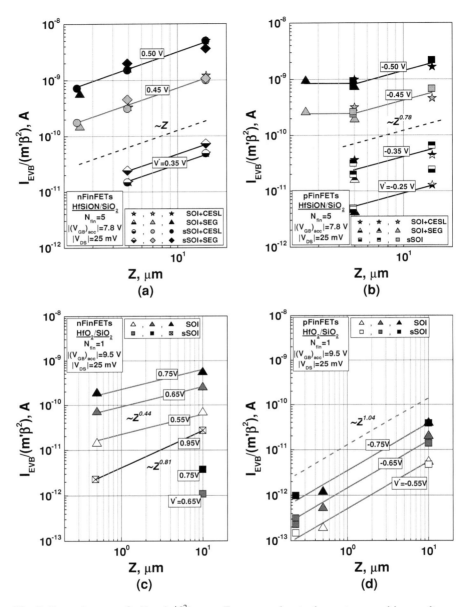

**Fig. 7** Dependences of $[I_{EVB}/m'\beta^2]$ on $Z$ measured at the gate overdrive voltages $|V^*| = 0.25$–$0.95$ V and accumulation back-gate voltage for $n$-type (**a**, **c**) and $p$-type (**b**, **d**) SOI and sSOI FinFETs with HfSiON/SiO$_2$ (**a**, **b**) and HfO$_2$/SiO$_2$ (**c**, **d**) gate dielectrics

dependence $[I_{EVB}/m'\beta^2] \propto Z$ is typical for the $p$-channel devices while sub-linear increase of $[I_{EVB}/m'\beta^2]$ with increasing $Z$ is observed for the $n$-channel FinFETs where $[I_{EVB}/(m'\beta)^2] \propto Z^{0.43}$ and $[I_{EVB}/(m'\beta)^2] \propto Z^{0.81}$ for the devices with SOI and sSOI substrates, respectively.

If one takes into account that the following relation can be used for $I_{EVB}$

$$I_{EVB} = j_{EVB}ZL_{eff} = N_{fin}\left(W_{eff}j_{EVBtop} + 2h_{fin}j_{EVBside}\right)L_{eff} \tag{15}$$

where $j_{EVB}$ is the average density of the EVB current, $j_{EVBside}$ and $j_{EVBtop}$ are the densities of the EVB currents flowing through the side and top interfaces, respectively, then the sub-linear dependences of $I_{EVB}$ on $Z$ observed for the $p$FinFETs with the HfSiON/SiO$_2$-dielectric and $n$FinFETs with the HfO$_2$/SiO$_2$-dielectric can be explained by the inequality $j_{EVBside} > j_{EVBtop}$. The smaller exponent in the dependence $[I_{EVB}/(m'\beta)^2]$ vs. Z in the latter case than in the former one at $W_{eff} \geq 0.87$ μm points to the fact that at $W_{eff} \geq 0.87$ μm the contribution of the EVB current passing through the side interfaces to the value of $I_{EVB}$ in the case of the $n$FinFETs with the HfO$_2$/SiO$_2$-dielectric is higher than in the case of the $p$FinFETs with the HfSiON/SiO$_2$-dielectric. However, the opposite situation takes place at $W_{eff} < 0.87$ μm.

Another difference between the LKE Lorentzians for the HfSiON/SiO$_2$- and HfO$_2$/SiO$_2$-devices is demonstrated in Fig. 8 where the dependences $[S_I(0)]_{LKE}$ vs. $\tau_{LKE}$ are shown for the devices of both types.

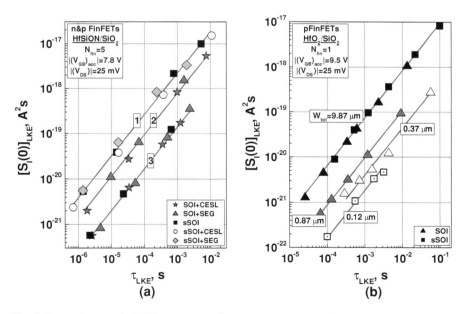

**Fig. 8** Dependences of $[S_I(0)]_{LKE}$ on $\tau_{LKE}$ for $n$-type (*curves 1 and 2*) and $p$-type (*curve 3*) HfSiON/SiO$_2$-FinFETs of $W_{eff} = 2.87$ μm with sSOI (*curves 1 and 3*) and SOI (*curves 2 and 3*) substrates (**a**) and $p$-type SOI and sSOI HfO$_2$/SiO$_2$-FinFETs of $W_{eff} = 0.12$–9.87 μm (**b**)

It is seen from Fig. 8a that $[S_I(0)]_{LKE} \propto (\tau_{LKE})^\alpha$ where $\alpha = 0.93$ is typical for the HfSiON/SiO$_2$-FinFETs from which it follows (see Eq. 3) that $[m'(\beta\mu)^2/C_{eq}] \propto [\tau_{LKE}(|V^*|)]^{-0.07}$. Figure 9 shows that $[\tau_{LKE}(|V^*|)]^{-0.07}$ and, hence, $[m'(\beta\mu)^2/C_{eq}]$ increases with increasing $|V^*|$. Since the LKE Lorentzians are observed at sufficiently high $|V^*|$ where $\mu$ can only decrease with increasing $|V^*|$ we conclude that $[m'\beta^2/C_{eq}]$ increases with increasing $|V^*|$.

At the same time, it is seen from Fig. 8b that for the HfO$_2$/SiO$_2$-FinFETs the relation $[S_I(0)]_{LKE} \propto (\tau_{LKE})^\alpha$ where $\alpha = 1.05$ is valid. This means that $[m'(\beta\mu)^2/C_{eq}]$ does not increase with increasing $|V^*|$. The dependences of $[S_I(0)/\tau]_{LKE}$ on $|V^*|$ are shown in Fig. 10 for the HfO$_2$/SiO$_2$-FinFETs. The small decrease of $[S_I(0)/\tau]_{LKE}$ with increasing $|V^*|$ can be explained by the decrease of $\mu$ (see Eq. 7). The difference between the data shown in Fig. 10a and b is explained by the following relations valid for $\mu$ for the devices studied: $(\mu_n)_{sSOI} > (\mu_n)_{SOI} > (\mu_p)_{sSOI} \approx (\mu_p)_{SOI}$.

Consider now the results of investigations of the FinFETs by the 1/$f$ noise methods. The noise measurements in this case were carried out at $V_{GB} = 0$ V.

Figure 11 shows the dependences $[f \times S_I(f)]$ vs. $f$ typical for the noise spectra of the $n$FinFETs with the HfSiON/SiO$_2$-dielectric. It is seen that the main feature of the spectra of Fig. 11 is the decrease of $[f \times S_I(f)]$ with decreasing $f$ from $f = (10-30)$ kHz down to sufficiently low $f$ (note that for those devices it was difficult to measure the value of $S_I(f)$ for the 1/$f^\gamma$ noise at $f > 10$ kHz). One finds for this decrease: $(f \times S_I) \propto f^{1-\gamma}$ where $\gamma = 0.83-0.75$.

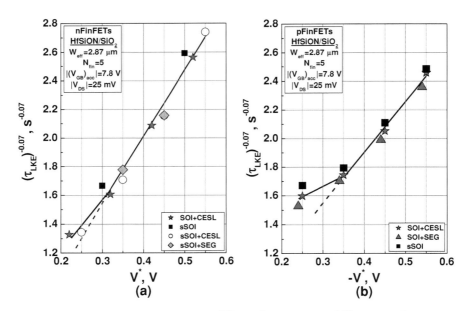

**Fig. 9** Dependences of the value of $(\tau_{LKE})^{-0.07}$ on $|V^*|$ where $(\tau_{LKE})^{-0.07} \propto [m'(\beta\mu)^2/C_{eq}]$ for $n$-type (**a**) and $p$-type (**b**) SOI and sSOI HfSiON/SiO$_2$-FinFETs

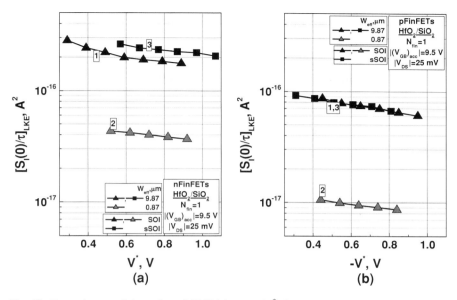

**Fig. 10** Dependences of the value of $[S_I(0)/\tau]_{LKE}$ on $|V^*|$ for $n$-type (**a**) and $p$-type (**b**) HfO$_2$/SiO$_2$-FinFETs with SOI (*curves 1 and 2*) and sSOI (*curves 3*) substrates for devices of $W_{eff} = 9.87$ μm (*curves 1 and 3*) and 0.87 μm (*curves 2*)

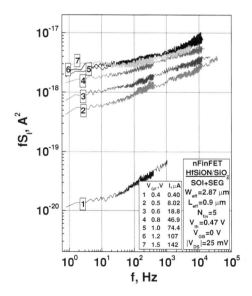

**Fig. 11** Drain current noise spectra multiplied by the frequency for the SOI + SEG $n$-type HfSiON/SiO$_2$-FinFET measured at $V_{GB} = 0$ V and different gate voltages

Figure 12 represents the dependences $[f \times S_I(f)]$ vs. $f$ typical for the noise spectra of the $n$FinFETs with the HfO$_2$/SiO$_2$-dielectric. The spectra shown in Fig. 12a appear to be typical for the SOI FinFETs as well as for the sSOI devices

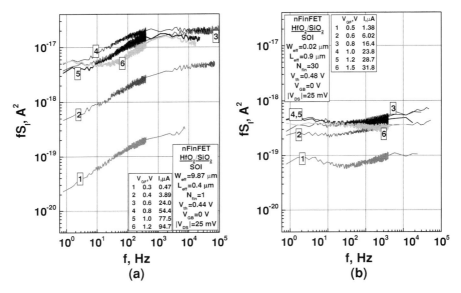

**Fig. 12** Drain current noise spectra multiplied by the frequency for the SOI $n$-type HfO$_2$/SiO$_2$-FinFETs of $W_{eff} = 9.87$ μm (**a**) and $W_{eff} = 0.02$ μm (**b**) measured at $V_{GB} = 0$ V and different gate voltages

of $W_{eff} \geq 0.12$ μm. Figure 12b shows the results typical for the SOI FinFETs of $W_{eff} = 0.02$ μm. It is seen that the shapes of the spectra $(f \times S_I)$ vs. $f$ shown in Fig. 12 differ significantly from those shown in Fig. 11. Indeed, it follows from Fig. 12a that the decrease of $(f \times S_I)$ with decreasing $f$ which corresponds to the $1/f^{0.7}$ noise takes place only at $f < 400$ Hz while the high-frequency plateau corresponding to the $1/f$ noise is observed at $f > 400$ Hz. As to the very narrow SOI FinFETs, it follows from Fig. 12b that the main component of their noise spectra is the $1/f$ noise which is observed from $f = 1$ Hz up to $f = 5$ kHz.

It has been found that the value of $S_{VG}$ for the devices considered is independent of the gate overdrive voltage $V^*$ in a sufficiently wide interval of $V^*$ and this takes place for the whole $[f \times S_I(f)]$ vs. $f$ curves (Fig. 13a). This is in support of the number fluctuations or trapping origin of the flicker noise observed. Therefore the distribution of $N_{ot}$ over $x$ can be estimated with the help of Eq. 14.

Figure 13b demonstrates that the value of $(S_I)_{1/f}$ is practically independent of $V^*$ at $0.45$ V $\leq |V^*| < 1.8$ V for the $n$FinFETs with HfSiON/SiO$_2$ gate dielectric. Taking into account Eq. 13, one can conclude that $\mu \neq \mu(V^*)$ at sufficiently high values of $V^*$. Moreover, the fact that $(S_I)_{1/f}$ is independent of $V^*$ means that the series resistance does not influence the value of $(S_I)_{1/f}$ and, hence, the dependence $I(V^*)$. The same effect was previously detected for the $n$-channel planar devices with SiON gate dielectric and has been used to explain the effect of the high gate voltage drain current leveling off observed in such devices [9].

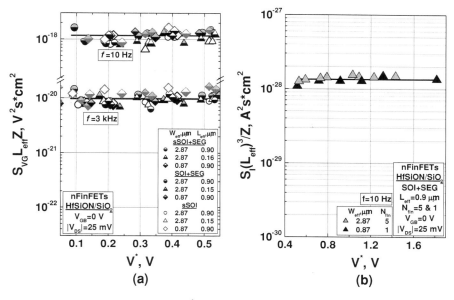

**Fig. 13** Dependences of $S_{VG}L_{eff}Z$ on $V^*$ for $n$-type HfSiON/SiO$_2$-FinFETs with sSOI and SOI substrates measured at $f = 10$ Hz and 3 kHz (**a**) and dependences of $S_I(L_{eff})^3/Z$ on $V^*$ for $n$-type SOI + SEG HfSiON/SiO$_2$-FinFETs measured at $f = 10$ Hz (**b**)

The distributions of $N_{ot}$ over $x$ obtained by Eq. 14 are shown in Fig. 14. The following conclusions can be drawn from Fig. 14:

1. The trap concentration $N_{ot}$ is not distributed homogeneously over $x$ in most types of devices studied but decreases from $(2-3.7) \times 10^{19}$ cm$^{-3}$eV$^{-1}$ at $x \approx 1$ nm to $(4-6.5) \times 10^{18}$ cm$^{-3}$eV$^{-1}$ at $x = 2.1$ nm. While in the devices with the HfSiON/SiO$_2$ dielectric this decrease begins immediately from $x \approx 1$ nm (curve 3 in Fig. 14a), in the case of the HfO$_2$/SiO$_2$ dielectric the high value of $N_{ot}$ remains constant up to $x \approx (1.2-1.4)$ nm (Fig. 14a, curves 1 and 2; Fig. 14b, curve 1). As a result, the values of $N_{ot}$ in the region 1.15 nm $\leq x \leq$ 2.1 nm are higher for the HfO$_2$/SiO$_2$ dielectric than for the HfSiON/SiO$_2$ one and the difference reaches a factor of 2 at $x = 1.4-1.7$ nm.
2. In very narrow ($W_{eff} = 0.02$ μm) SOI devices with the HfO$_2$/SiO$_2$ dielectric the values of $N_{ot}$ appear to be homogeneously distributed over the whole range of $x$ investigated and are relatively low ($2.5 \times 10^{18}$ cm$^{-3}$ eV$^{-1}$).
3. The global straining technique increases the trap concentration in the devices with the HfO$_2$/SiO$_2$ dielectric.

At the same time, it has been found that in the case of the HfSiON/SiO$_2$-dielectric there is no influence of the global straining technique on the trap distribution either in SEG- and in CESL-devices and that the distribution $N_{ot}(x)$ is practically not changed by using SEG drain and source regions.

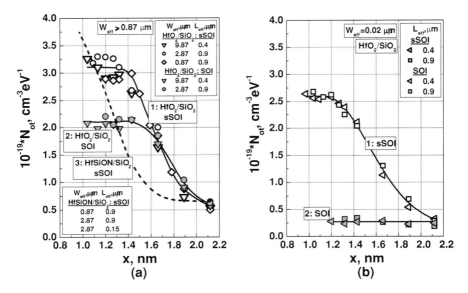

**Fig. 14** Distribution of noisy traps concentration $N_{ot}$ over distance $x$ from Si/SiO$_2$ interface for the sSOI (*curve 1*) and SOI (*curve 2*) n-type HfO$_2$/SiO$_2$-FinFETs of $W_{eff} \geq 0.87$ μm (**a**) and $W_{eff} = 0.02$ μm (**b**); *curve 3* presents the dependence $N_{ot}(x)$ for the sSOI n-type HfSiON/SiO$_2$-FinFETs of $W_{eff} \geq 0.87$ μm

A following remark with respect to the trap depths in Fig. 14 has to be made. If account is made of the position of the charge centroid in the silicon fin due to inversion layer quantization [17], a value in the range of 1 nm has to be subtracted.

It should be noted that the fact that the trap density in the interfacial SiO$_2$ layer of the *n*FinFETs studied appears to be higher than in the bulk high-$k$ layer is opposite to what has been found before in planar bulk transistors with high-$k$ [18]. This suggests a lower quality of the SiO$_2$ interface layer that seems to be a common feature here, irrespective of the process details, the main factor being the high-$k$ layer. The difference in profiles between the two gate dielectrics can be related to the different deposition method employed.

## 5 Conclusions

The results obtained by the methods based on the analysis of the BGI and LKE Lorentzian noise and $1/f^{\gamma}$ noise for investigation of the strained and non-strained fully depleted tri-gate FinFETs with different gate dielectrics are the following:

- The value of $(C_{eq}/\beta^2)$ is proportional to the fin width at $W_{eff} \geq 0.87$ μm and is determined by the width of the side fin interfaces at $W_{eff} \leq 0.37$ μm.
- An increase of $(m'\beta^2/C_{eq})$ with increasing $|V^*|$ is observed for the HfSiON/SiO$_2$-FinFETs under strong inversion conditions.

- The body factor $\beta$ found for the FinFETs investigated is higher than for their planar counterparts even in the case where $W_{eff} >> 2h_{fin}$.
- The EVB current density is much higher for the HfSiON/SiO$_2$ gate dielectric than for the HfO$_2$/SiO$_2$ one.
- The values of $[I_{EVB}/m'\beta^2]$ for the $n$-channel HfO$_2$/SiO$_2$-FinFETs are higher for SOI substrates than for sSOI ones, while the substrate type does not affect the value of $[I_{EVB}/m'\beta^2]$ for the $p$-channel HfO$_2$/SiO$_2$-FinFETs. Such a behavior of the EVB current correlates with the behavior of the ECB current in the HfO$_2$/SiO$_2$-devices. The effects are attributed to the influence of the strain on the barrier height at the Si/SiO$_2$ interface.
- The values of $[I_{EVB}/m'\beta^2]$ for the HfSiON/SiO$_2$-FinFETs do not change if one uses an SOI instead of an sSOI substrate and there is no correlation between the EVB and ECB currents in this case.
- A sub-linear dependence of $[I_{EVB}/m'\beta^2]$ on Z is typical for the HfSiON/SiO$_2$-$p$FinFETs and HfO$_2$/SiO$_2$-$n$FinFETs which is explained by the inequality $j_{EVBside} > j_{EVBtop}$.
- At sufficiently high values of $V^*$ the electron mobility in the channel no longer decreases with increasing $V^*$ and becomes independent of $V^*$.
- The bulk oxide trap density decreases with the distance $x$ from the Si/SiO$_2$ interface, opposite to what is commonly found on planar MOSFETs with high-$k$ gate dielectric. While in the HfSiON/SiO$_2$-FinFETs this decrease begins from $x \approx 1$ nm, in HfO$_2$/SiO$_2$-FinFETs the high value of $N_{ot}$ remains constant up to $x \approx (1.2–1.4)$ nm. In very narrow ($W_{eff} = 0.02$ μm) SOI devices with the HfO$_2$/SiO$_2$ dielectric the values of $N_{ot}$ are relatively low and homogeneously distributed over $x$. The global straining technique increases $N_{ot}$ in the HfO$_2$/SiO$_2$-devices and has no influence on $N_{ot}(x)$ in the HfSiON/SiO$_2$-devices.

## References

1. Van Der Ziel, A.: Fluctuation Phenomena in Semiconductors. Butterworths Scientific Publications, London (1959)
2. Lukyanchikova, N.: Noise Research in Semiconductor Physics. Gordon and Breach Science Publishers, London (1996)
3. Jayaraman, R., Sodini, C.G.: A 1/$f$ noise technique to extract the oxide trap density near the conduction band edge of silicon. IEEE Trans. Electron Devices **36**, 1773–1782 (1989)
4. Lukyanchikova, N., Garbar, N., Kudina, V., et al.: On the 1/$f$ noise of triple-gate field-effect transistors with high-$k$ gate dielectric. Appl. Phys. Lett. **95**, 032101–032103 (2009)
5. Lukyanchikova, N., Petrichuk, M., Garbar, N., et al.: Electron valence-band tunneling-induced Lorentzian noise in deep submicron silicon-on-insulator metal-oxide-semiconductor field-effect transistor. J. Appl. Phys. **94**, 4461–4469 (2003)
6. Lukyanchikova, N., Garbar, N., Smolanka, A., et al.: Origin of the front-back-gate coupling in partially depleted and fully depleted silicon-on-insulator metal-oxide-semiconductor field-effect transistors with accumulated back gate. J. Appl. Phys. **98**, 114506–114511 (2005)
7. Mercha, A., Rafí, J.M., Simoen, E., et al.: "Linear Kink Effect" induced by electron valence band tunneling in ultrathin gate oxide bulk and SOI MOSFETs. IEEE Trans. Electron Devices **50**, 1675–1682 (2003)

8. Sze, S.M.: Physics of Semiconductor Devices. A Wiley-Interscience Publication. John Wiley & Sons, New York (1981)
9. Lukyanchikova, N., Garbar, N., Kudina, V., et al.: High gate voltage drain current leveling off and its low-frequency noise in 65 nm fully-depleted strained and non-strained SOI nMOSFETs. Solid State Electron **52**, 801–807 (2008)
10. Ghibaudo, G.: New method for the extraction of MOSFET parameters. Electron Lett. **24**, 543–545 (1988)
11. Parton, E., Verheyen, P.: Strained silicon—the key to sub-45 nm CMOS. III-Vs Rev. Adv. Semicond. Mag. **19**, 28–31 (2006)
12. Ning, X.J., Gao, D., Bonfanti, P., et al.: Selective epitaxial growth of SiGe for strained Si transistors. Mater. Sci. Eng. B **134**, 165–171 (2006)
13. Frei, J., Johns, Ch., Vazquez, A., et al.: Body effect in tri- and pi-gate SOI MOSFETs. IEEE Electron Device Lett. **25**, 813–815 (2004)
14. Lee, W.-C., Hu, C.: Modeling CMOS tunneling currents through ultrathin gate oxide due to conduction- and valence-band electron and hole tunneling. IEEE Trans. Electron Devices **48**, 1366–1373 (2001)
15. Pantisano, L., Afanas'ev, V., Pourtois, G., et al.: Valence-band electron-tunneling measurement of the gate work function: application to the high-$k$/polycrystalline silicon interface. J. Appl. Phys. **98**, 053712–053718 (2005)
16. Zhao, W., Seabaugh, A., Adams, V., et al.: Opposing dependence of the electron and hole gate currents in SOI MOSFETs under uniaxial strain. IEEE Electron Device Lett. **26**, 410–412 (2005)
17. King, Y.-C., Fujioka, H., Kamohara, S., et al.: DC electrical oxide thickness model for quantization of the inversion layer in MOSFETs. Semicond. Sci. Technol. **13**, 963–966 (1998)
18. Nguyen, T., Savio, A., Militaru, L., et al.: Spatial distribution of electrically active defects in dual-layer ($SiO_2$/$HfO_2$) gate dielectric n-type metal oxide semiconductor field effect transistors. J. Vac. Sci. Technol. B **27**, 329–332 (2009)

# Mobility Characterization in Advanced FD-SOI CMOS Devices

G. Ghibaudo

**Abstract** A review of the main mobility results obtained in short channel devices (here GAA/DG, FD-SOI MOSFETs and FinFETs) are discussed for better understanding their transport limitations and performances. Regarding short channel GAA, FD-SOI and FinFET MOS devices, it has been shown that the mobility is strongly degraded at small gate length, whatever the architecture, the gate stack and the measurement method used. In particular, it has been found that, for FD-SOI, the mobility is more degraded at the top interface than at the bottom interface, revealing that defects are more numerous at the top channel region.

## 1 Introduction

The mobility and, more generally, the transport parameters of MOS devices are very important for the performance evaluation in advanced CMOS technologies. To this end, emphasis has been particularly put in the past years in transport improvement, aiming at enhancing the carrier mobility for gain in Ion, with no loss neither in leakage current nor load capacitance. Mechanical stress approach has already demonstrated ion [1] and mobility gains both for high gate length and short gate length transistors [2]. But mobility can also be improved by effective field reduction; indeed, large channel doping—especially for short gate transistors with pocket implants, which is necessary for short channel effects control—increases the effective field and degrades the mobility. So, thin film fully depleted single gate (SG) or double gate (DG) transistors, which allow low channel doping thanks

G. Ghibaudo (✉)
IMEP-LAHC, MINATEC/INPG, BP 257, 38016 Grenoble, France
e-mail: ghibaudo@minatec.inpg.fr

to their intrinsic short channel effects (SCE) control, should also benefit from a mobility gain [3]. However, if SOI fully depleted (FD) and DG type transistors have demonstrated high static performances, no actual gain was achieved due to channel doping suppression for short gates, both in literature and from experiments [4].

In this paper, a review of the main mobility results obtained in short channel devices (here GAA/DG, FD-SOI MOSFETs and FinFETs) are discussed for better understanding their transport limitations and performances.

## 2 Effective Mobility Experimental Results

We focus on low field mobility, although short channel transport is also ruled by drift velocity. It has been shown that these two parameters are well correlated [5], and thus mobility can be considered as a good indicator of the transport quality. We used Y-function technique [6], which allows to suppress the effect of series resistances, and to perform statistical extraction on wafers; the key point in short channel mobility computation being the extraction of the effective (electrical) length, performed by gate-to-channel split C(V) measurements (Fig. 1) [7]. The low field mobility, i.e. $\mu_0$ parameter is extracted from the gain parameter ($\beta = \mu_0 C_{ox} W/L_{eff}$) deduced from the slope of the Y(Vg) function at strong inversion [6].

Comparing in Fig. 2 the mobility at low gate length for heavily doped ($6 \times 10^{18}$ cm$^{-3}$) and undoped GAA transistors with equivalent Tsi, we notice that undoped channel shows as expected a high mobility for long (1 μm) transistors, but, for short gate lengths, doped and undoped channels have equivalent one. This feature explains the lack of performance increase for undoped DG devices [4].

The low field mobility was also extracted as a function of channel length for compressive and tensile stress N and P type FD-SOI MOS devices (Fig. 3). Note that in all cases there is a strong degradation of the mobility, by about a factor 2, as the channel length is reduced below 100 nm. This mobility behavior is therefore a general feature to thin film devices whatever their channel doping and stress level. As was already observed in bulk devices [2, 4, 8], the beneficial tensile stress effect on the mobility for 200–400 nm channel is cancelled at short channel length, especially for N type devices. As in bulk devices, this can be interpreted by a qualitative change in scattering process below 70–80 nm where neutral defects play an increasing role as in GAA devices [2, 4, 8].

In all cases, the ballistic contribution (see Sect. 3.3) has been found rather negligible, and, does not explain the strong mobility degradation below 80 nm. These results confirm that the mobility reduction in undoped thin silicon film SG or DG MOS devices with implanted source and drain junctions is a general trend in such CMOS technologies, which has to be overcome in order to fully take advantage of ballistic transport.

The backscattering coefficient (see Sect. 3.3) extracted for N and P type devices from the mobility data is given in Fig. 4 as a function of channel length [9].

**Fig. 1** Gate-to-channel capacitance measurements for various gate lengths (*top*). Calculation of ΔL based on gate-to-channel capacitance measurements (*bottom*) (after [4])

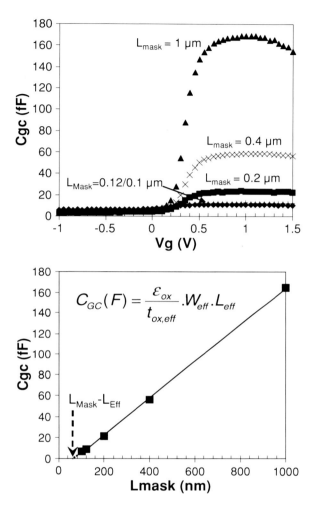

It decreases from 1 for long devices down to about 0.85 for the shortest ones. This means that, for the smaller devices, the ballisticity reaches at most 15%, indicating that in such FD-SOI undoped devices the transport is far from being ballistic.

The dependence of mobility on temperature is shown in Fig. 5 [10]. For room temperature and long channels, the electron mobility is slightly higher for $HfO_2$ than for HfSiON, but they become almost equal at 77 K. The general behavior of mobility increase with decreasing temperature is typical for acoustic phonon-dominated scattering. However, the slope is lower ($\mu \sim T^{-0.7}$) suggesting additional mechanisms. Figure 5 also shows a significant attenuation of mobility variation with temperature in short channels, below 100 nm, for both high-k architectures. The presence of implantation-induced neutral defects near S/D has been incriminated for additional scattering in short channels [4, 10].

**Fig. 2** Low field mobility compared for highly doped and undoped DG NMOSFETs (after [4])

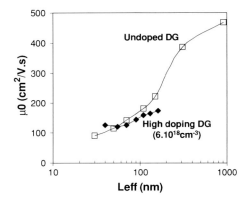

**Fig. 3** Low field mobility variations with channel length for N FD-SOI MOS devices with compressive and tensile CESL strain (after [9])

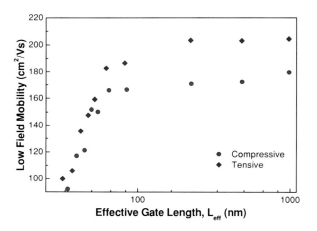

**Fig. 4** Variation of back-scattering coefficient $r$ with channel length for N and P type devices (tensile strain) (after [9])

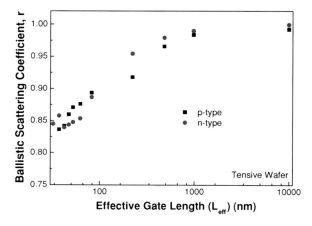

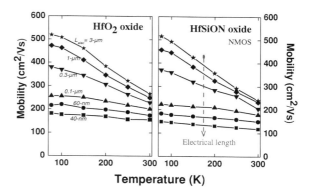

**Fig. 5** Comparison of mobility variation as a function of temperature for HfO$_2$ and HfSiON front-gate oxide (after [10])

In order to discriminate the role of the high-k gate stack used in these FD-SOI devices for the front gate, the effective mobility has also been investigated on the BOX back gate controlled via the substrate bias [10]. Typical mobility curves for the front and back channels of NMOS are shown in Fig. 6. From these figures several features are revealed:

1. The front-channel mobility (Si/High-K interface) is reasonable but consistently lower than at the back channel (Si/SiO$_2$), indicating that there are additional scatterings related to the front HK gate stack.
2. The back-channel mobility variation with temperature shows the dominant role of acoustic phonon scattering. The front-channel mobility improves less rapidly at low temperature, denoting the contribution of an extra mechanism (remote Coulomb scattering) at the Si/high-K interface.
3. In shorter channels, the mobility variation with temperature is strongly attenuated. As in GAA devices, implantation-induced neutral defects located close to source and drain are suspected to be responsible for such $\mu_{eff}(T)$ behavior [4, 10]. However their impact is lower on the back channel transport, which might indicate the presence of a vertical profile of implantation-induced neutral defects in the silicon film.

Low field mobility measurements were also performed on another FD-SOI lot featuring HfZrO$_2$ gate dielectric and TiN/TaN metal gate with various thicknesses and deposited with different techniques. Figure 7 shows the gate length dependence of the mobility for the various samples. The amplitude of the mobility with 10 nm TiN thickness is of similar value than that obtained with HfO$_2$ and HfSiON gate dielectric, revealing no specific impact of gate dielectric. However, it is clear from Fig. 7 that the TiN thickness has a strong impact on the mobility amplitude, especially for long gate lengths. Indeed, the mobility is degraded as the TiN thickness deposited by ALD is increased from 3 to 10 nm. This observation agrees with results obtained on long channel where mobility is reduced by the increasing amount of nitrogen diffusion at the channel-dielectric interface. However, it should be underlined that, for a given metal thickness, the deposition technique used for

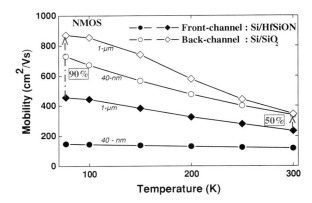

**Fig. 6** Comparison of mobility variation as a function of temperature for short/long and front/back channels in NMOS (after [10])

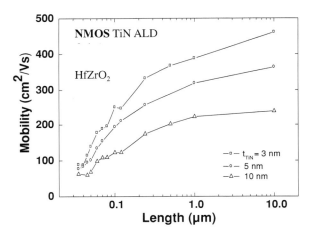

**Fig. 7** Variations of low field mobility with gate length for various TiN thicknesses (after [10])

the metallization (PVD vs. ALD) or metal species (TiN vs. TaN) do not change significantly the low field mobility versus gate length curves [10].

The low field mobility has also been extracted as a function of gate length on FinFET transistors (Fig. 8). The apparent effective mobility $\mu_{app}$ has been corrected for ballistic effect (see Eq. 3 in Sect. 3.3), allowing the drift–diffusion mobility $\mu_{dd}$ to be obtained for each gate length. In all cases, the mobility is found to decrease by about a factor 2 for the smaller gate length. This result confirms the general trend of mobility collapse observed at small channel length irrespective of the thin film technology studied (GAA, FD-SOI, DG-MOS). Similarly, the S/D implantation process is also suspected to be at the origin of such a collapse, originating from enhanced channel impurity scattering in channel regions close to source and drain [11].

In order to infer this analysis, the temperature dependence of the low field mobility has been studied for various gate lengths (Fig. 9). This figure indicates, as in Fig. 5 for FD-SOI devices, that there is a strong evolution of the scattering

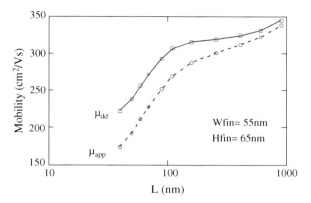

**Fig. 8** Variation of low field mobility with channel length for a FinFET. $\mu_{app}$ is the apparent mobility; $\mu_{dd}$ is the drift–diffusion mobility corrected for ballistic effect (after [11])

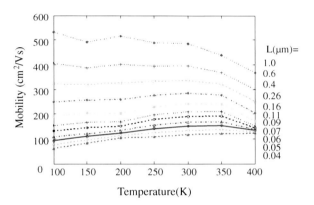

**Fig. 9** Variations of low field mobility with temperature for various channel lengths (1 μm down to 0.04 μm; $W_{fin}$ = 55 nm, after [11])

mechanism between long and short devices. For long transistor, the mobility is first increasing as the temperature is reduced due to phonon scattering reduction, whereas, for short ones ($L$ = 40–50 nm), the mobility is nearly constant or slightly decreasing as the temperature is lowered, revealing neutral/charged impurity scattering process.

## 3 Origin of Mobility Degradation at Short Channel Length

From Figs. 2, 3, 4, 5, 6, 7, 8 and 9, all the mobility results obtained in various technologies share several common features:
- The mobility is degraded at short gate length, whatever the device architectures (GAA, DG, FD-SOI, FinFET…) and whatever the measurement technique (Static, split CV, magneto-resistance).

- This degradation can be observed for rather large gate lengths (a few 100 nm). This characteristic length is not compatible with channel length values at which non-stationary and quasi-ballistic effects are expected to influence strongly the drain current (see below).
- The temperature dependence of the mobility indicates that the temperature deactivation observed for long channel devices (due to phonon freezing) is strongly attenuated for short channel devices. This is typical of scattering with charged or neutral defects.

Several mechanisms could be responsible for such local defect distribution and possible origins are discussed below. However, beforehand, we also review, as a possible mechanism, the ballistic effect and the remote Coulomb scattering due to the depletion in the source-drain regions. It should also be pointed out that the role of extra interface traps close to S/D regions has been ruled out as a possible reason for additional Coulomb scattering process in short channel, since it would imply a strong increase of threshold voltage for small gate length, not observed in all architectures [4, 10, 11].

## 3.1 Quasi-Ballistic Mobility Concept

Following [12], the quasi-ballistic drain current of a MOS transistor in linear operation can be expressed as [9]:

$$I_d = (1-r) \cdot \frac{W}{L} \cdot \mu_{bal} \cdot Q_i \cdot V_d = r \cdot \frac{W}{L} \cdot \mu_{dd} \cdot Q_i \cdot V_d = \frac{W}{L} \cdot \mu_{app} \cdot Q_i \cdot V_d \quad (1)$$

where $Q_i$ is the inversion charge, W and L the gate width and length, $V_d$ the drain voltage, $\mu_{bal}$ the ballistic mobility, $\mu_{dd}$ the drift–diffusion one (i.e. for long channel) whereas $\mu_{app}$ stands for the experimental or apparent mobility to be measured from drain current applying Eq. 1. The ballistic mobility $\mu_{bal}$ can be derived from Eq. 1 after considering the drain current expression in the ballistic limit (i.e. $r = 0$), giving,

$$\mu_{bal} = \frac{q \cdot v_T \cdot L}{2 \cdot kT} \quad (2)$$

where $v_T$ is the injection velocity at the virtual source ($\approx 1.2 \times 10^7$ cm/s for silicon). Eliminating the backscattering coefficient in Eq. 1 allows us to recover the Matthiessen-rule-like expression for $\mu_{app}$ [13]:

$$\frac{1}{\mu_{app}} = \frac{1}{\mu_{bal}} + \frac{1}{\mu_{dd}} \quad (3)$$

It is now easy to obtain the backscattering coefficient from Eq. 1 as [9],

$$r = 1 - \frac{\mu_{app}}{\mu_{bal}} \quad (4)$$

As it has been seen from Figs. 2, 4 and 8, the ballistic effect cannot fully explain the mobility degradation observed experimentally in GAA/FD-SOI and FinFET devices. It can only contribute up to 15–20% mobility reduction.

## 3.2 Long-Range Coulomb Interactions from Source and Drain

One possible explanation for the drift–diffusion mobility degradation at small channel length could be the long range remote Coulomb scattering (RCS) arising from the depletion charge in the source and drain regions similarly to what has been done for polysilicon gates [14]. Indeed, at the source–channel and drain–channel junctions, a depletion space charge is formed in the N+ doped source and drain electrodes (Fig. 10) resulting from the longitudinal control of the channel (Fig. 11).

This depletion charge that surrounds the channel can in principle contribute to a long range RCS at each point in the channel. For a given areal density of the space charge $Q_s$ at source and drain, as well as the remote efficiency function $G(x)$ [15] at a distance $x$ from source and drain, the mobility along the channel can be expressed as,

$$\frac{1}{\mu_{dd}(x)} = \frac{1}{\mu_0} + \alpha \cdot G(x) \cdot Q_s \quad (5)$$

where $\alpha$ is the Coulomb scattering coefficient ($\approx 10^4$ V s/C) [16] and

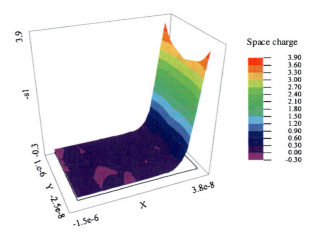

**Fig. 10** Surface plot of depletion charge (a.u.) in source (Nd = $5 \times 10^{19}$/cm$^3$) from 2D simulation

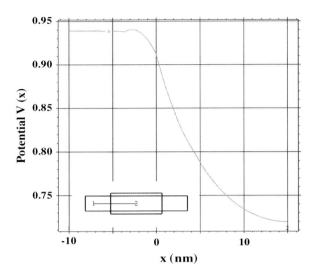

**Fig. 11** Variation of potential with space × around the source–channel junction from 2D simulation (tsi = 10 nm, tox = 1 nm, source doping Nd = 5 × 10$^{19}$/cm$^3$)

$$G(x) = \frac{1}{1 + x/\lambda c} + \frac{1}{1 + (L - x)/\lambda c} \quad (6)$$

The remote efficiency function G(x) has been derived from Sah et al. [15] original calculations and symmetrized to take into account the source and drain contributions. The parameter $\lambda c$ is a characteristic damping length of the order of 1.2 nm [15].

Figure 12 shows how the mobility calculated with Eq. 5 varies along the channel. The plot has been drawn using a reduced space coordinate (x/L) for various channel length (L) values. The amplitude of the S-D depletion charge has been computed by 2D simulation for a standard DGMOS structure (tsi = 10 nm, tox = 1 nm, S-D doping Nd = 5 × 10$^{19}$/cm$^3$). It lies around 1–2 × 10$^{12}$ q/cm$^2$ for the doping level used. As can be seen, the mobility is reduced by a factor 2 just at the junction and reaches a maximum in the middle of the channel, the maximum value being reduced as the gate length is shorter due to RCS effect. The global channel mobility has then been calculated in two manners (Fig. 13): (1) by full numerical simulation of Poisson and drift–diffusion transport equations and (2) by averaging the reciprocal mobility $\mu_{dd}(x)$ of Eq. 5 over the channel (this corresponds to adding in series the elementary resistance contributions).

Both results, which are close to one another, indicate that the effective mobility is degraded as the gate length is reduced below 100 nm. However, this degradation of 40% cannot fully explain the experimental results obtained on various devices for which a factor of 2–4 has been observed. It should also be mentioned that this phenomenon is certainly overestimated in this simulation since an abrupt doping profile has been used for the S-D junctions, leading to a larger space charge value.

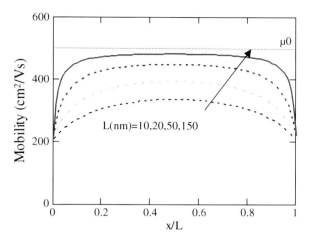

**Fig. 12** Variation of mobility along the channel between source and drain for various gate length $L$ as obtained from Eqs. 5 and 6 ($Q_s = 1.5 \times 10^{12}$ q/cm$^2$)

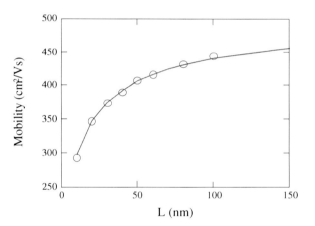

**Fig. 13** Variation of effective mobility with gate length $L$ as obtained from analytical formula (Eqs. 5, 6, *solid line*) and numerical 2D simulation (*symbols*) ($Q_s = 1.5 \times 10^{12}$ q/cm$^2$)

## 3.3 Scattering by Neutral Point Defects

The processing of source and drain regions may also be at the origin of additional scattering defects near the source and drain regions. Indeed, it is well known that ion implantation produces point defects such as vacancies, interstitial Si or clusters of defects that can be charged or neutral. The salicidation process can also generate local point defect supersaturation or depletion [17].

That is why the concentration and spatial extension of the point defects created by low energy dopant implantation used for S/D engineering into real devices (thin SOI with mask openings in the 10–50 nm range) have been evaluated by 2D Monte Carlo collision simulation [18]. In particular, the distribution of interstitial defects at the edges of the S/D and extending into the channel region of the devices and resulting from As implant (7 keV, $10^{15}$/cm$^2$) characteristic of LDD extension

**Fig. 14** 2D contour plot of Si interstitial concentration underneath the source after As implantation (7 keV, $10^{15}/cm^2$)

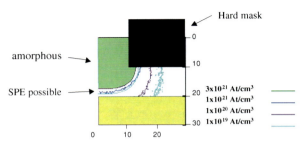

**Fig. 15** Lateral profiles of Si interstitial density in the channel at various depth from top interface

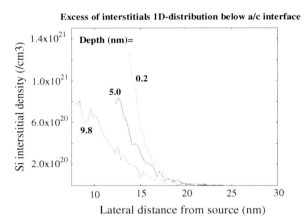

condition has been evaluated. As can be seen from Figs. 14 and 15, the Si interstitials injected underneath the hard mask (corresponding to nitride spacer and gate stack) exhibit a huge concentration larger than $8 \times 10^{20}/cm^3$ and with a lateral extension of the order of 10 nm.

In order to simulate the impact of such neutral defects on the channel mobility, we have approximated the interstitial lateral profile by an exponential function of the form $Nn(x) = Nmax.exp(-x/\lambda n)$ for each side of the channel and symmetrized it to account for both source and drain ends (Fig. 16). In such a way, the defect density from each side can merge for short channel devices, resulting in a huge defect concentration within a large portion of the channel.

The local mobility in the channel has been calculated using the Mathiessen rule by taking into account the neutral defect scattering law [19, 20] and a background long channel mobility value $\mu_{long}$ as:

$$\frac{1}{\mu(x,L)} = \frac{1}{\mu_{long}} + \frac{Nn(x,L)}{An} \quad (7)$$

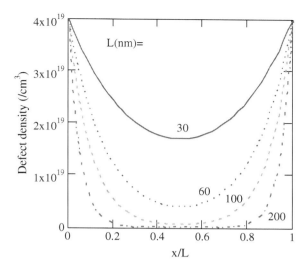

**Fig. 16** Neutral defect profiles along the channel for various $L$ as suggested from MC implantation simulations (Nmax $= 4 \times 10^{19}$/cm$^3$, $\lambda$n $= 10$ nm)

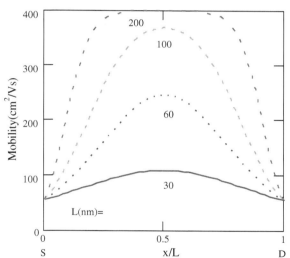

**Fig. 17** Local mobility profile $\mu(x,L)$ along the channel resulting from neutral defect profile of Fig. 16 ($\mu_{long} = 400$ cm$^2$/V s, An $= 2.5 \times 10^{21}$/cm V s)

where An $= 2.5 \times 10^{21}$/cm V s for silicon [19, 20].

The global effective mobility $\mu_{eff}$ of the device can be obtained by integrating over the channel the reciprocal local mobility such as:

$$\mu_{eff}(L) = \left[\frac{1}{L} \cdot \int_0^L \frac{1}{\mu(x,L)} dx\right]^{-1} \quad (8)$$

Figure 17 shows the local mobility profile along the channel for various gate lengths as obtained from Eq. 7 with typical parameter values. Note the strong

**Fig. 18** Variation of effective mobility $\mu_{eff}$ with gate length from local mobility profile of Fig. 17 (same simulation parameters)

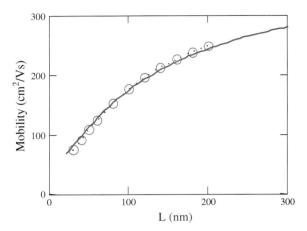

**Fig. 19** Variation of effective mobility $\mu_{eff}$ with temperature for various $L$ resulting from local mobility profile of Fig. 17

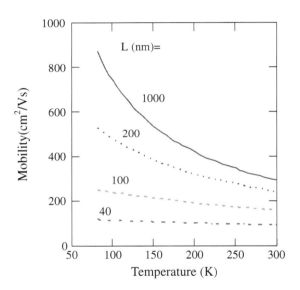

mobility degradation in the middle of the channel for gate length below 100 nm. The device global mobility obtained from Eq. 8 using the same parameters is reported as a function of gate length in Fig. 18. The symbols correspond to the effective mobility data obtained from TCAD simulations using the same local mobility profile and inferring the validity of the analytical model of Eq. 8. As can be seen, the mobility degradation given by this approach explains reasonably well the experimental results of Fig. 7 for parameters in agreement with the defect profiles found by MC implantation simulations.

The temperature dependence of the effective mobility have also been modeled by considering that the neutral scattering mobility is constant with temperature and that the long channel mobility which is limited by acoustic phonons varies as $T^{-1}$. This allows us obtaining the evolution of the effective mobility with temperature given in Fig. 19. The obtained behavior with temperature is in good agreement with the experimental data of Fig. 5, emphasizing the physical basis of the proposed explanation based on extra neutral defects in the channel.

## 4 Conclusions

Regarding short channel GAA, FD-SOI and FinFET MOS devices, it has been shown that the mobility is strongly degraded at small gate length, whatever the architecture, the gate stack and the measurement method used. In particular, it has been found that, for FD-SOI, the mobility is more degraded at the top interface than at the bottom interface, revealing that defects are more numerous at the top channel region. The negative role of the nitrogen diffusion from TiN/TaN metal gates has been confirmed by a significant reduction of low field mobility with the TiN thickness increase. Low temperature measurements of the mobility have indicated that the scattering processes are strongly modified for short channel devices, demonstrating that there is an increasing role of channel diffusion scattering centres, most likely neutral point defects, for gate length below 100 nm. These extra defects in the channel are likely suspected to be Silicon interstitials injected from the source and drain junction during the implantation process as confirmed by 2D Monte Carlo implantation simulations. The concentration and the lateral spatial extension of the generated defects have been implemented into a mobility model that explains reasonably well both the gate length dependence and the temperature variation of the mobility, emphasizing the physical merits of the proposed interpretation for the mobility collapse observed at small channel lengths.

**Acknowledgments** This work has been partially supported by European PULLNANO/FP6 integrated project and NANOSIL network of excellence.

## References

1. Ghani, T., Armstrong, M., Auth, C., Bost, M., Charvat, P., Glass, G., Hoffmann, T., Johnson, K., Kenyon, C., Klaus, J., McIntyre, B., Mistry, K., Murth, A., Sandford, J., Silberstein, M., Sivakumar, S., Smith, P., Zawadzki, K., Thompson, S., Bohr, M.: A 90 nm high volume manufacturing logic technology featuring novel 45 nm gate length strained silicon CMOS transistors. In: IEEE Proceedings of IEDM, pp. 978–980 (2003)
2. Andrieu, F., Ernst, T., Lime, F., Rochette, F., Romanjek, K., Barraud, S., Ravit, C., Boeuf, F., Jurczak, M., Casse, M., Weber, O., Brevard, L., Reimbold, G., Ghibaudo, G., Deleonibus, S.: Experimental and comparative investigation of low and high field transport in substrate- and process-induced strained nanoscaled MOSFETs. In: IEEE/VLSI Symposium, pp. 176–178 (2005)

3. Wong, H.-S.P., Frank, D.J., Solomon, P.M.: Device design considerations for double-gate, ground-plane, and single-gated ultra-thin SOI MOSFET's at the 25 nm channel length generation. In: IEEE Proceedings of IEDM, pp. 407–410 (1998)
4. Cros, A., Romanjek, K., Fleury, D., Harrison, S., Cerutti, R., Coronel, P., Dumont, B., Pouydebasque, A., Wacquez, R., Duriez, B., Gwoziecki, R., Boeuf, F., Brut, H., Ghibaudo, G., Skotnicki, T.: Unexpected mobility degradation for very short devices: a new challenge for CMOS scaling. In: IEEE Proceedings of IEDM, pp. 439–402 (2006)
5. Lochtefeld, A., Antoniadis, D.A.: Investigating the relationship between electron mobility and velocity in deeply scaled NMOS via mechanical stress. IEEE Electron. Device Lett. **22**, 591–593 (2001)
6. Ghibaudo, G.: New method for the extraction of MOSFET parameters. Electron. Lett. **24**, 543–545 (1988)
7. Romanjek, K., Andrieu, F., Ernst, T., Ghibaudo, G.: Improved split C-V method for effective mobility extraction in sub-0.1-µm Si MOSFETs. IEEE Electron. Device Lett. **25**, 583–585 (2004)
8. Lime, F., Andrieu, F., Derix, J., Ghibaudo, G., Boeuf, F., Skotnicki, T.: Low temperature characterization of effective mobility in uniaxially and biaxially strained nMOSFETs. Solid State Electron. **50**, 644–649 (2006)
9. Pappas, I., Ghibaudo, G., Dimitriadis, C.A., Fenouillet-Beranger, C.: Backscattering coefficient and drift-diffusion mobility extraction in short channel MOS devices. Solid State Electron. **53**, 54–56 (2009)
10. Pham-Nguyen, L., Fenouillet-Beranger, C., Vandooren, A., Skotnicki, T., Ghibaudo, G., Cristoloveanu, S.: In situ comparison of Si/high-kappa and $Si/SiO_2$ channel properties in SOI MOSFETs. IEEE Electron. Device Lett. **30**, 1075–1077 (2009)
11. Bennamane, K., Boutchacha, T., Ghibaudo, G., Mouis, M., Collaert, N.: DC and low frequency noise characterization of FinFET devices. Solid State Electron. **53**, 1263–1267 (2009)
12. Lundstrom, M.: On the mobility versus drain current relation for a nanoscale MOSFET. IEEE Electron. Device Lett. **22**, 293–295 (2001)
13. Shur, M.S.: Low ballistic mobility in submicron HEMTs. IEEE Electron. Device Lett. **23**, 511–513 (2002)
14. Yang, N., Henson, W.K., Hauser, J.R., Wortman, J.J.: Estimation of the effects of remote charge scattering on electron mobility of n-MOSFET's with ultrathin gate oxides. IEEE Trans. Electron. Devices **47**, 440–447 (2000)
15. Sah, C.T., Ning, T.H., Tschopp, L.L.: The scattering of electrons by surface oxide charges and by lattice vibrations at the silicon-silicon dioxide interface. Surf. Sci. **32**, 561–575 (1972)
16. Sun, S.C., Plummer, J.D.: Electron mobility in inversion and accumulation layers on thermally oxidized silicon surfaces. IEEE Trans. Electron. Devices **27**, 1497–1508 (1980)
17. Denorme, S., Mathiot, D., Dollfus, P., Mouis, M.: 2-Dimensional modelling of the enhanced diffusion in thin base N-P-N bipolar transistors after lateral ion implantations. IEEE Trans. Electron. Devices **42**, 523–527 (1995)
18. CEMES/CNRS: IPROS manual Monte Carlo simulation of ion implantation into real devices. Internal document, CEMES/CNRS (1994)
19. Erginsoy, C.: Neutral impurity scattering in semiconductors. Phys. Rev. **79**, 1013–1014 (1950)
20. Ouisse, T., Physica, B.: Neutral impurity scattering with electron screening, vol. 270, pp. 262–271 (1999)

# Special Features of the Back-Gate Effects in Ultra-Thin Body SOI MOSFETs

T. Rudenko, V. Kilchytska, J.-P. Raskin, A. Nazarov and D. Flandre

**Abstract** Ultra-thin body silicon-on-insulator (SOI) MOSFET is considered to be a strong candidate for ultimate scaling of CMOS technologies, because of its excellent suppression of the short-channel effects, even without the use of channel doping. Apart from undoped ultra-thin silicon body, nowadays SOI MOSFETs also feature ultra-thin gate high-k gate dielectrics and thin buried oxides. These innovating features bring about special electrical properties. In this work, we describe some of these properties revealed via the back-gate effects, including special behaviors of interface coupling, transport properties and gate tunneling currents, which may be beneficial for the back-gate control schemes.

## 1 Introduction

Ultra-thin body (UTB) silicon-on-insulator (SOI) MOSFET is recognized as promising candidate for scaling CMOS devices into nanometer region [1–7]. Thinning down the transistor body allows for suppressing the short-channel effects without the use of the channel doping, which is beneficial in the view of the carrier mobility and immunity to threshold voltage variability [1, 4, 6]. In addition to ultra-thin undoped body, the present-day SOI MOSFETs feature ultra-thin gate high-k gate dielectrics and thin buried oxides (BOX), which provide additional improvements of the short-channel characteristics and device performance [1, 4, 7].

---

T. Rudenko (✉) and A. Nazarov
Institute of Semiconductor Physics, NAS of Ukraine, Kiev, Ukraine
e-mail: tamara@lab15.kiev.ua

V. Kilchytska, J.-P. Raskin and D. Flandre
Université catholique de Louvain, Louvain-la-Neuve, Belgium

However, these innovating device features also bring non-conventional effects and unusual electrical properties [5, 7–9]. In this work, we describe some of these unusual properties revealed via the back-gate effects. Biasing the back gate in a fully-depleted (FD) SOI MOSFET enables to change the carrier and electric field distributions in the silicon film, which can influence the basic physical characteristics such as transport properties or gate tunneling currents. On the other hand, the back-gate effects are important from the practical viewpoint, because they can affect the operation of SOI devices. In this work, we focus on the special features of the interface coupling, back-gate effect on the effective mobility and gate tunneling currents inherent to SOI MOS structures with undoped ultra-thin bodies and ultra-thin gate dielectrics, which may be advantageous for the back-gate control schemes [10–12].

The experimental results presented in this paper are obtained on the n-channel SOI MOSFETs fabricated at CEA-LETI using UNIBOND (100) SOI wafers with two BOX thicknesses, namely, 145 and 11.5 nm. Details of fabrication process can be found in [13]. The thickness of the silicon body in the channel region is 11 nm. The devices have elevated source/drain regions to reduce parasitic resistance. No channel doping is used. The gate stack consists of ALD $HfO_2$ with equivalent oxide thickness (EOT) of 1.75 nm, and CVD TiN gate electrode. In this study, we use only long-channel devices with the gate length $L = 10$ μm and width $W = 10$ μm.

## 2 Unusual Features of Interface Coupling and their Physical Understanding

The effect of the charge coupling between the front and back SOI interfaces is a key property of any FD SOI MOSFET [14, 15]. This effect is typically characterized by the coupling curve, representing modulation of the threshold voltage at one gate by the opposite-gate bias, and it is usually described by the classical Lim–Fossum model [15]. Interface coupling is widely used for characterization purposes, namely, for the electrical determination of the Si film and BOX thicknesses [16]. Besides, the interface coupling is also used for adjusting the threshold voltage in SOI MOSFETs with thin BOX [8–10]. Recently, it has been demonstrated that UTB SOI devices exhibit non-conventional behaviors of interface coupling [7–9, 17]. Among them are the increased slope of the coupling curves extracted by the drain current and transconductance measurements [7, 9], impossibility of simultaneous achievement of strong inversion at one interface and accumulation at opposite interface in UTB SOI MOSFETs at reasonable oxide fields [8], and impact of quantum-mechanical (QM) effects [17].

The goal of this section is to gain an understanding of special features of interface coupling in UTB SOI MOSFETs. An analysis of interface coupling is performed by means of 1D numerical simulations in both classical and QM modes, using a self-consistent Schrödinger–Poisson solver (SCHRED [18]), and comparison of simulation results with the Lim–Fossum model and experimental data extracted from either the gate-to-channel capacitance or transconductance measurements of the long-channel devices. The threshold voltage was defined as the gate voltage where the second derivative of the inversion charge (or the derivative of the gate-to-channel capacitance) is at a maximum [19].

## 2.1 Extraction Procedure of the Coupling Curves from the Capacitance Data

The channel separation and extraction of the coupling curves have been performed using front-gate split C–V measurements for various back-gate biases (Fig. 1) [20]. The threshold voltage was determined from the position of the peak of the derivative of the gate-to-channel capacitance $C_{gc}$ with respect to the front-gate voltage $V_{gf}$. This method is theoretically equivalent to the transconductance change (or second derivative of the drain current $I_d$) method [21], however, in contrast to the latter, the capacitance derivative method is unaffected by the mobility degradation and series resistance effects, which facilitates the comparison between experiments and simulations. The procedure for extraction of the coupling characteristics is illustrated in Fig. 1.

For large positive $V_{gb}$, when the C–V curves exhibit a capacitance plateau related to the back channel, the derivatives reveal two clearly pronounced peaks: the first observed in the range of the capacitance plateau is due to activation of the back-channel, and its position at various $V_{gb}$ yields a relationship between $V_{gf}$ and the back-channel threshold voltage $V_{THb}$, while the second, whose position does not change with $V_{gb}$, corresponds to activation of the front channel, yielding the front-gate threshold voltage for inverted back interface $V_{THf\_back\_inv}$. For $V_{gb} \leq 0$, when the capacitance plateau on the C–V curves disappears, $dC_{gc}/dV_{gf}$ curves

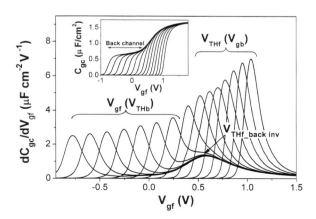

**Fig. 1** Experimental $dC_{gc}/dV_{gf}$ curves versus the front-gate voltage $V_{gf}$ for various back-gate voltages $V_{gb}$ illustrating the procedure for extraction of coupling characteristics. Inset shows the corresponding $C_{gc}(V_{gf})$ curves ($t_{BOX} = 145$ nm; $t_{Si} = 11$ nm; $V_{gb}$ from +40 to −30 V with step −5 V; $f = 100$ kHz)

reveal a single peak corresponding to the front-gate threshold voltage $V_{THf}$, and its shift with $V_{gb}$ gives the front-channel coupling curve. Similar procedure has been used for extraction of coupling curves from the transconductance measurements. The same $C_{gc}$ measurements have been used for channel separation and determination of the inversion-carrier density $N_{inv}$ at arbitrary $V_{gf}$ and $V_{gb}$ when extracting the front- and back-channel effective mobilities discussed in Sect. 3.

## 2.2 Analysis of Coupling Characteristics

Figure 2 presents the experimental coupling curves extracted from the peaks of $dg_m/dV_{gf}$ and $dC_{gc}/dV_{gf}$ curves for various $V_{gb}$ for a thick BOX device. Dashed lines indicate the slopes of the coupling curves predicted by the classical Lim–Fossum model [15]. As it is evident from Fig. 2, modulation of experimental front- and back-channel threshold voltages by the opposite-gate bias extracted by both methods is much stronger than expected from the Lim–Fossum model. Besides, it is clearly seen that there is a noticeable difference between results obtained by the two methods: the $dg_m/dV_{gf}$ method yields a higher slope of the front-channel coupling curve compared to the $dC_{gc}/dV_{gf}$ method, which is most likely related to the strong field-induced mobility degradation at the Si/HfO$_2$ front interface.

Figure 3 compares the coupling curves obtained by numerical simulations in both classical and QM modes [18] and experimental data extracted from the peaks of capacitance derivatives. It can be seen that at $V_{gb} = 0$ V for our 11-nm-thick silicon film, QM effect on the threshold voltage is negligible, in agreement with previously published studies asserting that QM effects give an observable impact on the threshold voltage of a SOI MOSFET if $t_{Si} < 10$ nm [22, 23]. However, biasing the second gate in opposite direction brings into appearance a noticeable difference between QM and classical threshold voltages, which moreover increases with the opposite-gate bias, resulting in distinctly different slopes of QM and classical coupling curves. In particular, QM simulations result in a higher

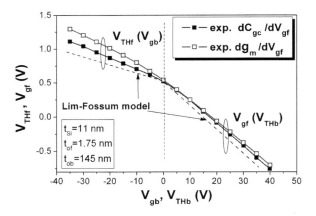

**Fig. 2** Experimental front-channel and back-channel threshold voltages versus opposite-gate bias extracted from the peaks of $d_{gm}/dV_{gf}$ and $dC_{gc}/dV_{gf}$ for the long-channel UTB SOI MOSFET with a thick BOX. *Dashed lines* show the slopes of the coupling curves predicted by the Lim–Fossum model

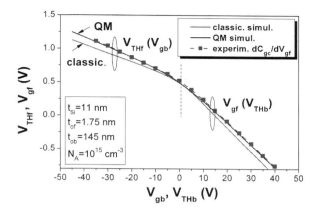

**Fig. 3** Comparison between coupling curves obtained by QM and classical numerical simulations (*lines*) and experimental data extracted from the peaks of the capacitance derivatives (*symbols*)

slope of the $V_{THf}(V_{gb})$-curve and a lower slope for $V_{gf}(V_{THb})$-curve (that is, higher slope of the $V_{THb}(V_{gf})$-curve) compared to classical simulations, which means that QM effects enhance modulation of $V_{THf}$ and $V_{THb}$ by opposite-gate bias. Note that experimental coupling curves extracted from the capacitance data differ significantly from the classical numerical simulation results, but are in excellent agreement with QM simulation results. The QM effect on the coupling characteristics of UTB SOI MOSFETs was first reported in [17]. It can be explained by the variation of the potential distribution and the transverse electric field in the silicon film at threshold conditions when biasing the second gate in the opposite direction. At low opposite-gate biases, when the potential distribution in the silicon film is nearly flat and the potential well has quasi-rectangular shape, QM effects are observable only if the silicon film is very thin (<10 nm) [22, 23]. However, when the second gate is biased in the opposite direction, the normal electric field in subthreshold conditions increases significantly, transforming the potential well into triangular shape and resulting in the shift of the ground state energy, thereby increasing the threshold voltage. In other words, the QM effect on the coupling characteristics is caused by the electrical confinement. This effect is important even for relatively thick SOI MOSFETs ($t_{Si} > 10$ nm), whose threshold voltage is usually considered to be unaffected by quantization effects.

The above QM effect cannot be properly described by the conventional "QM effective gate dielectric thickening" at the conduction side, because in low-doped UTB SOI devices at threshold conditions, the carrier confinement changes dramatically with opposite-gate bias, which is reflected by the strong variation of the position of the charge centroid, as shown by Fig. 4.

Figure 5 shows the coupling curves for different silicon film thicknesses and thick BOX ($t_{BOX} = 145$ nm), obtained by 1D numerical simulations in classical and QM modes [18]. It can be seen that for thick BOX, QM effect on the front-gate coupling curve does not depend on the film thickness, because when the capacitance of the silicon film considerably exceeds the capacitance of the BOX, the front-surface electric field at threshold conditions does not depend on the film thickness. However, for the back-channel, QM effect on the coupling curve

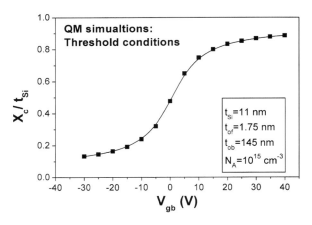

**Fig. 4** Normalized position (i.e. distance from top interface) of the QM inversion-charge centroid in the silicon film at the front-gate threshold conditions as a function of the back-gate bias, illustrating strong variation of carrier confinement in undoped UTB device with the back-gate bias at threshold

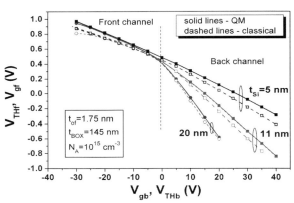

**Fig. 5** Coupling characteristics obtained by classical and QM numerical simulations for different silicon film thicknesses and thick BOX

increases with the film thinning. In the case of a thin BOX, QM effect on both front-gate and back-gate coupling curves increases with the film thinning [17].

Thus, an enhanced modulation of the front- and back-channel threshold voltages by opposite-gate bias (that is, increased slope of the coupling curves) observed in UTB SOI MOSFETs is caused mainly by quantization effects. It cannot be accounted by conventionally adding of the QM "darkspace" to the gate dielectric thickness because of the strong variation of the carrier confinement in undoped UTB devices at threshold conditions with the back bias. The described QM effect on interface coupling has to be taken into account in modeling and characterization of thin-film SOI MOSFETs.

## 3 Impact of the Back-Gate Bias on the Effective Mobility

The transport properties, and in particular the carrier mobility, in UTB SOI MOSFETs, depending on the quality of both silicon film interfaces, have been the

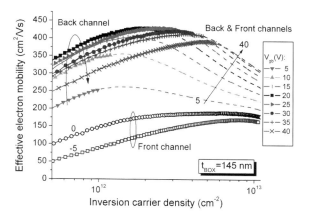

**Fig. 6** Effective mobility as a function of the inversion-carrier density for various $V_{gb}$. Open and full symbols correspond, respectively, to the front- and back-channel channel conductions. Dashed lines indicate regions where front and back channels co-exist

subject of much investigation in the recent years [23–27]. The most extensively studied issues have been the effect of the Si film thickness on the effective mobility $\mu_{eff}$ [24, 25] and the comparison between mobility in single-gate (SG) and double-gate (DG) modes [26, 27]. The effect of the back-gate bias on $\mu_{eff}$ in UTB SOI is of interest in view of the possibility to change the carrier concentration distribution and the position of the charge centroid in the Si film. In particular, adjusting the back-gate bias allows for realizing symmetric DG conditions in non-symmetric SOI MOS structure, which has been used in the comparison between the SG and DG mobilities [26, 27]. Then, the back-gate effect has been used to clarify the impact of the gate stack [20, 28, 29] as well as of the BOX thinning [20] on the effective mobility. Particularly, it has been found that in the case of ultra-thin gate oxides or high-k gate dielectrics, the front-channel mobility in UTB SOI MOSFETs is lower than the mobility in the back channel, which has been attributed to enhanced Coulomb scattering at the front interface [20, 28, 29]. It was found that BOX thinning does not deteriorate the quality of the silicon film and its interfaces, and thus the above effect of much higher back-channel mobility in UTB SOI MOSFETs with ultra-thin gate dielectrics holds for the thin BOX [20].

In this section, we focus on the evolution of the effective mobility in UTB SOI MOS structures with high-k gate dielectric when shifting the conduction channel in the film from front to back interface. The effective mobility was extracted by the front-gate split C–V measurements for various $V_{gb}$. Analysis of the C–V curves measured for various $V_{gb}$ and their derivatives was used for separating regions where front or back conduction dominates, as described in Sect. 2.1.

Figure 6 shows experimental $\mu_{eff}$ versus the inversion-carrier density $N_{inv}$ for various $V_{gb}$ obtained for a SOI MOSFET with 145-nm-thick BOX. Open and full symbols in Fig. 6 indicate regions where, respectively, front or back conduction dominates, and dashed lines indicate regions where front and back channels co-exist. It should be noted that in UTB SOI devices, the measured mobility involves the contributions from scattering at both silicon film interfaces, which cannot be completely separated, especially at low $N_{inv}$ and weak asymmetry of the

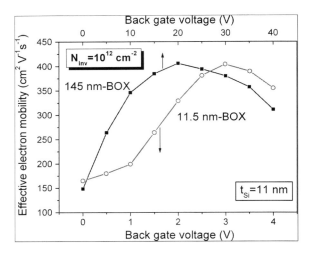

**Fig. 7** Experimental effective mobility as a function of $V_{gb}$ at constant $N_{inv} = 10^{12}$ cm$^{-2}$ for 145-nm-thick BOX (*full symbols, top axis*) and 11.5-nm-thick BOX (*open symbols, bottom axis*). (TiN/HfO$_2$, EOT = 1.75 nm, $W = 10$ µm, $L = 10$ µm)

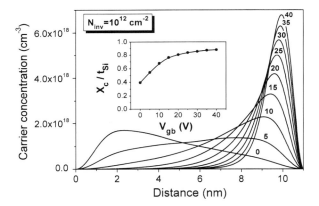

**Fig. 8** Carrier concentration distributions in the silicon film for a constant sheet inversion-carrier density Ninv = $10^{12}$ cm$^{-2}$ and different back-gate biases obtained by QM numerical simulations for $t_{Si} = 11$ nm, $t_{of} = 1.75$ nm, $t_{BOX} = 145$ nm, and $N_A = 10^{15}$ cm$^{-3}$. Shown in the *inset* is the normalized charge centroid versus back-gate bias

front and back surface potentials, when the conduction is determined by volume inversion. However, from Fig. 6 it is clearly seen that mobility is much higher when the back conduction dominates. The maximum back-channel mobility exceeds the front-channel mobility by more than factor of 2. With activation of the front channel the carrier density in the back channel remains essentially unchanged, so that further increase of $N_{inv}$ is due to the front channel, therewith the larger contribution of the front channel, the lower averaged mobility (dashed lines).

Figure 7 shows the variation of the experimental effective mobility with $V_{gb}$ at constant $N_{inv} = 10^{12}$ cm$^{-2}$ in undoped UTB devices with thick (top axis) and thin (bottom axis) BOX and TiN/HfO$_2$ gate stack. The corresponding carrier concentration distributions and the position of the charge centroid in the silicon film obtained by QM numerical simulations are presented in Fig. 8. It can be seen from Fig. 7 that, for given devices, application of positive $V_{gb}$, which shifts the inversion-

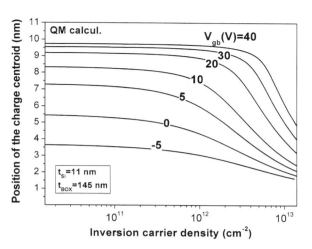

**Fig. 9** The position of the inversion-layer centroid obtained by QM numerical simulations as a function of $N_{inv}$ with $V_{gb}$ as a parameter ($t_{Si} = 11$ nm, $t_{of} = 1.75$ nm, $t_{BOX} = 145$ nm, $N_A = 10^{15}$ cm$^{-3}$)

charge centroid away from the front interface, allows to increase the value of $\mu_{eff}$ by 2.7 times compared to that at $V_{gb} = 0$ V. By correlating the data in Figs. 7, 8, we can conclude that, for given $N_{inv}$, the maximum mobility value is observed when the charge centroid is located approximately at a distance of $\sim 0.7\, t_{Si}$ from the front interface, that is, when the charge centroid is closer to the back interface. This indicates the presence of additional scattering defects at the front interface.

An important point in Fig. 7 is that the maximum $\mu_{eff}$ value related to the back channel is the same for thick and thin BOX, which demonstrates that BOX thinning does not deteriorate the quality of the back interface, being in agreement with results presented in [20]. Note that for the thick BOX, the steepest increase of $\mu_{eff}$ is observed at low $V_{gb}$ (<10 V), which can be attributed to a sharp variation of the position of the charge centroid in the range of small $V_{gb}$ when conduction is determined by volume inversion (Fig. 8). The gentle initial slope of the $\mu_{eff}(V_{gb})$ curve observed at small $V_{gb}$ (<1 V) for the thin BOX in Fig. 7 is due to substrate depletion effect; further increase of positive $V_{gb}$ results in a sharp rise of $\mu_{eff}$, similar to that for thick BOX.

Figure 9 shows the inversion-layer centroid (the average inversion-layer distance from the front interface) versus $N_{inv}$ for different $V_{gb}$ obtained by QM numerical simulations [18]. By combining the data presented in Figs. 6 and 9, we plotted in Fig. 10 the experimental $\mu_{eff}$ as function of the normalized position of the charge centroid in the silicon film for various values of $N_{inv}$ extracted from QM simulations. It is clearly seen that when the charge centroid moves from the front interface to the back interface, $\mu_{eff}$ increases. The maximum value of $\mu_{eff}$ is observed when the charge centroid lies in the bottom half of the film, namely, at a distance of 0.73 $t_{Si}$ from the front interface, and not in the center of the film, when symmetrical DG-like carrier distributions are provided. In this context, it is interesting to compare mobility in SG and DG modes in the case of different scattering rates at the front and back interfaces.

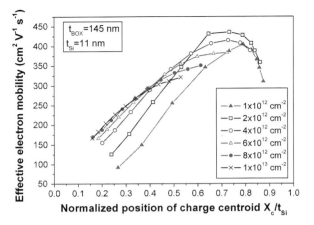

**Fig. 10** The experimental $\mu_{eff}$ as function of the normalized position of the charge centroid in the film for various values of $N_{inv}$ ($W/L = 10/10$ μm, TiN/HfO$_2$, EOT = 1.75 nm)

So far the comparison between mobility in DG and SG operating modes has been performed only for SOI devices featuring identical quality of the front and back interfaces [26, 27]. Below we will demonstrate that in the case of the devices with distinctly different two interfaces, the mobility improvement in the DG mode may be much stronger than for the devices with identical two interfaces. In the case of UTB SOI devices with thick BOX, the DG-like conditions (providing symmetrical carrier concentration and electric field distributions) are realized when $N_{inv\_DG} = 2 \times N_{inv\_SG}$ [26], that is, when $N_{inv\_front} = N_{inv\_back}$. The experimental procedure for finding the appropriate $V_{gf}$ and $V_{gb}$ values that provide such conditions is illustrated in Fig. 11. The comparison between mobility measured in the DG mode and SG front- and back-channel operating modes is presented in Fig. 12.

For the DG mode, the inversion-carrier density in Fig. 12 corresponds to 1/2 of the total inversion density which provides the same $E_{eff}$ as in the SG mode [26]. It can be seen from Fig. 12, that mobility in DG mode is noticeably lower than in SG back-channel operating mode. However, mobility in DG mode is significantly (more than factor of 2) higher than in SG front-channel operating mode, which differs from the case of SOI films with identical quality of the two interfaces, when for the films with thicknesses around 10 nm only a modest (<5%) mobility improvement is observed in DG mode at low $N_{inv}$ [26, 27]. Thus, in the case of SOI films with a higher scattering rate at the front interface, a significant mobility improvement can be obtained in DG mode for a wide range of $N_{inv}$: for low $N_{inv}$, when conduction is determined by volume inversion, this improvement is due to shift of the charge centroid further from the front interface featuring higher scattering rate, while for high $N_{inv}$, when coexisting front and back channels are almost independent, this improvement is due to the fact that half of the carriers (back-channel carriers) feature much higher mobility.

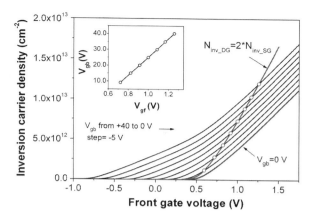

**Fig. 11** The experimental procedure for finding $V_{gb}(V_{gf})$ values giving DG-like conditions ($t_{Si} = 11$ nm, $t_{BOX} = 145$ nm)

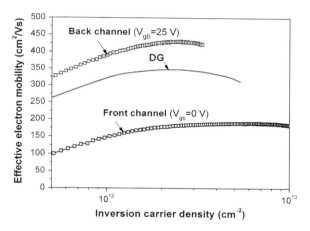

**Fig. 12** The comparison between mobility measured in the DG mode and front and back-channel mobilities measured in SG modes ($N_{inv\_DG}/2 = N_{inv\_SG}$, $t_{Si} = 11$ nm, $t_{BOX} = 145$ nm)

The presented mobility results allow for better understanding the transport properties of UTB SOI MOSFETs with ultra-thin and high-k gate dielectrics and provide an additional motivation for the employment of the back-gate control schemes.

## 4 Impact of the Back-Gate Bias on the Gate-to-Channel Tunneling Current and Drain-to-Gate Current Ratio

An important parameter of the advanced UTB SOI CMOS devices, featuring ultra-thin gate dielectrics, is gate tunneling leakage current which determines power consumption and the limits for the gate dielectric scaling. Previous experimental studies have shown that gate leakage current in a SOI device is affected by

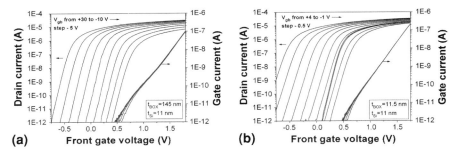

**Fig. 13** The gate tunneling (*left axis*) and drain (*right axis*) currents measured as a function of the front-gate voltage at various back-gate voltages in thick BOX **a** and thin BOX **b** UTB SOI devices (TiN/HfO$_2$, EOT = 1.75 nm, $V_d$ = 50 mV, $W/L$ = 10/10 μm)

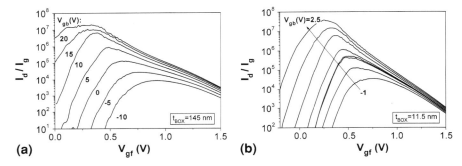

**Fig. 14** Experimental drain/gate current ratio as a function of $V_{gf}$ for various $V_{gb}$: **a** thick BOX; **b** thin BOX. It is clearly seen that application of $V_{gb}$ to lower the threshold voltage significantly improves the $I_d/I_g$ ratio in the subthreshold region and near threshold. (TiN/HfO$_2$, CET = 1.9 nm, $V_d$ = 50 mV, $W/L$ = 10/10 μm)

back-gate bias due to the change of the electric field distribution in the body [30]. However, in the case of undoped UTB SOI films, no change in the gate-to-channel tunneling current $I_g$ is observed when the back gate is positively biased up to large values of $V_{gb}$, though it strongly increases the drain current (Fig. 13). A noticeable variation of $I_g$ in the subthreshold region is observed only for negative $V_{gb}$; in strong inversion $I_g$ remains nearly unchanged. As a result, the $I_d/I_g$ ratio significantly increases with application of $V_{gb}$ lowering the threshold voltage, especially, in the subthreshold region and near threshold (Fig. 14).

Basically, the tunneling current is determined by the product of the charge available for tunneling by tunneling probability. So invariability of $I_g$ with variation of $V_{gb}$ resulting in strong variation of the drain current may appear to be surprising. Insensitivity of gate tunneling current to application of positive $V_{gb}$ can be qualitatively explained as follows. The gate tunneling current is observed only in the range of $V_{gf}$ where the front channel is activated. At high positive $V_{gb}$, in the above range of $V_{gf}$ the front and back inversion channels co-exist, and the potential

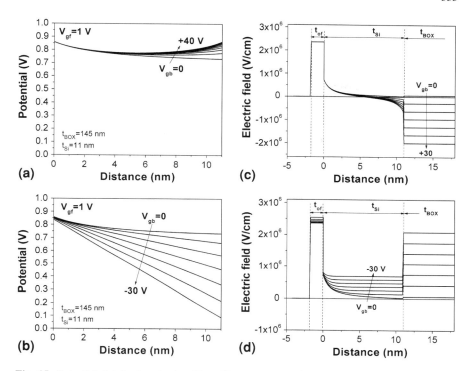

**Fig. 15** Potential distributions in the silicon film upon strong inversion at the front interface and various **a** positive $V_{gb}$ and **b** negative $V_{gb}$ and the corresponding electric field distributions for **c** positive $V_{gb}$ and **d** negative $V_{gb}$, obtained by classical numerical simulations [18] for $t_{Si} = 11$ nm, $t_{of} = 1.75$ nm, $t_{BOX} = 145$ nm, $N_A = 10^{15}$ cm$^{-3}$

distribution in the silicon film exhibits a minimum, where the electric field is zero, as illustrated by Fig. 15a, c. In the point of the potential minimum, the electric field changes its direction, so that the electric fields in the front and back channels have the opposite direction. Thus, only the front-channel inversion charge gives a contribution to the gate current. Since the front-channel threshold voltage for the inverted back interface $V_{THf\_back\_inv}$ is independent on $V_{gb}$, the carrier density in the front channel $N_{inv\_front}$ at certain $V_{gf}$ is the same for different positive $V_{gb}$. In other words, application of positive $V_{gb}$ increases the total inversion charge contributing to the drain current, while the inversion charge contributing to $I_g$ remains unchanged. Furthermore, when both interfaces are biased in inversion such that a point of the potential minimum or zero electric field exists in the film, the electric field at the front interface in the silicon film $E_{sf}$ is determined only by the front-channel inversion charge $qN_{inv\_front}$ because the contribution of the doping charge in undoped UTB SOI films is negligible $E_{sf} \approx qN_{inv\_front}/\varepsilon_{Si}$. Since for positive $V_{gb}$, $N_{inv\_front}(V_{gf})$ does not change with $V_{gb}$, the front-surface electric field $E_{sf}(V_{gf})$ and thus the electric field in the front-gate dielectric also does not change with $V_{gb}$. This is confirmed by the electric field distributions obtained by classical numerical simulations for strongly inverted front interface and various $V_{gb}$ presented in Fig. 15c. It can be seen that for $V_{gb} \geq 0$ V, the electric field at the front interface and the electric field in the front-gate dielectric, and thus tunneling probability, are unaffected by $V_{gb}$. Therefore, since both the carrier density available for tunneling and tunneling probability are unchanged by $V_{gb} \geq 0$ V, the gate tunneling current $I_g$ also must remain unchanged.

It is more difficult to explain the behavior of the gate tunneling current for negative $V_{gb}$. As can be seen from Fig. 13, negative $V_{gb}$ decreases the gate current in weak inversion, while in strong inversion, $I_g$ remains almost unaltered. For negative $V_{gb}$, the electric field in the film is unidirectional, so that the total carrier density can contribute to $I_g$. In this case, both the carrier density available for tunneling and the electric field in the gate dielectric change with negative $V_{gb}$, featuring, however, the opposite trends in respect of $V_{gb}$, namely: at the same $V_{gf}$, $N_{inv}$ decreases with negative $V_{gb}$ due to increase of $V_{THf}$, while $E_{of}$ increases due to increase of the electric field in the body (Fig. 15b, d). Whether or not the above opposite trends allow for explaining the observed invariability of the gate-to-channel tunneling current with negative $V_{gb}$ at high $V_{gf}$ is unclear. Thus, an understanding of this observation requires further investigations, involving systematic QM simulations of gate tunneling current in undoped UTB SOI MOS devices for various back-gate biases.

## 5 Conclusions and Possible Implications of the Discussed Effects in the Back-Gate Control Schemes

Based on the analysis of coupling characteristics, it has been shown that advanced UTB SOI MOSFETs with ultrathin gate dielectrics feature enhanced modulation of the threshold voltage by the opposite-gate bias (or increased slope of coupling characteristics) compared to the classical Lim-Fossum model, due to QM effects caused by the electrical confinement. This QM effect is important even for relatively thick-film SOI MOSFETs (with $t_{Si} > 10$ nm), whose threshold voltage is usually considered to be unaffected by quantization effects. Furthermore, this effect cannot be properly described by the conventional "QM effective gate dielectric thickening" at the conduction side because of the strong variation of the carrier confinement in low-doped ultra-thin films at threshold conditions.

It has been found that in advanced UTB SOI MOSFETs featuring ultra-thin high-k gate dielectrics and both thick and thin BOX, mobility at the front interface is significantly (more than twice) lower than at the back interface, indicating additional scattering defects at the front interface. It was shown that in UTB SOI MOSFETs with strongly different scattering rates at the two interfaces, mobility at low and moderate inversion (being of particular interest for low-power devices) is highly sensitive to the position of the charge centroid in the silicon film, and thus to bias at the second gate (more sensitive than in the case of identical interfaces).

Based on the comparison of mobilities for thick and thin BOX, it has been found that BOX thinning does not degrade the quality of the silicon film and its interfaces.

The special features of the back-gate effects in UTB SOI MOSFETs discussed in this paper are useful not only for deep understanding of coupling and transport behaviors of these advanced devices, but they may be of advantage in the back-gate control schemes.

In the back-gate control schemes, back-gate bias is used to modulate the front-gate threshold voltage through interface coupling [10–12]. From this viewpoint, an increased slope of the coupling characteristics inherent to UTB SOI MOSFETs with ultra-thin-gate dielectrics is beneficial.

Then, as follows from the mobility results, in the case of ultra-thin gate dielectrics, application of the back-gate voltage to increase the drive current in the active mode should also bring about the mobility enhancement due to shift of the charge centroid away from the front interface, featuring high Coulomb scattering.

Furthermore, from the analysis of the behavior of the gate tunneling current with back-gate bias, one can expect a large dividend in the drain/gate current ratio in the active mode. This should provide additional improvements in the device performance.

Thus, special features of the back-gate effects in UTB SOI MOSFETs discussed in this paper may be of advantage in the back-gate control schemes.

**Acknowledgments** This work has been partly funded by the European Commission under the frame of the Network of Excellence "NANOSIL" (Silicon-based Nanostructures and Nanodevices, No. 216171) and "EuroSOI+". The devices have been fabricated within the frame of SOITEC/LETI Nanosmart research program and authors thank CEA-LETI and SOITEC for their support.

## References

1. Omura, Y., Nakashima, S., Izumu, K., Ishii, T.: 0.1-μm-gate, ultrathin-film CMOS devices using SIMOX substrate with 80-nm thick buried oxide layer. IEDM Technical Digest, pp. 675–678 (1991)
2. Su, L., Jacobs, J., Chung, J., Antoniadis, D.: Deep-submicrometer channel design in silicon-on-insulator (SOI) MOSFET's. IEEE Electron Dev Lett **15**, 183–185 (1994)
3. Suzuki, S., Ishii, K., Kanemaru, S., Maeda, T., Tsutsumi, T., Sekiwaga, T., Nagai, K., Hiroshima, H.: Highly suppressed short-channel effects in ultrathin SOI n-MOSFETs. IEEE Trans Electron Dev **47**, 354–359 (2000)
4. Wong, H.-S.P., Frank, D.J., Solomon, P.M.: Device design considerations for double-gate, ground-plane, and single-gated ultra-thin SOI MOSFET's at the 25 nm channel length generation. IEDM Technical Digest, pp. 407–410 (1998)
5. Ernst, T., Munteanu, D., Cristoloveanu, S., Quisse, T., Hefyene, N., Horiguchi, S., Ono, Y., Takahashi, Y., Murase, K.: Ultimately thin SOI MOSFETs: special characteristics and mechanisms. Proceedings of IEEE International Conference, pp. 92–93 (1999)
6. Weber, Q., Faynot, O., Andrieu, F., Buj-Dufournet, C., et al.: High immunity to threshold voltage variability in undoped ultra-thin FD SOI MOSFETs and its physical understanding. IEDM Technical Digest, pp. 245–248 (2008)

7. Pretet, S., Ohata, A., Dieudonne, F., Alliber, F., Bresson, N., Matsumoto, T., et al.: Scaling issues for advanced SOI devices: gate oxide tunnelling, thin buried oxide, and ultra-thin films. The Electrochemical Society Proceedings 2003-02, pp. 476–487 (2003)
8. Eminente, S., Cristoloveanu, S., Clerc, R., Ohata, A., Ghibaudo, G.: Ultra-thin fully-depleted SOI MOSFETs: special charge properties and coupling effects. Sol State Electron **51**, 239–244 (2007)
9. Ohata, A., Cristoloveanu, S., Vandooren, A., Cassé, M., Daugé, F.: Coupling effect between the front and back interfaces in thin SOI MOSFETs. Microelectron. Eng. **80**, 245–248 (2005)
10. Tsuchiya, R., Horiuchi, M., Kimura, S., Yamaoka, M., Kawahara, T., Maegawa, S., Iposhi, T., Ohji, Y., Matsuoka, H.: Silicon on thin BOX: A new paradigm of the MOSFET for low-power and high-performance applications featuring wide-range back-bias control. IEDM Technical Digest, pp. 631–634 (2004)
11. Yang, I.J., Vieri, K., Chandrakasan, A., Antoniadis, D.A.: Back gated CMOS on SOIAS for dynamic threshold voltage control. IEDM Technical Digest, pp. 877–879 (1995)
12. Hiramoto, T.: Low power and low voltage MOSFETs with variable threshold voltage controlled by back-bias. IEICE Transactions on Electronics E83-C, pp. 161–169 (2000)
13. Andrieu, F., Faynot, O., Garros, X., Lafond, C., et al.: Comparative scalability of PVD and CVD TiN on $HfO_2$ as a metal gate stack for FDSOI cMOSFETs down to 25 nm gate length and width. IEDM Technical Digest, pp. 641–644 (2006)
14. Colinge, J.P.: Silicon-on-insulator technology: materials to VLSI, 3rd edn. Kluwer, Boston (2004)
15. Lim, H.K., Fossum, J.G.: Threshold voltage of thin-film silicon-on-insulator (SOI) MOSFETs. IEEE Trans. Electron. Dev. **30**, 1244–1251 (1983)
16. Cristoloveanu, S., Li, S.: Electrical Characterization of Silicon-On-Insulator Materials and Devices. Kluwer, New York (1995)
17. Poiroux, T., Widiez, J., Lolivier, J., Vinet, M., Cassé, M., Prévitali, B., Deleonibus, S.: New and accurate method for electrical extraction of silicon film thickness on fully-depleted SOI and double-gate transistors. IEEE Int SOI Conf Proc, pp. 73–74 (2004)
18. Schred Simulation Tool: http://nanohub.org (2010)
19. Park, C.-K., Lee, C.-Y., Lee, K., Moon, B.Y., Byun, Y.H., Shur, M.: A unified current-voltage model for long-channel MOSFETs. IEEE Trans. Electron. Dev. **38**, 399–406 (1991)
20. Rudenko, T., Kilchytska, V., Burignat, S., Raskin, J.P., Andrieu, F., Faynot, O., Tiec, Y., Landry, K., Nazarov, A., Lysenko, V.S., Flandre, D.: Experimental study of transconductance and mobility behaviors in ultra-thin SOI MOSFETs with standard and thin buried oxides. Solid State Electron. **54**, 164–170 (2010)
21. Wong, H.-S., White, M.H., Krutsick, T.J., Booth, R.V.: Modeling of transconductance degradation and extraction of threshold voltage in thin oxide MOSFETs. Solid State Electron. **30**, 953–968 (1987)
22. Omura, Y., Horiguchi, S., Tabe, M., Kishi, K.: Quantum-mechanical effects on the threshold voltage of ultrathin-SOI n MOSFETs. IEEE Electron. Dev. Lett. **14**, 569–571 (1993)
23. Uchida, K., Koga, J., Ohba, R., Numata, T., Takagi, S.I.: Experimental evidences of quantum-mechanical effects on low field mobility, gate-channel capacitance, and threshold voltage of ultrathin body SOI MOSFETs. IEDM Technical Digest, pp. 29.4.1–29.4.4 (2001)
24. Esseni, D., Mastrapasqua, M., Celler, G.K., Fiegna, C., Selmi, L., Sangiorgi, E.: Low field electron and hole mobility of SOI transistors fabricated on ultrathin silicon films for deep submicrometer technology application. IEEE Trans. Electron. Dev. **48**, 2842–2850 (2001)
25. Uchida, K., Watanabe, H., Kinoshita, A., Koga, J., Numata, T., Takagi, S.: Experimental study on carrier transport mechanism in ultrathin-body SOI n- and p-MOSFETs with thickness less than 5 nm. IEDM Technical Digest, pp. 47–50 (2002)
26. Esseni, D., Mastrapasqua, M., Celler, G.K., Fiegna, C., Selmi, L., Sangiorgi, E.: An experimental study of mobility enhancement in ultrathin SOI transistors operated in double-gate mode. IEEE Trans. Electron. Dev. **50**, 802–807 (2003)

27. Uchida, K., Koga, J., Takagi, S.: Experimental study on carrier transport mechanisms in double-gate and single-gate ultrathin-body MOSFETs—Coulomb scattering, volume inversion, and $\delta T_{SOI}$-induced scattering. IEDM Technical Digest, pp. 805–808 (2003)
28. Ohata, A., Cristoloveanu, S., Cassé, M.: Mobility comparison between front and back channels in ultra-thin silicon-on-insulator transistors by the front-gate split capacitance-voltage method. Appl. Phys. Lett. **89**, 032104 (2006)
29. Ohata, A., Cassé, M., Cristoloveanu, S.: Front- and back-channel mobility in ultrathin SOI-MOSFETs by front-gate split CV method. Solid State Electron. **51**, 245–251 (2007)
30. Majkusiak, B., Badri, M.H.: Semiconductor thickness and back-gate voltage effects on the gate tunneling current in the MOS/SOI system with ultrathin oxide. Trans. Electron. Dev. **47**, 2347–2351 (2000)

# Part IV
# Sensors and MEMS on SOI

# SOI Nanowire Transistors for Femtomole Electronic Detectors of Single Particles and Molecules in Bioliquids and Gases

V. P. Popov, O. V. Naumova and Yu. D. Ivanov

**Abstract** The need for high-throughput, label-free multiplexed sensors for chemical and biological sensing has increased in the last decade in the newer applications like healthcare, genomic and proteomic diagnostics, environmental and industrial monitoring; quality control, core defense and security areas, etc. Human genome decoding was prolonged during 13 years and has shown a complex relation between genes, proteins and illnesses. But the amount of genes is only $3 \times 10^4$, while the amount of proteins is as high as $5 \times 10^6$. To decode this amount of proteins a massive set of samples with biological solutions should be analyzed using fast and massively parallel analyzing tools.

## 1 Introduction

Only parallel collection and analysis with a matrix of sensor elements will provide the "bandwidth" necessary for protein decoding. The task is complicated by the large differences for different protein contents in bioliquids ranged from micro- to attomole amounts [1]. High sensitivity and selectivity combined with wide range of analyte concentration measurements in bioliquids and gases are needed.

In order to make these challenging sensor elements, nanowire (NW) field-effect transistors (FETs) were nanostructured using a "top-down" approach in high-quality monocrystalline silicon-on-insulator (SOI). The SOI film has a thickness of a few tens of nanometers. Electronic detection of biochemical interactions or label-free physical adsorption of particles or molecules on the open surface

V. P. Popov (✉) and O. V. Naumova
A.V. Rzhanov Institute of Semiconductor Physics, SB RAS, Novosibirsk, Russia
e-mail: popov@isp.nsc.ru

Yu. D. Ivanov
V.N. Orekhovich Institute of Biomedical Chemistry, RAMS, Moscow, Russia

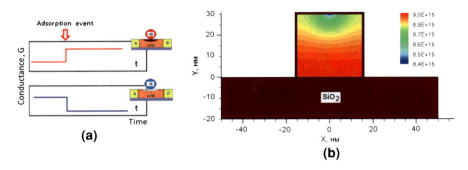

**Fig. 1 a** N-type Si nanowire conductance change due to a change of the surface potential $\varphi S$ at the moment of particle absorption of a positive (*top*) or negative (*bottom*) effective charge; **b** DESSIS simulation of electron density in a nanowire with a 30 × 30 nm cross-section and a single negative charge placed at the surface of the 1 nm native oxide covering the NW. The NW is doped with phosphorus to a concentration of $N_o = 10^{16}$ cm$^{-3}$

channel of NW FETs is possible in an electrolytic environment (Fig. 1). These immobilized substances serve as a virtual gate for FETs changing channel conductivity from one to ten orders of magnitude depending on their concentrations and the effective charges they carry. Even a single charge located at the surface of a 30 × 30 nm lightly doped Si NW can cause more than a 10% change in carrier concentration across the channel (Fig. 1b).

High sensitivity to chemical and biological substances is usually achieved in sensor elements by having a large ratio between surface area and volume. This requirement is automatically satisfied in nanosized NW sensors. In addition, there is a good size compatibility between the sensor and the sensed species. In the present work, the nanostructuring of silicon on insulator layers with a thickness of a few tens of nanometers was carried out at the last stage of electronic nanochip fabrication by etching in a fluorine and chloride plasma mixture. Electrical characterization confirmed that the fabrication approach produces a high-quality SOI nanowire transistor matrix with open channels operating in wide temperature range down to temperatures close to liquid helium. The sensitivity of NW FETs to negative Cl$^-$ ions in aqueous solution (pH 6) was shown to be as high as 10 femtomoles. The sensitivity of NW FETs to negative protein BSA molecules in a pH 7.4 buffer solution was shown to be much better than 1 femtomole and single-cell organisms such as viruses and bacteria were within the detection limit. Using NW FET matrix it is also possible to provide the sensing for toxic gases in the ppb–ppm range.

## 2 Optimization of the Fabrication Technology for NW Sensors

To optimize device sensitivity, a lot of efforts were made to minimize the serial resistance of the SOI NW FETs and to lower the noise from the interface states and metal contacts. Si nanowire transistors are the most promising sensor element for electronic sensing of biochemicals, chemical toxins and pathogens [2, 3] due to the compatibility of silicon and $SiO_2$ with body tissues and liquids. For clinical or even home use the sensors should be integrated with CMOS electronics. In our approach nanometer sizes were obtained using electron beam lithography and etching in a fluorine–chloride gas–plasma mixture after standard CMOS processes. To study the prospects of SOI-nanowire transistors (NW FETs) as sensing elements for electronic biochips (using open channels in nanowire transistors) and, in particular, organic molecules in bioliquids, microfluidic biochips [4] were manufactured with the following parameters: silicon layer thickness, $t$, of 10–50 nm, a natural oxide thickness on the nanowire surface of less than 2 nm, a nanowire width, $w$, of 20–500 nm. The length of the nanowires is 10 µm, and there are 12 nanowires on each chip. The diameter of the sensitive zone is not less than 2 mm.

The NW FET are SOI structures with a 300-nm buried oxide layer thickness; the thickness of the top silicon layer was 10–50 nm. The SOI structures were fabricated by a technology developed at the Rzhanov Institute of Semiconductor Physics of the Siberian Branch of the Russian Academy of Sciences [5]. A specific feature of SOI structures used in the present work was the fact that the interface between the top silicon layer and the buried oxide was the bonding interface. The original SOI structures with the silicon layer thickness equal to 400–500 nm had the $n$-type of conductivity, which determined the regime of SOI NW FET operation in accumulation of electrons and, hence, ensured higher values of mobility than in a hole channel.

The mobility of the charge carriers in the used SOI structures was found on the basis of the gate characteristics of the test MOS transistors with a channel length of 2 µm, which were produced by a standard CMOS technology on SOI with a 400-nm top silicon layer. The substrate of the SOI structures was used as a gate for forming an inversion channel near the bonding interface. Using the gate characteristics $I_{DS}(V_G)$ ($I_{DS}$ is a drain–source current and $V_G$ is a gate voltage) and the algorithm proposed in [6, 7], we determined the mobility of the charge carriers as a function of the effective electric field $\mu_{eff}$ ($E_{eff}$).

## 3 SOI NW Fluid Cell

The top silicon layer thickness was reduced from the initial value to 10–50 nm in a consecutive cycle of operations: thermal oxidation and oxide removal in hydrofluoric acid. Applying optical lithography to nanometer silicon layers, we fabricated SOI transistors consisting of drain–source areas connected by a silicon ribbon (10 µm long and 3 µm wide). A SOI transistor image obtained by an optical microscope is shown in Fig. 2a. The linear NW sizes of the SOI transistor were decreased to 15–500 nm by the method of lateral nanostructuring of SOI with the use of electron lithography and gas–plasma etching. Electron lithography

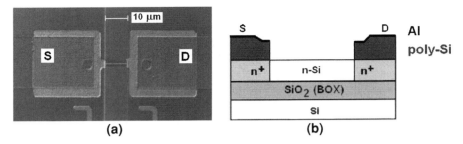

**Fig. 2 a** Optical microphotograph of single sensor element based on the nanowire (NW) field effect transistor FET with 10 μm open channel between source (S) and drain (D); **b** Schematic cross-section along the nanowire channel with raised highly doped S–D polysilicon regions and Al metal contacts

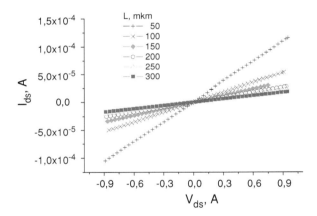

**Fig. 3** Ohmic contacts to 20 nm SOI layers are confirmed by the current slopes on test wire structures with different length $L$

(with a Raith-150 lithograph) combined with gas–plasma-enhanced chemical etching (in $SF_6:CFCl_3$ mixture) was applied at the final stage of SOI NWT fabrication (after the contacts were formed), which did not involve high-temperature annealing for eliminating defects generated during SOI nanostructuring. In the course of gas–plasma chemical etching, the substrate bias with respect to the plasma was within 10 V to minimize radiation damages. Thus, the conditions close to the gas etching process were provided.

Highly doped by phosphorous 250 nm source–drain raised regions were used for contact formation as standard steps in the CMOS technology. They provided good ohmic contacts to 10–50 nm SOI layers and low length of impurity diffusion near 0.2 μm, while for the thick SOI layer (200 nm) an initial channel concentration $(6/8) \times 10^{16}$ cm$^{-3}$ is reached at the distance 1.5/1.8 μm (Figs. 3, 4). The following technologies were applied to SOI layers to create ohmic contacts (drain–source areas):

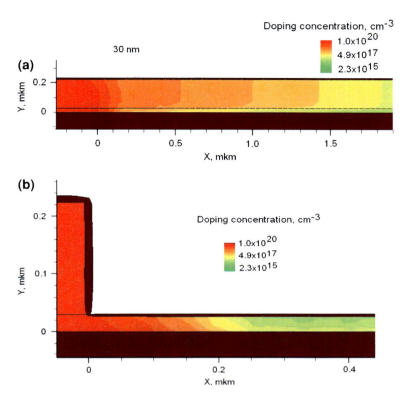

**Fig. 4** Simulation of donor distributions near the source and drain regions after P⁺ implantation and annealing at 950°C, 20 min in SOI layers with two thicknesses: **a** 200 nm, **b** 30 nm

1. deposition of a polycrystalline silicon film 250 nm thick by the low pressure chemical vapor deposition (LPCVD) method;
2. doping of the polycrystalline silicon film by phosphorus;
3. doping activation at 950°C;
4. deposition of an aluminum film;
5. post-metallization annealing at 420°C.

Individual SOI NWTs were insulated from each other by a silicon dioxide film obtained by tetraethoxysilane (TEOS) pyrolysis in the LPCVD process. Such film allows multiple exposure of biosensors in various solutions. A special fluid cell was developed for SOI NW FET tests to ensure a microflow of the biochemical fluid or gases to the SOI NW FET containing chip (Fig. 2).

Special treatment was used to stabilize the surface properties before and during the measurements in the atmosphere or/and in the different bioliquids. The free surface of NW FETs was treated by the following two types. In treatment 1, a solution $H_2O_2:H_2O = 1:4$ was used to remove organic contaminations from NW surface. The duration of this stage was a varied parameter. Then the native-oxide

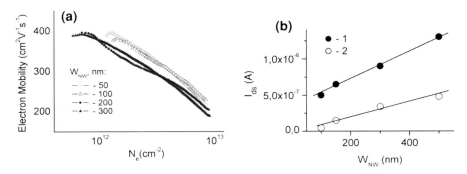

**Fig. 5 a** Mobility of electrons versus their concentration for different NW width, $W_{NW}$ (with permission from [8]); drain current dependence on the width $W_{NW}$ of NW channel: *b1* in the air; *2* into PBS solution

layer was removed from NW surface by treating this surface in aqueous solution of hydrofluoric acid, $HF:H_2O = 1:50$, during 30 s. After each stage in treatment 1, the surface was rinsed with deionized water. The second cleaning treatment, treatment 2, was as follows: first, the surface of NW FETs was treated during 4 min with Remover Shipley solution, conventionally used in microelectronics for removing organic contaminations from structure surfaces; this step was followed by treating the samples in acetone ($CH_3$–C(O)–$CH_3$) (for 1 min) and washing them in deionized water. The natural oxide was removed from the silicon surface by treating the samples for 30 s in $HF:H_2O$ solution ($HF:H_2O = 1:50$). Treatments of the NW surface in hydrogen peroxide for times not shorter than 3 min allow removal of organic contaminations after prolonged storage in air. With such treatments, reproducible stable states for subsequent treatment of this surface in hydrofluoric acid solutions can be achieved.

The drain-gate characteristics of the SOI NW FETs were measured using Lab-View© data acquisition board and software from National Instrument Corp. (USA). The measurements were performed either in air or in a liquid cell with deionized water or test buffer solutions with different pH values and protein contents. Other important details of measurements were presented already in our work [8].

The values of the effective low field mobility in nanowire structures with small thickness of silicon layer (25–50 nm) were calculated according to [7] and were equal to 400–600 $cm^2/(V\ s)$ for electrons in accumulation (Fig. 5). These values are comparable with the data, obtained for field effect transistor with high quality thermal oxide on volume silicon [7].

## 4 Operating Parameters of the Fluid Cell

The important parameters for the sensor operating are (1) the range of the operating temperatures, (2) the time of reading of signal, (3) the stability of signal and

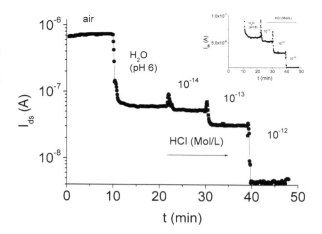

**Fig. 6** Time dependence of current (logarithmic and linear scale on the inset) for device with $w = 100$ nm and $t = 30$ nm measured on the air, into pure $H_2O$ (pH 6) and with the different content of HCl molecules (Mol/L)

(3) the sensitivity of detecting of the bio particles. Figure 6 shows the time dependence of current for devices with $w = 100$ nm and $t = 30$ nm measured in air, into $H_2O$ (pH 6) and the different HCl solution. The device demonstrates the high sensitivity to the environment ambient, to switching from one solution on another when system in air (equilibrium is installed during 2–2.5 min) and stability of characteristics.

The main problems in the formation of SNWT are to create high-quality-nanowires with low density of surface states, high mobility of charge carriers and low resistance ohmic contacts to ensure high sensitivity of the sensor elements. Novosibirsk Institute of Semiconductor Physics (ISP) has developed the process of SOI NW FET matrices with undoped silicon high mobility channels and the virtual gates of adsorbed substances for use them as a bio-chemical sensors with femto-mole sensitivity. The proposed routing using optical and electronic lithographies on few nanometer silicon layers based on the standard high temperature CMOS processes in the starting phase and the formation of perfect silicon nanowires without a silicon dioxide degradation parameters by RIE procedure at the last stage of the routing, provides low defect levels on surface and linear dependence of the conductance on the width of nanowire $W_{NW}$ even without additional thermal treatments (Fig. 5b).

The sensitivity of NW FETs to the HCl concentration (to the negative ions $Cl^-$) in aqua solution (pH 6) was shown to be as high as 10 femtomoles. Response of device to the BSA molecules is shown in Fig. 7.

The effective charge accumulated on the SOI NW surface ($\Delta Q_s$) after immobilization of BSA molecules was calculated from the flatband voltage shift ($\Delta V_{FB}$) of measured $I_{ds}$–$V_g$ curves (inset of Fig. 7) and the buried-oxide capacitance $C_{BOX}$ as $\Delta Q_s = \Delta V_{FB} C_{BOX}/q$ (here q is the electron charge).

The linear dependence between protein concentration and trapped charge is obtained at the lower concentration of BSA. The resolution on the voltage axis for $I_{ds}$–$V_g$ dependencies is much lower than 10 mV. It corresponds to the measured value of $\Delta Q_s = 10^{11}$ cm$^{-2}$ and determined by design parameters of SOI. Linear

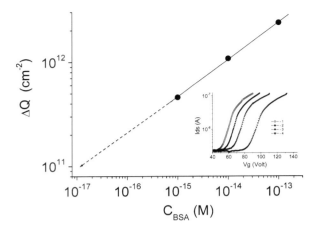

**Fig. 7** A charges accumulated on the SOI NW surface after immobilization of BSA molecules. The *inset* shows $I_{ds}$–$V_g$ curves measured for SOI NW FET with $w = 100$ nm, $t = 15$ nm at Vds = 0.15 V. Content of BSA molecules in the solution, $C_{BSA}$ (M/L): (1) 0 (before treatment), (2) $10^{-15}$, (3) $10^{-14}$, (4) $10^{-13}$

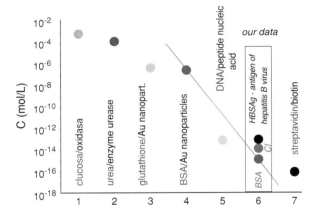

**Fig. 8** Sensitivity to analyte substances in bioliquids according to the published data in [1–7]

approximation of $\Delta Q_s$–$C_{BSA}$ curve on the $C_{BSA}$ axis allows to expect the limit of detecting as high as 10 attomoles for used NW FETs. Confirmed in our experiments sensitivity to the BSA molecules near 1 femtomole (about 600 molecules in 1 mm³ of liquid) is among the best world results in this area (Fig. 8). Our measurements show the ability to sense single cells like *E. coli*, or hepatitis B viruses, or liver cancer proteins in the specific reactions of immobilization by using this electronic biosensor. The last method is essential for early diagnosis and treatment of dangerous diseases. The operation of SOI nanowires from liquid He to 250°C temperatures allows to register adsorption–desorption kinetics for many gaseous substances too [9].

To increase the sensitivity and selectivity the sensor is covered (sensitized) by the special substance enlarging adsorbing rate of elected particle. An electronic chip with the array of elements, integrated control schema alerting sensitized

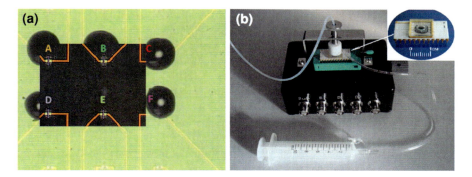

**Fig. 9** a Immobilization scheme of protein molecules in microdroplets (*inset* dark area with FETs activated by antibodies, see text); **b** microfluidic cell for delivering gases or bioliquids

management and microcell, ensuring gas or liquid flow serves as a universal electronic sensor platform, promising to use in various areas such as organic molecule detectors in bioliquid cell or toxic/explosives in the atmosphere. An example of using this biochip in clinic for multipurpose diagnostics by antibody/antigen specific reactions with immobilized proteins is presented in Fig.9, where each NW FETs are marked by A, B—antibodies for $\alpha$-fetoprotein detection (anti-AFPim); C—reference point; D, E—antibodies for surface antigen of hepatitis B virus HBsAg (anti HBS NE3im); F—antibodies for troponin (a marker of myocardial infarction). The results of measured sensitivity are also presented in Fig. 8. Similar biochips can be used for developing the universal platform of electronic biosensors for chemical and biological analysis for environmental, medical and private purposes.

## 5 Advantages of the SOI NWs Fluid Cell

The main advantage of the nanowires manufactured by lateral nanostructuring on the structure of silicon on insulator (SOI) is the use of the buried oxide (BOX) structures as a gate dielectric and the silicon substrate as a back-gate (BG) resulting in SOI nanowires the functions of nanowire field effect transistors (NW FETs). Accordingly, when registering adsorbed particles not only the conductance modulation of SOI nanowires is recorded, but their subthreshold transistor characteristics are matched to ensure exponential dependency of conductivity on the surface potential instead of linear one at simple surface conductor modulation. Another important advantage of NW FETs is their compatibility with standard CMOS technology that allows to create a lab-on-chip (LoC), integrating sensory elements with control and processing information schemes.

Developed routing has allowed creating transistor like nanostructures with undoped channel and high charge carrier mobility and heavily doped contact areas

(Fig. 2b). Minimum values of effective mobility in the nanowire structures with the thickness 10–50 nm and the width 20–100 nm are restricted by 400 and 100 cm$^2$/(V s) for electrons in accumulation and the holes in inversion, respectively. These values are comparable with the data for FETs with thermal oxide on bulk silicon.

The SOI nanowire transistors isolated in the chip matrix using protective TEOS allow multiple exposing of these SNWTs in the various solutions and buffer liquids and reproducibility of the measurements. In particular, we showed, that negatively charged HCl$^-$ ions in demonized (DI) water is detected in the concentration as low as $10^{-14}$ M with stationary conductivity established in the real time interval <2 min. Sensor design provides a reproducibility of the measurements for protein molecules of bovine serum albumin (BSA) in PBS solutions (KH$_2$PO$_4$) with pH = 7.4 (5 mM) in the interval of the concentrations $10^{-15}$–$10^{-4}$ M when nonspecific hostage molecules trap to the open silicon channel.

It means that electron lithography and gas–plasma RIE etching are suitable for manufacturing the silicon nanowire transistors, which are sensitive to the seizure–release of single organic molecules even without additional thermal treatments to remove the process violations. Confirmed sensitivity to BSA molecules near 1 femtomole (about 600 molecules in 1 microliter of liquid) is among the best world developments in this area and provides detecting a single viral particles and cancer cells, as well as toxic and pathogen substances in the specific reactions of immobilization by using this electronic nanosensor, that is essential for the safety, early diagnosis and treatment of dangerous diseases.

Thus, the developed in the present work technological process of SOI NW FET nanostructuring provides a maximum compatibility with the standard CMOS technology using nanowire formation on the last stage. The low-temperature and "low-defect" method of SOI nanostructuring based on gas–plasma etching and raised polysilicon contact formation with subsequent high dose ion doping allow to get the charge transport characteristics of SOI NW FETs like in the bulk silicon. On the basis of the elaborated processes the biochips with matrices of SOI nanowire transistors were manufactured and their femtomole sensitivity to the test organic and inorganic molecules was confirmed as powerful as the world's best results.

**Acknowledgments** The authors would like to thank M. A. Ilnitskii for Synopsis TCAD NW FET simulations. This work was supported in part by FCNTP (grant no. GK 02.740.11.0791).

# References

1. Road Map "Russian Human Proteome Project". http://www.ibmc.msk.ru/content/intelligence/RHUPO_RoadmapRus.pdf (2010)
2. Li, C., Curreli, M., Lin, H., et al.: Complementary detection of prostate-specific antigen using in o nanowires and carbon nanotubes. J. Am. Chem. Soc. **127**, 12484–12485 (2005)
3. Stern, E., Klemic, J.F., Routenberg, D.A., et al.: Label– free immunodetecting with CMOS compatible semiconducting nanowires. Nature **445**, 519–522 (2007)

4. Naumova, O.V., Safronov, L.N., Fomin, B.I., et al.: Silicon-on insulator nanowire transistors for medical biosensors. In: International Conference and Proceedings of EuroSOI-2009, pp. 69–70 (2009)
5. Popov, V.P., Antonova, A.I., Frantsuzov, A.A.: Properties of silicon-on-insulator structures and devices. Semiconductors **35**, 1030–1037 (2001)
6. Liang, M.S., Choi, J.Y., Ko, P.K., et al.: Inversion-layer capacitance and mobility of very thin gate-oxide MOSFET's. IEEE Trans. Electron. Dev. **33**, 409–413 (1986)
7. Chen, K., Wann, H.C., Dunster, J., et al.: MOSFET carrier mobility model based on gate oxide thickness, threshold and gate voltages. Solid State Electron. **39**, 1515–1518 (1996)
8. Naumova, O.V., Fomin, B.I., Nasimov, D.A., et al.: SOI nanowires as sensors for charge detection. Semicond. Sci. Technol. **25**(8), 055004 (2010)
9. Naumova, O.V., Popov, V.P., Safronov, L.N., et al.: Ultra-thin SOI layer nanostructuring and nanowire transistor formation for femtomole electronic biosensors. ECS Trans. **25**, 83–87 (2009)

# Sensing and MEMS Devices in Thin-Film SOI MOS Technology

J.-P. Raskin, L. Francis and D. Flandre

**Abstract** Silicon-on-Insulator (SOI) technology is emerging as a major contender for heterogeneous microsystems applications. In this work, we demonstrate the advantages of SOI technology for building thin-film field-effect biosensors and optical detectors, physical and chemical sensors on thin dielectric membrane as well as three-dimensional (3D) microelectromechanical (MEMS) sensors and actuators. The flatness and robustness of the thin membrane as well as the self-assembling of 3D microstructures rely on the chemical release of the microstructures and on the control of the residual stresses building up in multilayered structures undergoing a complete thermal process. The deflection of multilayered structures made of both elastic and plastic thin films results from the thermal expansion coefficient mismatches between the layers and from the plastic flow of a metallic layer. The proposed CMOS-compatible fabrication processes were successfully applied to suspended sensors on thin dielectric membranes such as gas-composition, gas-flow and pressure sensors and to 3D self-assembled microstructures such as thermal and flow sensors.

## 1 Introduction

Silicon-on-Insulator (SOI) technology has now been widely demonstrated and recognized to be a mature and viable alternative to mainstream bulk silicon (Si) for the integration of high-speed, low-power digital [1] and analog/RF [2] CMOS

---

J.-P. Raskin (✉), L. Francis and D. Flandre
Institute of Information and Communication Technologies, Electronics and Applied Mathematics (ICTEAM), Université catholique de Louvain (UCL), Place du Levant, 3, B-1348 Louvain-la-Neuve, Belgium
e-mail: jean-pierre.raskin@uclouvain.be

circuits, as well as niche applications under extreme high temperature [3] or radiation operating conditions [4]. Indeed, the use of SOI substrate is likely the best way to overcome the limitations of bulk technology. The insulating layer blocks charges transport between the active layer and the substrate, positioning this technology as the best one for radiation and high-temperature environments, like outer space, military or industrial and automotive sectors. In addition, the reduction of parasitic capacitances and leakage currents allow better high-frequency performances and lower power consumption when compared to silicon bulk process.

Recently, SOI technology has also been applied to Micro-Electro-Mechanical Systems (MEMS) [5, 6]. The buried oxide (BOX) present in the SOI material is successfully used as an etch-stop layer for wet or dry bulk micromachining, either from the back or from the front side of the wafer. For instance, a stacked BOX/silicon membrane can be obtained by releasing the silicon substrate from the backside of the wafer. The high quality of the top crystalline silicon film enables semiconductor devices such as temperature sensors, microheaters, metal oxide semiconductor field-effect transistors (MOSFETs), microwave circuits to be built within the membrane in order to improve thermal or electrical insulation. The membrane can also play the role of a structural support for mechanical sensors, such as pressure sensor for instance.

A brief overview concerning the most recent achievements with respect to SOI-CMOS devices and MEMS structures co-integration is given hereafter in this section. The combination of SOI-CMOS with MEMS is ideal in terms of lower power consumption of analog interface circuits and faster response of MEMS sensors. From the technology point of view, the BOX offers a natural etch-stop layer or intrinsic sacrificial layer when designing membranes, cantilevers or interdigitated capacitances. A wide variety of processes have been proposed for monolithic integration of electronics with micromechanical devices, either the MEMS process first, last or intertwined with the CMOS process. As it remains rare to find a fully integrated process at a single site, SOI-CMOS-MEMS are most of the time sequentially fabricated through separate foundries. This allows a selection of standard or tailored SOI substrates, whereas specific and tailored steps have to be performed in order to prevent the CMOS circuits to be damaged by the MEMS release or to prevent high aspect ratio MEMS to impair the planar processing of CMOS devices [7].

The MEMS processing of SOI-CMOS transistors provides a thermal insulation beneficial to uncooled infrared (IR) and TeraHertz (THz) sensors, to chemical gas sensors operating at high temperature and to flow measurements [8–12]. The BOX is a natural etch-stop for the backside removal of the silicon substrate by Deep Reactive Ion Etching (DRIE), thus SOI technology is preferred over regular CMOS technology since thermal insulation is achieved easily [9]. Transistors thermally insulated by MEMS processing become highly sensitive active bolometers while the performances of as-processed transistors are not degraded by the post-CMOS-MEMS processing [8]. Uncooled IR sensors and THz imagers have higher performances at a lower cost compared to state-of-the-art uncooled sensors based on bolometers implemented in non-CMOS materials like vanadium oxide or amorphous silicon [8, 9]. Subthreshold operation of the MOS transistors is preferred since the conduction mechanism at that regime is based on diffusion, which is very sensitive to temperature variations [9]. Still based on DRIE, obtained thin membranes can be used for smart gas sensors operating at high temperatures (>300°C) [10, 11]. The high temperatures are obtained by microheaters embedded in the membrane using p-type MOSFETs, polysilicon or Pt resistors. The in situ temperature measurement is then done with differential transducing circuits [10]. A fully integrated gas sensor with automatic temperature reading was demonstrated with a power consumption of 57 mW at 300°C and a thermal response of 10 ms [10]. The same sensing platform integrated with zinc oxide nanowires was recently demonstrated for hydrogen sensing with a power consumption of 16 mW at 300°C [11]. An array of SOI-CMOS micro-thermal shear stress sensors was used for flow measurements at macro-scale [12]. The flow sensor was fabricated with a commercial 1-μm SOI-CMOS process and a post-CMOS DRIE back etch, the CMOS aluminum metallization served as sensing material. Small sensors ($18.5 \times 18.5$ μm$^2$ sensing area on $266 \times 266$ μm$^2$ oxide membrane) and large sensors ($130 \times 130$ μm$^2$ sensing area on $500 \times 500$ μm$^2$ membrane) feature low power (100°C temperature rise at 6 mW) and a small time constant of only 5.46 and 9.82 μs for the small and large sensing areas, respectively. These devices are robust under transonic and supersonic flow conditions and successfully identified laminar, separated, transitional and turbulent boundary layers in a low speed flow [12].

The monocrystalline layer of the SOI substrate offers a structural layer for mechanical transduction, such as cantilevers, with performances independent from the CMOS technology. Resonant cantilevers integrated in a CMOS chip have a high mass sensitivity reported in the range of 8 ag/Hz, used to determine e.g. the time-resolved evaporation rate of glycerine drops in the attoliter range [13, 14].

The monolithic co-integration of SOI-CMOS with MEMS is also of interest for accelerometers using interdigitated capacitive transducers with a thickness of 10 μm and above [15–17]. Analog devices demonstrated a monolithic high-g acceleration sensor (ADXSTC3-HG) to measure in-plane acceleration up to $10,000g$ while subjected to $100,000g$ in the orthogonal axes, and packaged in an eight-pin leadless chip carrier [16]. Special tricks have to be adopted in the processing of both parts when a thick SOI layer is used, such as trench isolation [15] or micro-bridge interconnections [17]. Such examples show that after adaptations the fabrication technology can be applicable widely to high-dense integration of monolithic CMOS-MEMS on thick SOI wafers with capacitive transducers, or similarly that Multi-Users Multi-Chips modules for SOI-CMOS-MEMS can be developed successfully for a wide set of applications [17, 18].

Besides sensing applications, SOI RF-MEMS compatible with CMOS and high-voltage devices have been proposed for system-on-a-chip applications integrating for instance RF-MEMS switches with driver and processing circuits for single-chip communication applications. SOI high-voltage, CMOS devices and RF-MEMS capacitive switches were fully integrated with high-resistivity SOI substrates, showing an insertion loss of 0.14 dB and isolation of 9.5 dB at 5 GHz [19]. Still in the signal filtering domain, 22–25 MHz bulk lateral resonators with 100-nm-wide air gap on partially depleted SOI with quality factor as high as 120,000 were demonstrated under vacuum at room temperature [20].

In this chapter, based on several experimental research results on the material, process and component properties, we aim at demonstrating in Sects. 2 and 3, 4 and 5 how field effects into thin monocrystalline silicon film, bulk and surface micromachining techniques, respectively, can be used for the fabrication of fully integrated microsensors and actuators in SOI-CMOS technology. The co-integration issue of the microstructure with its associated electronics will also be addressed. Through this chapter the residual stress in thin films is a key parameter not only associated to the reliability of suspended sensors on thin membrane but also as a tool for assembling original three-dimensional (3D) MEMS.

## 2 Nano-Interdigitated Array Gate Thin-Film SOI MOSFET for Biosensors

Label-free electronic biosensors are crucial components for the seamless integration of proteomics with information technology. Electronic biosensors have strongly benefited from the development and application of nanotechnologies. Among recent proposed devices, nanowire sensors have attracted much attention because the large surface-to-volume ratio of nanowires potentially increases sensor sensitivity to the single-molecule detection level [21–29]. Nano-interdigitated arrays (nIDAs), employed as capacitance and impedance sensors, also show promises of high sensitivity [30–34]. However, their reliability is limited by any conduction path (for example a nanoscale dot) between two combs, and their sensitivity is deteriorated by high ion concentrations, such as typically found in biological solutions. On the other hand, ion-sensitive field-effect transistor, in which the gate of a MOSFET is replaced by an ion-sensitive membrane, serves as more macroscopic biosensors. But this type of sensor suffers from a low sensitivity and needs to integrate a reference electrode [35].

In [36], we propose and demonstrate a new protein sensor by opening nano-trenches in the gate electrode of a conventional SOI MOSFET to allow biomolecules to get access to the surface of the gate oxide, thereby effectively creating a nIDA-gate MOSFET. The binding reaction of antibody Ixodes ricinus immuno suppressor (anti-Iris) protein [37, 38] to its corresponding antigen (Iris) is used as test system to establish the sensitivity and specificity of our sensor. When the anti-Iris antibody binds the Iris antigen attached to the fingers of the nIDA, the same as the case of the nIDA capacitance sensors, the total gate capacitance is increased, which is monitored by the enhancement of the drain current in the MOSFET. Most importantly, the increase of net positive charges existing in the protein layer results in a negative shift in the threshold voltage, which also gives rise to the significant enhancement of the drain current. This makes the present sensor more sensitive than nIDA capacitance sensors. The nanotailored gate provides more available binding sites for the antibody molecules compared to dielectric-modulated field-effect transistors with a vertical or planar nanogap [39, 40]. Besides, for a given antibody–antigen pair, the sensitivity of the present sensor can be improved by optimizing the geometrical parameters of the MOSFET. Moreover, this sensor with real-time and label-free capabilities can easily be used for the detection of other proteins, DNA, viruses and cancer markers.

## 2.1 Sensing Principle

The nIDA-gate MOSFET sensor has to be seen as an integrated transducer which converts the protein binding process directly into a measurable current without any labeling. The sensor is composed of a source and a drain connected by a silicon channel capacitively coupled to a nIDA-gate (Fig. 1a). For the sake of clarity, we first focus on two fingers only (Fig. 1b). One is used as the control gate while the other remains floating. The drain current in the silicon channel linearly depends on the total gate capacitance, which consists of two capacitances in series: the gate oxide capacitance ($C_{ox}$) and the coupling capacitance ($C_c$). The coupling capacitance is the sum of the inter-finger capacitance ($C_{fin}$) and the fringing capacitance ($C_{frin}$). When antibody binds to a specific antigen attached to the finger's surface of the nIDA-gate, the $C_c$ enhancement caused by the variation of the dielectric parameters (such as dielectric permittivity and protein layer thickness) promotes an increase of the drain current in the sensor. On the other hand, it is known that the protein carries positive charges [41, 42], thus with the binding of the antibody,

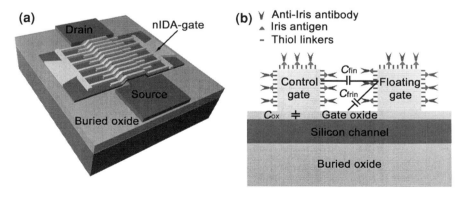

**Fig. 1** Schematic nIDA-gate MOSFET sensor. **a** Three-dimensional structure showing source and drain regions connected by a silicon channel (*blue*) capacitively coupled with a nano-interdigitated array gate (*yellow*). **b** Cross-section of two fingers, showing the successive binding of the thiol linkers, Iris and anti-Iris biomolecular layers, the gate oxide capacitance ($C_{ox}$), finger capacitance ($C_{fin}$) and fringing capacitance ($C_{frin}$)

the net positive charge attached on the surface of the nIDA-gate is increased. This is analogous to increasing charges in the oxide of conventional nMOSFET, which results in a negative shift in the threshold voltage, thereby increasing the drain current as well. Conclusively, the drain current increase, caused by the enhancement of the total gate capacitance and the threshold voltage shift, provides a direct detection for the binding of complementary antibody–antigen pairs.

## 2.2 Sensor Fabrication Process

The sensors are fabricated on p-type SOI wafers. The silicon body of the sensor, including source, drain and channel regions, is defined by UV lithography and Reactive Ion Etching (RIE). The source and drain regions are heavily doped by Arsenic. A 33-nm-thick $SiO_2$ is thermally grown to form the gate oxide and isolate the sidewalls of the silicon body. Then, the gold nIDA-gate is built by e-beam lithography and lift-off process. Generally, any metal, even polysilicon, can be used as gate material. However, gold is selected here because it allows using self-assembled monolayer (SAM) thiol as a linker [43, 44] to anchor proteins. The bare nIDA-gate has a finger width of 100 nm and repeat period of 200 nm. Finally, to eliminate any interference and protect the electronic leads from being chemically etched or damaged during sensor tests, the area outside of the nIDA-gate is covered with a layer of PolyMethylMethAcrylate (PMMA). Thus, only the nIDA-gate is exposed to the environment during sensor testing.

After fabrication, the cleaned sensor is immediately immersed in a 2 mM ethanol solution of 11-mercaptoundecanoic acid (11-MUA, 95%) for 24 h to form SAM linkers over the surface of the gold nIDA-gate. The carboxylic acid groups of the SAM are then activated by immersion in an aqueous solution of N-hydroxysuccinimide (0.05 M NHS, 98+%) and 1-(3-dimethylaminopropyl)-3-ethylcarbodiimide hydrochloride (0.2 M EDC, 98+%) for 30 min, followed by rinsing in water [45]. The antigen is then covalently grafted onto the sensor by immersing the activated surfaces for 1 h in an aqueous phosphate-buffered solution of the Iris antigen (1 mg/ml in PBS). At that stage, the sensor is ready to detect the targeted anti-Iris antibody. To ensure saturation of the binding biomolecules, the sensor is immersed in anti-Iris solutions of varying concentrations for 1 h. The sensor is subsequently rinsed with distilled water and blown dry with nitrogen.

## 2.3 Current Response

The drain current as a function of gate voltage ($I_{DS}$–$V_{GS}$) characteristics of the sensors are then measured by using HP4156C in air. Figure 2a plots the $I_{DS}$–$V_{GS}$ characteristics of a typical sensor, in the bare state, after grafting the thiol linkers,

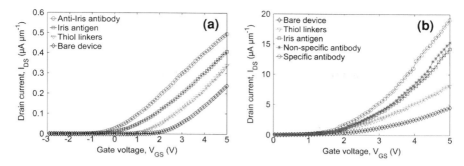

**Fig. 2** Experimental current responses during protein detection. $I_{DS}$–$V_{GS}$ characteristics for **a** a typical sensor, in the bare state, after grafting the thiol linkers, anchoring the Iris receptor, and binding the antibody (0.36 ng ml$^{-1}$), and **b** in the cases of non-specific and specific antibody bindings

anchoring the Iris receptor, and binding the antibody from a 0.36 ng/ml anti-Iris solution. For a given gate voltage, the drain current in this sensor is markedly increased after each step. Particularly, after grafting the thiol linkers, a −0.6 V shift in the threshold voltage ($V_{th}$) is observed, followed by a shift of −1.4 V when the Iris receptor is anchored with the thiol linkers and finally $V_{th}$ shifts by −0.3 V after binding the antibody. According to the simulation results presented in [36], we suggest that the high enhancement of the drain current is mainly attributed to the net positive charge increase in the biomolecular layers, which results in a high current response in the present sensor. This is an advantage of our sensor over the nIDA capacitive sensors.

The selectivity of the sensor is assessed by replacing the anti-Iris antibody solution with a highly concentrated (2 mg/ml) solution of non-specific antibody 286F7, an antibody directed against the thyroid stimulating hormone (thyrotropin) [46]. A very small current change is observed in Fig. 2b. Then a droplet of anti-Iris antibody with a concentration of 360 ng/ml is positioned onto the same device, a significant current change is obtained. This sensor has 45-nm-high, 50-nm-wide floating gate, 80-nm-wide control gate, 80-nm-wide inter-finger distance, and 30 μm × 25 μm detection area. The significant current change is attributed to the specific binding between the anti-Iris antibody and its antigen. However, it should be noted that the small current change in the case of the non-specific antibody is due to the fact that a few thiol linkers without binding antigen may attract the positive charges existing in non-specific protein or the positive ion present in the buffer solution. Even so, the concentration of the non-specific antibody is much higher than that of the specific antibody. Therefore, we can conclude that the present sensor has a good selectivity.

Those experimental results show that conventional devices can be turned into high sensitivity nano-biosensors simply by redesigning their components at the nanometer scale. The reason for the high sensitivity of the sensor is the modulation of the drain current not only by the dielectric parameter changes of the biological layer but also by the changes of charge concentration originating from the biological layer. The sensor is able to detect the binding reaction of a typical antibody (anti-Iris) at a concentration lower than 1 ng/ml. The detection limit can be strongly improved by optimizing the geometrical parameters of the nIDA-gate MOSFET, such as finger width and height, inter-finger distance, detection area, channel width and thickness as well as gate oxide thickness. The sensor exhibits a reproducible specific detection for anti-Iris antibody and a high selectivity. Furthermore, the sensor fabrication is mostly compatible with CMOS technology as the nIDA-gate is added by using post-processing steps. This novel nano-biosensor is thus expected to find widespread use for the integration of biotechnology with information technology.

## 3 Blue/UV Sensors in Thin-Film SOI

Optical detection at low wavelengths close to blue and ultraviolet (i.e. for $\lambda < 480$ nm) is commonly used in biomedical and environmental fields. For example, DNA concentration measurement, bacteria or protein detection require optical light whose emission wavelength is below $\lambda = 300$ nm. On the other hand, UV and ozone rates are measured in upper wavelengths close to $\lambda = 400$ nm. So, there is a real need for miniaturized sensors able to detect in this wavelength range. In addition, such applications require low levels of detection, due to the low power of the emitted optical signals to monitor. It is thus of importance to minimize photodiode dark current and to achieve high optical responsivity in order to reach a detection sensitivity as high as possible.

The optical responsivity of a photodevice can be defined as:

$$R = \frac{I_D - I_{Dark}}{P_{opt}} \qquad (1)$$

where $I_D$ is the total current flowing through the device, $I_{Dark}$ is the dark current and $P_{opt}$ is the incident optical power.

Whereas photodetectors in (Al)GaN-based technologies can target high optical responsivity below $\lambda = 300$ nm thanks to their appropriate bandgap [47], silicon devices absorb light as a function of their thickness up to wavelengths close to 1,200 nm. A solution to specifically and efficiently absorb low wavelengths with silicon-based technologies is to implement the sensors in thin silicon layers based on SOI technology [48]. Thin-film SOI structures are more and more used due to their insulating properties for integrated circuits (ICs). This provides the opportunity to implement a complete system-on-a-chip and is a way to fabricate a low-wavelength photodetector with a low-cost process. As this type of microsystem can be useful for complex manipulations in biomedical labs, there is an opportunity to replace existing large-sized and expensive apparatus with portable photosensors.

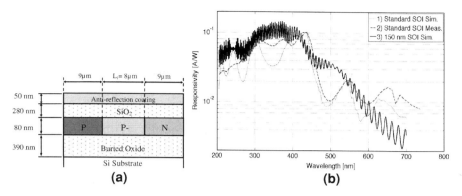

**Fig. 3 a** Cross-section view of an interdigitated PIN lateral photodiode with typical dimensions for the SOI technology at UCL. **b** Simulations and measurements of spectral responsivity of the lateral PIN photodiodes for two different SOI technologies

In [49], the influence of the fabrication process parameters on light absorption of lateral PIN photodiodes designed and characterized in various SOI technologies is discussed. The standard SOI-CMOS technology available at Université catholique de Louvain (UCL) features a 80-nm-thick top silicon layer which is optimum for the absorption of light emission below $\lambda = 400$ nm. The thin Si film is separated from the silicon substrate by a 400-nm-thick BOX. A CMOS process silicon oxide of 280 nm, which consists in a gate oxide of 30 nm and an interdielectric oxide of 250 nm, covers the photodiode. In addition, an anti-reflection coating (ARC) of 50 nm is deposited on the top of the wafer. This ARC can be either silicon nitride ($Si_3N_4$) or alumina ($Al_2O_3$) since their refraction index is quite similar. The design of a PIN diode always faces the trade-off between a small intrinsic length providing an increase of the photogenerated electron–hole pairs collection but also a considerable increase of the dark current and a larger intrinsic length that reduces this dark current but increases electrons–holes recombination and thus decreases the photogenerated current through the device. Based on the specification of the UCL SOI technology, an intrinsic length of $L_i = 8$ μm is chosen. The other lateral dimensions, i.e. anode and cathode lengths, respectively, $L_n$ and $L_p$, have to be as short as possible to maximize the photosensible area of the device but their size is fixed by the technology, which is in our case: $L_n = L_p = 9$ μm taking into account the contact and metal interconnects.

The responsivity of the SOI PIN photodetector is measured by steps of 10-nm wavelength with optic fiber direct illumination from a high power light source. The optical characterizations of the fabricated devices presented in Fig. 3 (curve 2) agree quite well with the simulation results (curve 1). The differences below $\lambda = 400$ nm may be explained by process variations (e.g. slight thickness variations of the oxide and ARC over the total area of the photodiode). High responsivity values in UV, close to $R = 0.1$ A/W with a spectral cut-off wavelength of $\lambda_c = 450$ nm and an attenuation of 90% over 50 nm, are obtained. Measured dark current is as low as $I_{Dark} = 10$ pA and corresponds to the simulated value for a photodiode with a total surface area of 1 mm$^2$ and a reverse bias of $V_d = -0.5$ V.

Based on the same multilayered structure (i.e. using similar material layers stacked in the same order) of PIN diode, a new design based on an industrial 150 nm fully depleted (FD) SOI-CMOS technology [50] has been investigated. Featuring a Si film thickness of 50 nm, our simulations (curve 3 in Fig. 3) indicate higher responsivity values thanks to an increase of the sensitive area (the anode and cathode lengths are reduced to less than 1 μm) and the cut-off wavelength lower than $\lambda_c = 400$ nm due to the reduction of the silicon film thickness and the attenuation of 90% is now achieved over 200 nm because of the thicker top oxide for that industrial process. Another important improvement would be achieved by reducing the front silicon oxide thickness while keeping the 40-nm-thick silicon nitride ARC. It would reduce the oscillations in the bandwidth and considerably increase the responsivity values.

Starting from a standard SOI wafer, the responsivity of the photodetector can be improved by adding specific layers (anti-reflection) as mentioned above but also by etching parts of detector stacked layers (BOX) as proposed in [49]. Indeed, SOI technology can be extended to the so-called Silicon-on-Nothing (SON) technology [51] by etching a part of the buried silicon dioxide. Etching an air cavity of 250-nm-deep below the intrinsic channel of the SOI PIN photodiode improves its responsivity at $\lambda = 400$ nm [49]. Moreover, the air cavity can also be used for special applications, i.e. in introducing particles or gases in the cavity for a detection based on the refraction index variation.

The designed and fabricated SOI PIN photodetectors have been implemented with a current-to-frequency converter [52] and in a biomedical DNA detection system [53]. The aim of the second system is to measure DNA, in closed tube containers, at very low concentration levels. The photodiodes are directly illuminated with a 280 nm ray of light and the tube container is placed along the direct optical path, between the light source and the photodiodes. With natural DNA absorption peak at around 280 nm wavelength [54], we can expect a monotonic relation between the DNA concentration and the induced photocurrent (e.g. a lower photocurrent for larger DNA concentration, and a large photocurrent for the smallest DNA concentrations). Experimental results presented in [53], using a wide range of DNA concentration (from 400 to pg/μl), demonstrate higher efficiencies with respect to commonly used methods or apparatus in a biology and clinical labs.

In [55], the temperature influence on the performance of a lateral SOI PIN photodiode when illuminated by low-wavelength optical illumination, in the range of blue and UV, has been investigated. Experimental measurements were performed from 100 to 400 K and two-dimensional numerical simulations were conducted. Experimental results show that the current generated by these wavelengths incidence reduces both at cryogenic and moderately high temperatures, reaching a maximum value around 250–300 K while further temperature increase causes another current rise. Two-dimensional numerical simulations show that the current reduction observed at low temperatures is caused by the bandgap narrowing, whereas its increase at high temperatures is related to the doping and temperature dependence of carrier lifetime.

## 4 Bulk Micromachining: SOI-MEMS Sensors on Thin Membranes

In the recent years, dielectric membranes have been used in microsensors fabrication. The main interests of such membranes come first from their thermal insulating properties which are particularly adequate to reduce the electrical consumption in gas and flow sensors [56–58] or increase the sensitivity in IR sensing applications [59]; second, they take part in the sensing mechanism such as for ultrasonic transducers [60] or pressure sensors applications.

Furthermore, co-integrated sensors appear being particularly advantageous. Indeed, the sensor and its associated electronics can be placed on the same die increasing the signal-to-noise ratio and the integration level of the sensor, its miniaturization and drastically reducing its packaging costs. Beyond its well known electrical advantages, SOI technology appears particularly suitable for MEMS applications and especially for membranes fabrication. Indeed, the BOX constitutes a natural back-etching stop layer, allowing the release of thin and stress-controlled membranes. Moreover, the released thin monocrystalline silicon–BOX stack corresponds to the mechanical membrane with the possibility to co-integrate the transducer part into the active thin silicon layer. The interest for SOI technology in the fabrication of suspended sensors on thin dielectric membranes is demonstrated at the end of this section.

### 4.1 Fabrication of Thin and Flat Dielectric Membranes

The membrane supporting suspended sensors has to combine low thermal conductivity (i.e. small thickness) with high mechanical strength (i.e. large thickness) [61]. While insuring compatibility with a FD SOI-CMOS process, of interest for micropower or high-temperature applications [62], the 400-nm-thick buried thermal oxide of the SOI substrate can advantageously constitute the first part of the membrane. A second part of the membrane stack is composed of the densified Plasma Enhanced Chemical Vapor Deposition (PECVD) oxide layer related to the interconnect dielectric between the polysilicon and aluminum layers of the CMOS process. For mechanical robustness, a nitride layer must be added and the thickness must be chosen such as to compensate for the high residual stresses present in the deposited films. The thermal oxide usually shows a compressive residual stress ($\sigma_{ox}$) equal to 250 MPa while the film of nitride undergoes a tensile residual stress ($\sigma_{nit}$) equal to about 1.2 GPa. The total stress of the sandwich layer is calculated by:

$$\sigma_{tot} = \frac{t_{ox}\sigma_{ox} + t_{nit}\sigma_{nit}}{t_{ox} + t_{nit}} \quad (2)$$

where $t_{ox}$ and $t_{nit}$ are the thicknesses for the oxide and nitride, respectively.

Equation 2 shows that the ratio between the two thicknesses should be equal to 4 to generate a flat membrane with zero bending moment. However, in practice, the best choice is to take a ratio equal to about 2 to have a slightly tensile membrane and to take into account its high critical lengthening [63]. In practice, based on stress measurements [64], we selected thicknesses of 300 nm for the Low Pressure Chemical Vapor Deposition (LPCVD) nitride layer and 290 nm for the PECVD oxide layer for a total thickness of around 1 μm with low overall residual stress (80 MPa) and low stress gradient (0.78%).

The sensor fabrication process is optimized in order to be integrated in intermediate and post-processing of a standard IC SOI-CMOS fabrication. As presented at the end of this section for the case of a pressure sensor, in comparison with the standard SOI-CMOS process, only two more masks are needed for the intermediate processing: the first one for patterning silicon on the membrane location and the second one for etching the nitride layer.

## 4.2 Integrated Microheater

As will be shown later in this section, thin dielectric membranes can be useful in the design of high performance gas and flow sensors based on a suspended microheater. The design chosen for the heater is a loop-shaped phosphorous doped polysilicon resistor (Fig. 4) optimized based on finite element simulations (ANSYS software) in order to achieve low-power consumption and high thermal uniformity. A 340-nm-thick phosphorous doped polysilicon is used for the heater

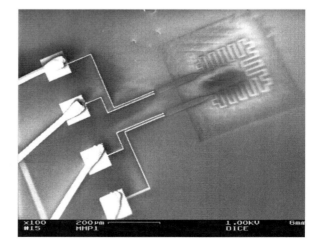

**Fig. 4** Packaged polysilicon micro-heater on 440 × 440 μm² membrane. On-chip aluminum connections (*bright lines*) and gold wires to the package can be seen

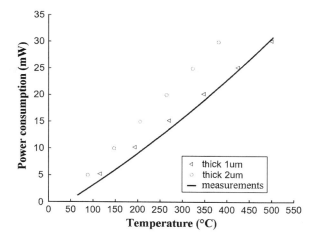

**Fig. 5** Influence of the membrane thickness on the microheater power consumption versus membrane temperature (*open triangles* and *open circles* are simulations results)

because it is the common material and thickness for a gate in CMOS and SOI-CMOS fabrication. The doping level is imposed by the IC process to get a resistivity of around 25 Ω/square. This value is a good trade-off between thermal uniformity and low-power consumption [65].

The membrane dimensions have also a strong influence on the microheater performance. Increasing its thickness by a factor of 2 increases the power consumption by about 30% at 400°C (Fig. 5), whereas using a membrane thinner than 1 μm leads to a decrease in power, thermal uniformity and also in membrane robustness. On the other hand, large surface membranes lead to high thermal insulation but are also detrimental to robustness. The active area of our membranes can be heated to working temperatures equal to 400°C with 25 mW only, which is a much better result in comparison with other published microheaters [66, 67]. Thermal uniformity was confirmed by thermo-reflectometry measurements [68]. Thermal ageing and reliability tests (deformation, high-frequency switching and high temperature) were performed in order to validate the high reliability of the membrane.

## 4.3 Gas Sensors

This last decade, most of the commercially available metal oxide sensors are based on screen-printing onto ceramic substrates. As the sensing mechanism of these devices relies on chemisorption reactions that take place at the surface of the metal oxide, the devices must be heated at temperatures of approximately 300°C, in order to enhance the reaction. As a result, the power consumption of commercially available devices is too high for portable or hand-held applications. Additionally, ceramic substrates have a high thermal inertia, which makes it difficult to modulate the operating temperature of the sensors. Attempts to join these high-temperature

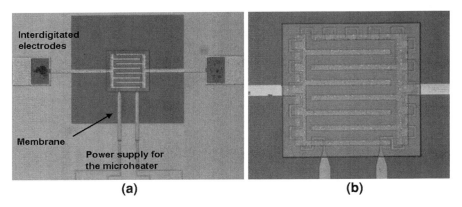

**Fig. 6** Top view images of a gas sensor integrated on top of thin membrane using a SOI wafer as starting substrate: **a** the sensing area and connections for the loop-shaped polysilicon microheater and the interdigitated Ti-Pt electrodes, **b** zoom-in on the gas-sensor active area

materials with a standard silicon process have only been moderately successful [69]. Silicon resistive sensors operating at these temperatures generally suffer from reduced thermal stability and durability. Gardner et al. [70] have designed sensors using composite structures with a platinum resistive heater embedded in low stress silicon nitride with electrodes exposed for metal oxide deposition. Other sensors based on nitride or oxy-nitride, membranes including micro-calorimeters have been reported [71–75]. Although these sensors have shown exceptional thermal stability as well as the durability to withstand thermal cycling, the process is not fully CMOS compatible, and so suffers from a higher cost to produce and does not present the possibility of circuit integration. Resistive heaters based on a polysilicon resistor have been reported, even though these devices have a tendency to suffer from two major short falls. Specifically, the high doping level required for the polysilicon resistors puts significant stress into the membrane, and they show poor long-term thermal stability. Other type CMOS-compatible sensors have been successfully fabricated by Suehle et al. [76] through a commercial foundry (MOSIS). These sensors are based on aluminum oxide micro-hotplates and polysilicon heaters. Aluminum tends to limit the operation temperature due to electromigration and this leads to increased power losses.

In [77], the design of a micro-hotplate gas sensor based on SOI technology has been demonstrated. After the complete compatible SOI-CMOS process to fabricate the loop-shaped polysilicon microheater (Fig. 4) and closed electronics, titanium (10 nm)–platinum (200 nm) interdigitated electrodes are evaporated and patterned by lift-off on oxide film (Fig. 6). Evaporation followed by lift-off is an easier technique and more economical than sputtering but produces X-rays which shift the threshold voltage of the CMOS transistors. By annealing the wafer at 432°C for 30 min under forming gas, the trapped charges in the gate oxide are removed and the threshold voltage returns to normal. The gap width between interdigitated electrodes is critical to achieve a good sensitivity. The height of the electrical field lines will depend on this gap width: a thicker sensitive layer will require higher electrical field lines to sense up to the layer surface, and therefore a wider gap. The gap width can also have an impact on the selectivity between gases. Finally, high resistance values (i.e. wide gap and low fingers number) are required when high temperature need to be reached since the resistivity of the gas sensitive layer decreases with the temperature. The membrane is finally released in a single post-processing step, consisting of a backside bulk micromachining with a tetramethylammonium hydroxide (TMAH)-based etching solution.

Two kinds of sensitive layers are deposited using two different techniques. Well-known sputtering is the more common technique but allows to deposit only very thin layers (around 0.6 μm). Drop coating and screen-printing techniques are more interesting since they offer a better sensitivity to gases thanks to their higher thicknesses (between 2 and 20 μm for screen-printing and around 10 μm for drop coating).

Using a standard RF sputtering technique [78], sensitive layers of tin ($SnO_2$) and tungsten ($WO_3$) oxides are deposited by lift-off on the membranes after the wafer backside etching, as the last step before the dicing and encapsulation. As described in Sect. 4.1, the membranes yield is not damaged by the photolithography thanks to the high membrane robustness. The drop coating is performed on other sensors after their packaging and wire bonding using an electronically controlled system formed by a pneumatic injection part connected to a syringe. The pastes are prepared from tin and tungsten nano-particles and an organic solvent. Once the deposition is made the sensors are left for 10 min for leveling and after that are dried in an oven for 5 min at 150°C. The firing process is performed in situ using the sensors heating element at 400°C during 24 h in order to completely eliminate the organic vehicle and to obtain adhesion of the active layer to the substrate.

A series of tests are conducted by introducing all contaminants into the test chamber using dry air as carrier gas and monitored the resistance of the sensors during the measurement. The gases under test are carbon monoxide, nitrogen dioxide, ammonia, ethanol, acetone, and methane. During the measurements the relative humidity is kept between 10 and 15%. The sensors detect the volatiles (observe the sharp decrease in sensor resistance represented in Fig. 7) and soon after the exposure, their resistance increases towards the baseline value. This shows that after the volatiles have crossed the sensor chamber, the sensors are in the presence of clean air again. Sensors operate at three different temperatures (150, 200 and 250°C) set by applying a current to their heating resistor (polysilicon microheater). The sensitivity of the sensors is defined as $R_a/R_g$, where $R_a$ is the resistance in dry air and $R_g$ is the resistance in presence of gas. The results show that the sensors sensitivity increases with the operating temperature. The tested sensors show high selectivity to ethanol, ammonia and $NO_2$. Both type of active layers ($SnO_2$ and $WO_3$) deposited with either RF sputtering or drop-coating, are sensitive to ethanol vapor (Fig. 8) [77]. At the same time,

**Fig. 7** Time response to 10 ppm ethanol of the SnO2 drop-coated sensors

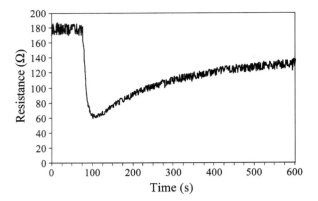

**Fig. 8** Sensitivity of sputtered and drop-coated $SnO_2$ sensors to ethanol at various working temperatures from 150 to 250°C

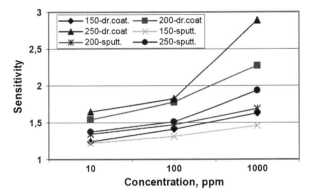

the drop-coated tungsten trioxide sensors show good response to ammonia and nitrogen dioxide [77].

The sensor response time ranges between 30 and 45 s, depending on its operating temperature. It is defined as the time it takes to reach 90% of its steady value after exposure to a gas. Generally, the sensor response is faster when operated at a higher temperature. The response of the sensors to variations in the ambient moisture was also studied in [77]. It was observed that the sensor resistance decreases when the relative humidity is increased. Furthermore, when the sensors operate at higher temperatures, the change in their resistance caused by a change in the moisture level is higher. It has been demonstrated that the devices with drop-coated active layers are more affected by changes in the moisture level than the sputtered ones. These results are in accordance with previous studies on gas sensors with thin and thick active layers [79]. However, the response to moisture can be considered as moderate, when compared to the sensitivities found for the different vapors studied.

## 4.4 Flow Sensors

Based on the suspended microheater presented in Sect. 4.2 a calorimetric-based flow velocity sensor can be designed by integrating thermopiles (such as Al/PolySi) close by the heating area along the gas flow (Fig. 9). Many gas-flow sensors reported in literature are said "CMOS compatible" [80] but most of them use layers or processing steps which are not inherent in standard ICs fabrication. Furthermore, only a few have been actually produced following standard CMOS processes. The originality of our design is that it is compatible with a standard SOI-CMOS process, on the same wafer than ICs. On the other hand, many different flow sensing principles can be found in the literature, leading to a difficult choice between low-power consumption, high airflow rate range, sufficient flow rate sensitivity, and compatibility with IC processes. Our sensor challenges most recent realizations by providing attractive trade-offs between these parameters for a large range of applications.

Measurements in the presence of nitrogen flow reveal a good sensitivity on a large airflow rate range (0–8 m/s) for the flow transducer (Fig. 10). In comparison with published results, our design offers the advantage to reach high temperatures at very low-power consumption, as well as a good response time (25 ms), which contributes to provide a high sensitivity to flows [81]. In comparison with the CMOS-compatible sensors published in the literature, our devices based on a simple design and process provide good results owing to the high performance of the 1 μm thin dielectric membrane.

## 4.5 Pressure Sensors

While thin ONO membrane from a SOI substrate was first designed in the gas and flow sensing context for its effective thermal isolation, it provides quite interesting mechanical properties for building pressure sensors as demonstrated in this section.

In order to sense the mechanical deflection of the membrane under pressure, beside the more classical passive resistors mounted in Wheatstone bridges [82–85], novel active structures have been developed and already reported [86–89]. These inherently co-integrated solutions are using transistors mounted in ring oscillators placed on thick silicon membranes, as pressure sensitive elements. Offering a frequency output, such systems give an advantageous alternative to analogue output signal sensors, by easing their interfacing, reducing their noise sensitivity and allowing easy non-linearity compensations by the use of a microprocessor. In this section, we present Voltage Controlled Oscillators (VCOs) in SOI technology, directly implemented on a thin and released dielectric membrane for sensing mechanical pressure.

The samples followed a standard SOI-CMOS process with a very few extra fabrication steps for the definition and building-up of the released dielectric

**Fig. 9** Calorimetric flow sensor with one direction of detection

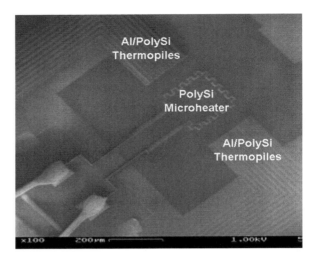

**Fig. 10** Measured differential voltage across the thermopiles as a function of the applied flow velocity for two membrane sizes

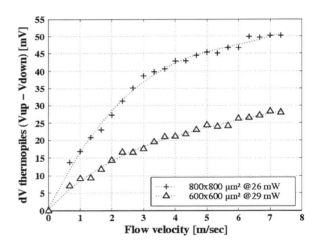

membrane. Figure 11 presents the different fabrication steps for building a thin SOI pressure sensor with its electromechanical transducers and ICs directly integrated on top of the released thin membrane. A $\langle 100 \rangle$ Unibond SOI wafer, with a 100 nm top thin silicon film and a 400-nm-thick BOX, is 25 nm oxidized to bring a protection to the Si film during the first steps of the process. A photolithography is performed and the active film is removed from the future membrane areas with a $SiCl_4$-based RIE etching (Fig. 11a). Next, the active zones are defined by a patterned LPCVD nitride film which acts as a mask for locally wet thinning the top silicon film and for the LOCOS oxidation (Fig. 11b, c). This mask is removed by wet etching. Following that, another LPCVD nitride layer is deposited

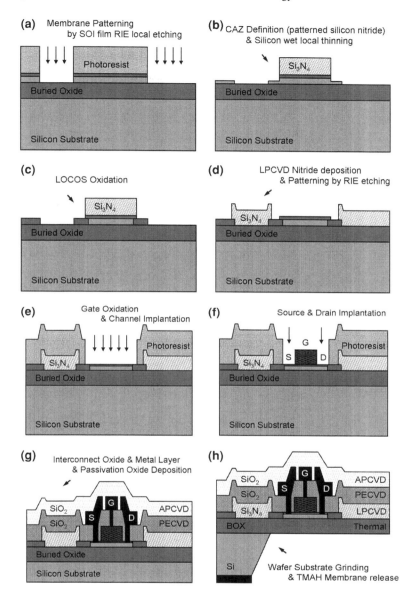

Fig. 11 Schematic illustration of the process flow of the multilayered membrane with lying MOS device: **a** SOI thin film locally removed from oxidized starting SOI wafer on future membrane areas, **b** LPCVD silicon nitride film deposition and patterning according to the active zones, Si local wet thinning **c** LOCOS oxidation, **d** membrane's nitride film deposition and RIE patterning, **e** gate oxidation and channel implantation, **f** source and drain implantation, **g** deposition of interconnect oxide, aluminum s-gun evaporation and passivation layer deposition, **h** wafer's substrate grinding, metal layer evaporation and patterning, TMAH membrane release; the final system configuration

and patterned, which constitutes the second layer of the membrane stack and covers all membrane areas except active devices locations. The etching of this latter film is performed by a combination of $SF_6$- and $CHF_3$-based plasmas. Although $SF_6$ provides an interesting 2.5:1 selectivity between nitride and oxide as well as a end-of-attack detection, $CHF_3$ is adopted to terminate the nitride film etching for precaution reasons: in case of oxide overetch the silicon film would be very rapidly etched with $SF_6$. From this point, wafers follow a standard SOI-CMOS process with gate oxidation, channel implantation, polysilicon gate deposition and patterning, source and drain implantation, interconnect oxide deposition, metal layer evaporation and passivation oxide deposition (Fig. 11e–g). Finally, at the back end of the process, the wafers are thinned by grinding from 725 μm down to 200 μm and the membranes are released by back-etching (Fig. 11g). This release is performed in TMAH-10% solution at 90°C with addition of 35 g/l of dissolved silicon powder and 15 g/l of APS (ammonium persulfate), ensuring a nearly infinite selectivity to aluminum used as back mask.

Several devices lying along the border of the released thin membrane are measured: n- and p-transistors, with or without channel doping and two different channel sizes ($L \times W$: $3 \times 3$ μm$^2$, $20 \times 20$ μm$^2$). Figure 12a shows representative $I_{DS}$–$V_{DS}$ characteristics with and without applied force on the membrane for a 20 μm $\times$ 20 μm nMOSFET. Figure 12b presents the current variation $I_{DS}$ related to the induced strain in the transistor channel. These measurements are in good agreement with values obtained by Gallon [90].

A smart pressure transducer can be obtained by integrating CMOS ring oscillators on the membrane. This configuration gives a direct digital output, avoiding external analogue circuitry and thus is less sensitive to noise compared to the conventional Wheatstone bridge. A pressure change on the membrane induces a variation of the oscillator frequency. This variation is related to the delay variation of each inverter due to the piezoresistive effect induced by the stress in the channel resistivity of the MOSFETs:

$$-\frac{\Delta f}{f} \approx \frac{\Delta \tau_{\text{inverter}}}{\tau_{\text{inverter}}} \approx \frac{1}{2}\left(\left(\frac{\Delta R}{R}\right)_{\text{NMOS}} + \left(\frac{\Delta R}{R}\right)_{\text{PMOS}}\right) \quad (3)$$

Referring to the SOI transistor piezoresistive coefficients [90], designing inverters with n- and p-MOSFETs channels, respectively, parallel and perpendicular to the strain will increase the VCO frequency. Choosing the opposite configuration (nMOSFET$_{\parallel}$ and pMOSFET$_{\perp}$) should lead to a decrease of this frequency. Ring oscillators are very sensitive to temperature because of the dependence of the carrier mobility, the threshold voltage (second-order effect), and the piezoresistive coefficients (first-order effect). By considering the frequency ratio of these two oscillators, first-order effects due to temperature and supply voltage variations can be suppressed and pressure sensitivity is amplified. Each ring oscillator is made of seven CMOS inverters. Channel dimensions are Lp = Ln = 12 μm, Wp = Wn = 24 μm and the ring oscillator is 150 μm $\times$ 150 μm in size. The VCOs are located at the borders of the membrane where the stress is

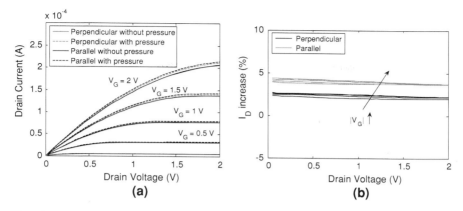

**Fig. 12** Output characteristics of perpendicularly oriented (*gray lines*) and parallel (*black lines*) intrinsic nMOSFETs at various gate biases: **a** $I_{DS}$–$V_{DS}$ curves with (*dashed lines*) and without (*solid lines*) applied stress on the membrane, **b** the corresponding $I_{DS}$ increase as a function of the drain voltage. $L = 20$ μm, $W = 20$ μm

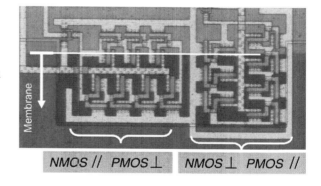

**Fig. 13** Optical photograph of an oscillator pair at the border of a membrane. *Left oscillator* the nMOS devices are parallel to the stress and pMOS are perpendicular, and the devices are inversely oriented for the oscillator on the *right*

larger (Fig. 13). When $V_{DD} = 2$ V, the oscillator frequency is 5.55 and 4.96 MHz for the configuration with nMOSFET$_\perp$ and nMOSFET$_\parallel$, respectively. Measurements under pressure are presented in Fig. 14. As predicted, we have a positive and negative shift of the VCOs oscillation frequencies according to their orientation. The positive shift is more important due to the higher piezoresistive coefficients sum for this configuration. Eventually, the difference of resonance frequency without applied pressure can be attributed to the internal residual stress of the membrane.

Sensitive edgeless transistors placed at the centre of a 250-μm-side membrane (Fig. 15a) are characterized in dynamic regime. For dynamic measurements, the devices are mounted in series with a resistor and placed close by the edge of a bladed rotor, which can rotate up to 135 rpm. The spectral response of the system

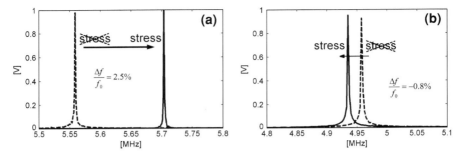

**Fig. 14** VCOs resonance frequency shift according to their orientation when a force is applied at the centre of the membrane: **a** the channel of nMOS devices are parallel to the stress while the pMOS are perpendicular, **b** the channel of nMOS devices are perpendicular to the stress while the pMOS are parallel

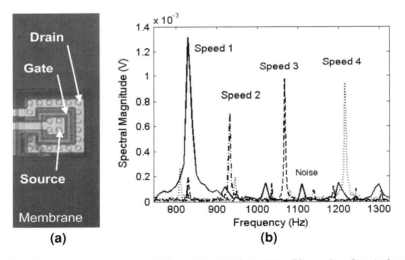

**Fig. 15 a** Optical photograph of a nMOS edgeless transistor ($W = 72$ μm, $L = 2$ μm) placed at the center of a membrane. **b** Spectral dynamic response of a nMOS edgeless transistor excited at various rotor blade speeds: 830, 921, 1,067 and 1,215 Hz, respectively

is given in Fig. 15b at various blade rotation frequencies, from 830 and 1,215 Hz. It is worth noting that the output signal is clear and accurate with respect to the rotor speed over this range of speeds, revealing a real opportunity of application of this kind of active element sensors. Some low amplitude noisy peaks are distinguishable in the graph, which can be attributed to the basic character of the packaging technology in use for these first experiments. Hence, the compactness, the sensitivity and the dynamic response of these systems allow us to figure out a wide range of possible applications, such as propeller shaft speed measurements [91].

## 5 Surface Micromachining: 3D Thin-Film SOI-MEMS Sensors and Actuators

This section aims at demonstrating the major interests of using SOI wafers for the fabrication of surface micromachined MEMS sensors and actuators. The surface micromachining techniques can be defined as the release of a material layer (structural layer) by the selective underetch of a thin sacrificial layer. Starting with a SOI wafer, the BOX can be used as a sacrificial layer for the release of a freestanding thin silicon layer which acts as the structural layer.

### 5.1 3D Surface Micromachined Microstructures

Microcantilevers-based sensors are silicon beams that are typically less than 1-μm thick. The term 3D is used here to design a structure presenting a non-flattened geometry, i.e. an out-of-plane curvature when the reference plane is the silicon substrate wafer. 2D microcantilevers are widely used in atomic force microscopy (AFM), mass sensing, contact sensing and force measurements. Classically with MEMS-based cantilevers, a change in surface tension or surface stress, due to interfacial interactions between the surface and the environment or intermolecular interactions on the surface [92], is detected electrically by a piezoresistive gauge or a piezoelectric resonance frequency shift (static or dynamic mode). 3D microcantilevers have an out-of-plane movable part, which can be viewed as the artificial copy of short cat whiskers or hairs which are sensitive to air flow impinging on it or to temperature variations.

Building true 3D geometries is a major challenge for the MEMS community, most mature MEMS being bi-dimensional microstructures involving only lateral movements (except perhaps micro-mirrors). Self-assembling microstructures are mainly studied for the development of out-of-plane inductors which allow isolation from the lossy silicon substrate, improving quality factor and self-resonance frequency [93]. Various techniques, such as melt hinge pad [94] or plastic deformation magnetic assembly [95] are used to generate out-of-plane deflections. Recently, new processes and applications have been investigated for 3D MEMS. The fabrication of a tunable capacitor based on the reflow of photosensitive cyclotene resist has been proposed in [96]. Sasaki et al. report on a SOI-MEMS process based on 3D multilayered structures for the fabrication of thick 3D buckled structures [97]. These techniques add extra and tricky fabrication steps to the classical CMOS process.

Providing solutions for the co-integration of MEMS or NEMS with state-of-the-art thin-film SOI MOS technology can then be seen as a mainstream technology for not only the fabrication of highly integrated low-power MOS circuits but also as a technology platform for the fabrication of high quality MEMS and NEMS sensors and actuators.

In this section, we present a simple one-mask SOI-MEMS process, compatible with a classical CMOS technology, for the fabrication of 3D self-assembled microstructures. Our process relies on the control of the chemical release and on the control of stresses appearing in multilayered structures which originate from the thermal expansion mismatch between the material layers as well as the control of the plastic yielding of a metallic layer. Similar methods, using stresses as the self-assembling technique, have been proposed in the literature [93, 97] but the thermal and plastic properties of metallic films were never analyzed or systematically used in the fabrication of such structures.

## 5.2 Design Principle

The control of stresses in thin films is a major issue for microelectronics and micromachined structures. Residual stresses are, for many reasons, considered as detrimental and huge efforts have been made to mitigate their effect. On the other hand, in some cases such as in self-assembling microsystems, use can be made of those stresses to generate a desired geometry. In our case, the stress gradient building up in multilayered systems is used as the assembly mechanism.

The residual stresses in a micro- or nanostructure can have different origins: (a) *intrinsic stresses* are those resulting from the growth or deposition of a film, and (b) *extrinsic stresses* are the ones induced by external factors. The main source of extrinsic stress is the temperature. Those stresses arise from the difference in thermal expansion coefficients ($\alpha$) of the materials composing the multilayered structure. Since materials are deposited or grown at high temperature and have different $\alpha$, the deformation upon cooling of each layer is different, resulting in a stress gradient across the multilayer.

When dealing only with elastic films such as $Si_3N_4$, $SiO_2$ and Si, a thermal cycle (heating up followed by a cooling down) has no effect on the stress as long as chemical stability of the material is ensured, e.g. no diffusion takes place. On the other hand, working with plastic films makes the stresses dependent on thermal history. The room temperature tensile stress of a metallic layer deposited on an elastic substrate usually increases after a thermal treatment [98, 99]. This stress increment results from the plasticity of the Al layer, and is affected also by the hardening capacity and temperature dependence of the flow properties. As explained later, the use of a thermal treatment to enhance stress gradient across the multilayered structures is crucial for successful assembling of our microstructures.

## 5.3 3D MEMS Fabrication Process

A SOI wafer with a100-nm-thick monocrystalline silicon thin film on top of a 400-nm-thick BOX lying on a 780-µm-thick bulk silicon handling substrate is

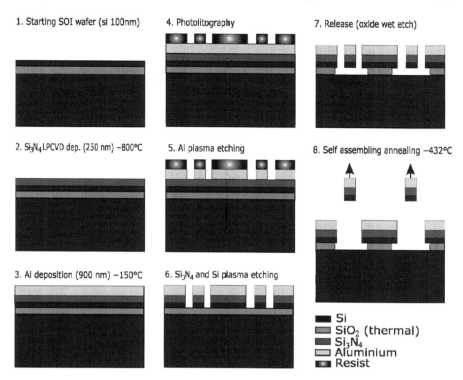

**Fig. 16** One-mask process flow for the fabrication of self-assembled 3D SOI-MEMS

considered as the starting material. The fabrication of 3D capacitive sensors requires the deposition of two layers on top of the silicon film forming the active layer of the starting SOI wafer. As a result of thermal expansion mismatch between layers, important stresses will then be induced in such multilayered structures that are then controlled a posteriori by monitoring the stack thickness. The aluminum layer is then deposited at a temperature of 150°C.

As illustrated in Fig. 16, the multilayered structure is made out of three layers. Starting with a SOI substrate, a 250-nm-thick layer of silicon nitride is coated using LPCVD silicon nitride. Then, a process known as "loading" is implemented which consists in cooling the system to room temperature (20°C) from the thin-film deposition temperature (i.e. 800°C for silicon nitride). In a standard CMOS process, LPCVD nitride is used to define the active zone of SOI MOSFETs or the n- and p-type doped zones in a bulk CMOS process. Thirdly, a third layer of 900-nm thick of aluminum is then evaporated at 150°C and represents a new loading phase. During these two cooling processes, stress builds up in the layers as a result of the thermal expansion coefficient mismatch between the two materials.

After a single photolithographic step to pattern the structures as desired, the aluminum film is firstly etched by a chlorine-based plasma. Subsequently, a $SF_6$- and $SiCl_4$-based plasmas are used to etch, respectively, $Si_3N_4$ and Si, while aluminum is used as a masking material due to its lack of chemical reactivity with the considered plasma.

The release of the microstructure must then be performed by withdrawing the buried silicon dioxide ($SiO_2$) located below the thin silicon layer by using a mixture of concentrated hydrofluoric acid (HF 73%) and isopropanol 1:1. After further rinsing in pure isopropanol, the samples are dried in a critical point dryer (Tousimis 915B) in order to avoid adhesion of the beams to the substrate. The critical point dryer operates by replacing isopropanol after the wet release by liquid $CO_2$ and vaporizing it by applying temperature and pressure above the critical point, thus avoiding the formation of liquid drops in between the beam and the substrate. Finally, the last step of the process consists in a thermal annealing after release, in order to promote the assembling of the microstructures by modifying the stress state in the upper plastic layer. The material grain size and morphology and therefore the electrical and mechanical properties of aluminum depend strongly on the annealing temperature. The duration of the thermal treatment necessary to achieve assembling depends on the annealing temperature. The choice of the annealing method is a trade-off between high tensile stress and preserving aluminum quality. In [100], it has been demonstrated that a thermal annealing at 432°C in a forming gas (95% nitrogen, 5% hydrogen) for 30 min is sufficient to self-assemble the microstructures and thus to obtain the desired 3D MEMS. This thermal treatment is similar to the last step of our SOI-CMOS process used to improve the metal-(poly)silicon contacts as well as to reduce the gate oxide interface traps. It could then be performed without modifying the thermal budget of our standard CMOS process thus insuring CMOS compatibility of the proposed 3D MEMS process.

SEM images of several out-of-plane microstructures obtained by this process are given in Fig. 17, showing the design flexibility offered by this process.

Since the CMOS circuits and the 3D MEMS are made with the same materials, our team has been focusing these last years on the co-integration of surface micromachined sensors with their associated electronics on the same chip. Two main process flows have been proposed in [101] for the release of the 3D MEMS co-integrated with their associated SOI-CMOS electronics: either the BOX is used as sacrificial layer (etch in HF-based solution) and the top thin silicon film is part of the multilayered stack 3D MEMS, or the top silicon layer itself is used as sacrificial layer (etched by $SF_6$ plasma) in the MEMS areas. Figure 18 shows the main process steps for this later case. The co-integration of the CMOS electronics and 3D microbeams requires three extra lithographic steps in comparison with the complete CMOS process. One photolithography must be performed to protect the MEMS areas when processing the IC (implantation, etching, etc.) and two additional photolithography steps are required to protect the IC when processing only the MEMS devices (etching and release). As presented in the next section, this

Sensing and MEMS Devices in Thin-Film SOI MOS Technology 381

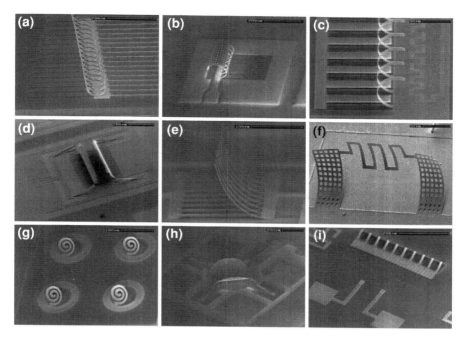

**Fig. 17** SEM images gallery: **a** spiral beams, **b** capacitive plate, **c** thermal actuator (zoom), **d** rectangular bolometer, **e** interdigitated capacitor, **f** inductor, **g** spirals, **h** circular bolometer, **i** thermal actuator

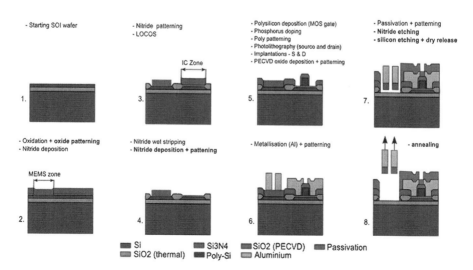

**Fig. 18** SOI MOS-MEMS co-integration process flow

process flow has been successfully used to build 3D MEMS flow sensors with their associated SOI-CMOS electronics.

## 5.4 Flow Sensor

The sensing mechanism of our flow sensors is based on interdigitated structures where half of the fingers are self-assembled while the others remain in the substrate plane. If air flows parallel to the substrate surface, the assembled fingers bend upwards or downwards, depending on flow direction resulting in a lower or higher capacitance between out-of-plane and flat fixed fingers. From a process point of view, the release of only half of the fingers is simply obtained by designing the fixed fingers wider than the others as can be seen in Fig. 19a. The 3D capacitive cantilevers are connected at the internal nodes of CMOS ring oscillators made of five inverters in order to perform a capacitance-to-frequency conversion. In order to get rid of electronics drift due to temperature or humidity variations, the absolute resonance frequency of the capacitive 3D MEMS ring oscillator is continuously compared to its of an identical ICs loaded with in-plane interdigitated capacitors built on the same chip. The frequency shift between both ring oscillators is directly related to the gas velocity flowing at the surface of the chip. Measurements under air flow are presented in Fig. 19b. The flow bends downwards the cantilevers, increases their capacitance and, therefore, lowers the oscillating frequency. For a range of flow velocities from 0 to 20 m/s, there is a measurable change in capacitance (3D MEMS) over a range of 75–275 fF which translates to relative frequency shift of more than 11%.

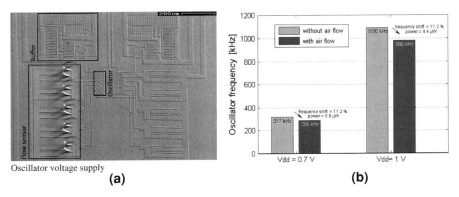

**Fig. 19** **a** SEM picture of a co-integrated 3D MEMS capacitive sensor with its associated SOI-CMOS circuit, a ring oscillator with an output buffer. **b** Frequency response of the 3D capacitive MEMS oscillator with and without applied air flow [109]

**Table 1** Review of some published flow sensors results

| Ref. | Technology | Static power consumption | Flow range (m/s) | IC compatibility |
|---|---|---|---|---|
| [105] | Si substrate | Yes | 2–18 | Yes |
| [106] | Membrane | Yes | 0.01–200 | No |
| [107] | Bridge | Yes | 0–4 | No |
| [108] | Membrane | Yes | 0–8 | Yes |
| Our work | Surface machining | No | 0–20 | Yes |

The three first are on bulk micromachining techniques and composed of a microheater and thermopiles

These sensors consume extremely low power, present high sensitivity and large sensing range as well as occupy low chip area (Table 1). The main advantage of this simple architecture, out of the process simplicity, is the absence of static power consumption when compared to classical solutions proposed to sense flow or temperature in the literature using thermopiles [81], piezoresistive transducer [102], hot-wire anemometer [103], etc.

Another advantage is the possibility to change the sensitivity or sensing range by changing the sensors geometry. Lower or higher speed flow can be sensed by decreasing or increasing, respectively, the cantilever stiffness. Stiffer (softer) cantilevers can be obtained either by reducing (increasing) the length or by increasing (decreasing) the thickness. Practically, flow sensors arrays composed of cantilevers characterized by various lengths have been considered in order to generate a wide range flow sensing.

To assist with the design of flow sensors, the mechanical and electrical behaviours of the microsystems can be predicted using analytically simplified cases or finite element analysis (FEA) simulations. As well as electrical simulations, the mechanical behaviour under various stimuli is also a key parameter for optimization. Given the thermal budget and elastic multilayer properties of the flow sensor, FEA simulations and analytical models can be used to optimize sensor design. Using the classical curvature method to extract the materials properties, simulations and simple analytical modelling of the 3D elements generate results in very good agreement with measurements of the fabricated microstructures [104].

## 5.5 Thermal Sensor

Under a substantial increase in temperature, the out-of-plane component of these interdigitated capacitive thermal sensors (bolometers in Fig. 17) bends downwards because this part is made up of three different layers (Si, $Si_3N_4$ and Al from bottom to top). Each layer is characterized by its own thermal expansion coefficient (around $2 \times 10^{-6}$, $3 \times 10^{-6}$ and $20 \times 10^{-6} \circ C^{-1}$, respectively). As aluminum

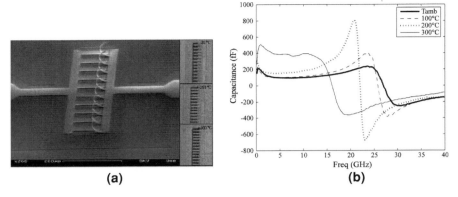

**Fig. 20** a Built capacitive thermal sensor (*insets* cantilevers position at different temperatures). b Response of thermal sensors composed of 200-μm-long fingers (2-μm-thick top Al layer) with a 1.5 μm spacing

expands 10 times more than Si and is 1-μm thick, a large bending motion downwards occurs, resulting in temperature detection. The structure shown in Fig. 20a is composed of 200-μm-long interdigitated fingers with a gap of 1.5 μm between adjacent fingers. The insets in Fig. 20a present the position of the assembled cantilevers as a function of temperature, for temperature ranging from 20 to 300°C.

Those structures were then electrically measured at different temperatures, allowing us to relate a resonance frequency shift or capacitance value shift to a given temperature. A shift of around 17 GHz was observed for a temperature ranging from 20°C (45 GHz) to 300°C (28.3 GHz) with a large shift between 200°C (38 GHz) and 300°C. Capacitance values at 9 GHz (plateau in measurements) range from 66.4 to 82.9 fF with values in the same range for temperatures between 20 and 100°C. The best sensitivity both for capacitance and self-resonance frequency is thus in the 200–300°C range.

## 5.6 Magnetic Sensor

The out-of-plane magnetic flux is converted into a mechanical force by Lorentz force $F$ on the M-shaped $Si_3N_4$–Al bilayered cantilever (Fig. 21):

$$F = BIL \sin(\theta) \qquad (4)$$

$I$ is the half-loop current, $L$ the top beam length, $B$ the magnetic field flux density across this beam and $\theta$ the angle between the magnetic field and the current flowing into the top beam. A piezoresistive gauge located at the M-shaped cantilever anchor converts the bending motion of the beam into an electrical signal.

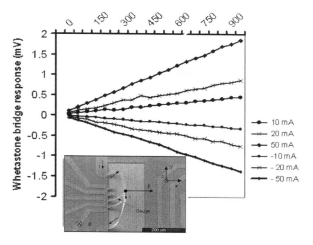

**Fig. 21** Si-Lorentz magnetometer with its integrated Wheatstone bridge. Sensor response in mV for different rectangular currents at 4 kHz

**Fig. 22** Si$_3$N$_4$–Ni micro-tweezers

The piezoresistor is then incorporated into a Wheatstone structure with three other fixed resistors. As shown in Fig. 21, the response is linear over a large range of the magnetic field (from 1 to 1,000 G) with a small offset at 0 V. The output voltage signal for the Lorentz force-based device for different values of current through the bridge shows a sensitivity of 0.2 mV/G [110].

## 5.7 Thermal Actuators

Under a substantial and controlled local heating by Joule effect in thin conductive layer, the out-of-plane cantilevers bend downwards because of different thermal expansion coefficients. Building circular cantilevers assemblies (in place of straight in line), a Si$_3$N$_4$–Ni bilayered micro-tweezer for the release or capture of micro-compounds (cells, proteins, microbes, yeasts, etc.) is shown in Fig. 22.

## 6 Conclusions

The great potential of SOI technology for MEMS fabrication has been described through several sensing applications.

It has been demonstrated that the electronic transport into the thin top silicon channel of a SOI wafer is quite sensitive to the binding of biomolecules at its surface. nIDA-gate MOSFET biosensors have been designed and fabricated by opening nanotrenches in the gate electrode of a conventional SOI MOSFET to allow biomolecules to get access to the surface of the gate oxide. Those integrated SOI biosensors are highly sensitive, compact and label-free.

Thin SOI PIN photodiodes exhibit an intrinsic remarkable optical responsivity in blue and ultraviolet wavelengths which are commonly used in biomedical and environmental fields.

Starting from a SOI wafer, the sacrificial and the structural layers are already available. Indeed, in the case of bulk micromachined sensors, the handling silicon substrate is the sacrificial layer and the BOX is the structural layer (part of the thin membrane) and, in the case of surface micromachined 3D sensors, the BOX is the sacrificial layer and the top thin silicon layer is the structural one. The CMOS compatibility issue has also been addressed.

Suspended sensors on thin membranes as well as 3D self-assembled micromachined interdigitated sensors were successfully built in thin-film SOI-MEMS technology thanks to the control and use of residual stresses building up in multilayered structures.

The proposed processes require only steps compatible with a classical SOI-CMOS process and constitute thus a solution for the co-integration of sensors with their associated electronics— read-out circuit. Moreover, thanks to the simplicity of the proposed processes, reliability from die to die and from wafer to wafer is quite high.

**Acknowledgments** We would like to thank all the PhD students, senior researchers, and professors who have actively participated to the simulation and experimental results presented in this chapter: Mr. N. André, Mr. B. Olbrechts, Mr. B. Rue, Mr. Olivier Bulteel, Mr. R. Pampin, Dr. L. Moreno Hagelsieb, Dr. X. Tang, Dr. P. Ivanov, Dr. J. Laconte, Dr. F. Iker, Dr. G. Rinaldi, Prof. S. Demoustier-Champagne, Prof. I. Stiharu and Prof. T. Pardoen. We would like to also thank Mr. P. Simon (Welcome) for performing some of measurements, the UCL clean rooms technicians and engineers (Winfab) for their precious support during the processing of the SOI-MEMS. This research has been financially supported by the European Commission through Networks of Excellence: SINANO, NANOSIL and EuroSOI+, by Walloon Region: MEMSACOM, CAVIMA, NANOTIC, MINATIS, and the Communauté française de Belgique: Action Concertée de Recherche, ARC no. 05/10-330, Innovative technologies for physical and (bio)chemical nano-sensors.

# References

1. Bernstein, K., Rohrer, N.: SOI circuit design concepts. Kluwer Academic Publishers, Dordrecht (2000)
2. Flandre, D., Raskin, J.-P., Vanhoenacker, D.: SOI CMOS transistors for RF and microwave applications. In: Dean, M.J., Fjeldly, T.A. (eds.) CMOS RF Modeling, Characterization and Applications. World Scientific Publishing Co., London (2002). ISBN 981-02-4905-5
3. Flandre, D.: Silicon-on-Insulator technology for high temperature metal oxide semiconductor devices and circuits. In: Kirschman, R. (ed.) High temperature electronics. IEEE Press, Piscataway (1998)
4. Leray, J.L., Dupont-Nivet, E., Musseau, O., Coic, Y.M., Umbert, A., Lalande, P., Pere, J.F., Auberton-Herve, A.J., Bruel, M., Jaussaud, C., Margail, J., Giffard, B., Truche, R., Martin, F.: From substrate to VLSI: investigation of hardened SIMOX without epitaxy, for dose, dose rate and SEU phenomena. IEEE Trans. Nucl. Sci. **35**, 1355–1360 (1988)
5. Diem, B., Rey, P., Renard, S., Bosson, S.V., Bono, H., Michel, F., Delaye, M., Delapierre, G.: SOI SIMOX: from bulk to surface micromachining, a new age for silicon sensors and actuators. Sens. Actuators A **46**, 8–16 (1995)
6. Mokwa, W.: Advanced sensors and microsystems on SOI. Int. J. High Speed Electron Syst. **10**, 147–153 (2000)
7. Kiihamaki, J., Ronkainen, H., Pekko, P., Kattelus, H., Theqvist, K.: Modular integration of CMOS and SOI-MEMS using "plug-up" concept. In: Proceedings of the 12th International Conference on TRANSDUCERS, Solid-State Sensors, Actuators and Microsystems, vol. 2, pp. 1647–1650 (2003)
8. Gitelman, L., Stolyarova, S., Bar-Lev, S., Gutman, Z., Ochana, Y., Nemirovsky, Y.: CMOS-SOI-MEMS transistor for uncooled IR imaging. IEEE Trans. Electron. Devices **56**, 1935–1942 (2009)
9. Corcos, D., Goren, D., Nemirovsky, Y.: CMOS-SOI-MEMS transistor (TeraMOS) for TeraHertz imaging. In: IEEE International Conference on Microwaves, Communications, Antennas and Electronics Systems, COMCAS 2009, pp. 1–5 (2009)
10. Lu, C.-C., Liao, K.-H., Udrea, F., Covington, J.A., Gardner, J.W.: Multi-field simulations and characterization of CMOS-MEMS high-temperature smart gas sensors based on SOI technology. J. Micromech. Microeng. (2008). doi:10.1088/0960-1317/18/7/075010
11. Ali, S., Santra, S., Haneef, I., Schwandt, C., Kumar, R., Milne, W., et al.: Nanowire hydrogen gas sensor employing CMOS micro-hotplate. In: Proceedings of 2009 IEEE Sensors, pp. 114–117 (2009)
12. Haneef, I., Coull, J.D., Ali, S.Z., Udrea, F., Hodson, H.P.: Laminar to turbulent flow transition measurements using an array of SOI-CMOS MEMS wall shear stress sensors. In: Proceedings of 2008 IEEE Sensors, pp. 57–61 (2008)
13. Gaudo, M.V., Abadal, G., Verd, J., Teva, J., Perez-Murano, F., Costa, E.F., Montserrat, J., Uranga, A., Esteve, J., Barniol, N.: Time-resolved evaporation rate of attoliter glycerine drops using on-chip CMOS mass sensors based on resonant silicon micro cantilevers. IEEE Trans. Nanotechnol. **6**, 509–512 (2007)
14. Villarroya, M., Figueras, E., Pérez-Murano, F., Campabadal, F., Esteve, J., Barniol, N.: SOI-silicon as structural layer for NEMS applications. Proc. SPIE **5116**, 1–11 (2003)
15. Chen, T.D., Kelly, T.W., Collins, D., Berthold, B., Brosnihan, T.J., Denison, T., Kuang, J., O'Kane, M., Weigold, J.W., Bain, D.: The next generation integrated MEMS and CMOS process on SOI wafers for overdamped accelerometers. In: Proceedings of the 13th International Conference on Solid-State Sensors, Actuators and Microsystems, TRANSDUCERS'05, vol. 2, pp. 1122–1125 (2005)
16. Davis, B.S., Denison, T., Kaung, J.: A monolithic high-g SOI-MEMS accelerometer for measuring projectile launch and flight accelerations. In: Proceedings of 2004 IEEE Sensors, pp. 296–299 (2004)

17. Takao, H., Ichikawa, T., Nakata, T., Sawada, K., Ishida, M.: Post-CMOS integration technology of thick-film SOI MEMS devices using micro bridge interconnections. In: Proceedings of the 21st International Conference on Micro Electro Mechanical Systems, MEMS 2008, pp. 359–362 (2008)
18. Takahashi, K., Mita, M., Nakada, M., Yamane, D., Higo, A., Fujita, H., Toshiyoshi, H.: Development of multi-user multi-chip SOI CMOS-MEMS processes. In: Proceedings of the 22nd International Conference on Micro Electro Mechanical Systems, MEMS 2009, pp. 701–704 (2009)
19. Guan, L., Sin, J.K.O., Liu, H., Xiong, Z.: A fully integrated SOI RF MEMS technology for system-on-a-chip applications. IEEE Trans. Electron. Devices **53**, 167–172 (2006)
20. Badila-Ciressan, N., Mazza, M., Grogg, D., Ionescu, A.: Nano-gap micro-electro-mechanical bulk lateral resonators with high quality factors and low motional resistances on thin Silicon-on-Insulator. Solid-State Electron. **52**, 1394–1400 (2008)
21. Cui, Y., Wei, Q.Q., Park, H.K., Lieber, C.M.: Nanowire nanosensors for highly sensitive and selective detection of biological and chemical species. Science **293**, 1289–1292 (2001). doi:10.1126/science.1062711
22. Hahm, J.-I., Lieber, C.M.: Direct ultrasensitive electrical detection of DNA and DNA sequence variations using nanowire nanosensors. Nano Lett **4**, 51–54 (2004)
23. Li, Z., Chen, Y., Kamins, T.I., Nauka, K., Williams, R.S.: Sequence-specific label-free DNA sensors based on silicon nanowires. Nano Lett **4**, 245–247 (2004)
24. Patolsky, F., Zheng, G.F., Lieber, C.M.: Fabrication of silicon nanowire devices for ultrasensitive, label-free, real-time detection of biological and chemical species. Nat Protocols **1**(4), 1711–1724 (2006)
25. Talin, A.A., Hunter, L.L., Léonard, F., Rokad, B.: Large area, dense silicon nanowire array chemical sensors. Appl. Phys. Lett. (2006). doi:10.1063/1.2358214
26. Elfström, N., Juhasz, R., Sychugov, I., Engfeldt, T., Kalström, A., Linros, J.: Surface charge sensitivity of silicon nanowires: size dependence. Nano Lett **7**, 2608–2612 (2007)
27. Lin, M.C., Chu, C.J., Tsai, L.C., Lin, H.Y., Wu, C.S., Wu, Y.P., Wu, Y.N., Shieh, D.B., Su, Y.W., Chen, C.D.: Optical switching of porphyrin-coated silicon nanowire field effect transistors. Nano Lett **7**, 3656–3661 (2007)
28. Elfström, N., Karlström, A.E., Linnros, J.: Silicon nanoribbons for electrical detection of biomolecules. Nano Lett **8**, 945–949 (2008)
29. Maki, W.C., Mishra, N.N., Cameron, E.G., Filanoski, B., Rastogi, S.K., Maki, G.K.: Nanowire-transistor based ultra-sensitive DNA methylation detection. Biosens. Bioelectron. **23**, 780–787 (2008)
30. Finot, E., Bourillot, E., Meunier-Prest, E., Lacroute, Y., Legay, G., Cherkaoui-Malki, M., Latruffe, N., Siri, O., Braunstein, P., Dereux, A.: Performance of interdigitated nanoelectrodes for electrochemical DNA biosensor. Ultramicroscopy **97**, 441–449 (2003)
31. Ueno, K., Hayashida, M., Ye, J.Y., Misawa, H.: Fabrication and electrochemical characterization of interdigitated nanoelectrode arrays. Electrochem. Commun. **7**, 161–165 (2005)
32. Zhu, X.S., Ahn, C.H.: On-chip electrochemical analysis system using nanoelectrodes and bioelectronic CMOS chip. IEEE Sens. J. **6**(5), 1280–1286 (2006)
33. Zou, Z.W., Kai, J.H., Rust, M.J., Han, J.Y., Ahn, C.H.: Functionalized nano interdigitated electrodes arrays on polymer with integrated microfluidics for direct bio-affinity sensing using impedimetric measurement. Sens. Actuators A **136**, 518–526 (2007)
34. Moreno-Hagelsieb, L., Foultier, B., Laurent, G., Pampin, R., Remacle, J., Raskin, J.-P., Flandre, D.: Electrical detection of DNA hybridization: three extraction techniques based on interdigitated Al/Al$_2$O$_3$ capacitors. Biosens. Bioelectron. **22**, 2199–2207 (2007)
35. Bergveld, P.: Thirty years of ISFETOLOGY: what happened in the past 30 years and what may happen in the next 30 years. Sens. Actuators B **88**, 1–20 (2003)
36. Tang, X., Blòndeau, F., Prevot, P.-P., Pampin, R., Godfroid, E., Jonas, A., Nysten, B., Demoustier-Champagne, S., Iñiguez, B., Colinge, J.-P., Raskin, J.-P., Flandre, D., Bayot,

V.: Direct protein detection with a nano-interdigitated gate MOSFET. Biosens. Bioelectron. **24**, 3531–3537 (2009)
37. Prevot, P.P., Adam, B., Zouaoui Boudjeltia, K., Brossard, M., Lins, L., Cauchie, P., Brasseur, R., Vanhaeverbeek, M., Vanhamme, L., Godfroid, E.: Anti-haemostatic effects of a serpin from the saliva of the tick *Ixodes ricinus*. J. Biol. Chem. **281**, 26361–26369 (2006)
38. Prevot, P.P., Couvreur, B., Denis, V., Brossard, M., Vanhamme, L., Godfroid, E.: Protective immunity against *Ixodes ricinus* induced by a Salivary serpin. Vaccine **25**, 3284–3292 (2007)
39. Poghossian, A., Cherstvy, A., Ingebrandt, S., Offenhäusser, A., Schöning, M.J.: Possibilities and limitations of label-free detection of DNA hybridization with field-effect-based devices. Sens. Actuators B **111–112**, 470–480 (2005)
40. Im, H., Huang, X.-J., Gu, B., Choi, Y.-K.: A dielectric-modulated field-effect transistor for biosensing. Nat. Nanotechnol. **2**, 430–434 (2007)
41. Park, K.Y., Kim, M.S., Choi, S.-Y.: Biosens. Bioelectron. **20**, 2111–2115 (2005)
42. Patolsky, F., Lieber, C.M.: Nanowire Nanosens. Mater. Today **8**, 20–28 (2005)
43. Flynn, N.T., Tran, T.N.T., Cima, M.J., Langer, R.: Long-term stability of self-assembled monolayers in biologically-related media. Langmuir **19**, 10909–10915 (2003)
44. Kim, D.-S., Park, J.-E., Shin, J.-K., Kim, P.-K., Lim, G., Shoji, S.: An extended gate FET-based biosensor integrated with a Si microfluidic channel for detection of protein complexes. Sens. Actuators B **117**, 488–494 (2006)
45. Gooding, J.J., Situmorang, M., Erokhin, F., Hibbert, D.B.: An assay for the determination of the amount of glucose oxidase immobilised in an enzyme electrode. Anal. Commun. **36**, 225–228 (1999)
46. Cecchet, F., Duwez, A.-S., Gabriel, S., Jérôme, C., Jérôme, R., Glinel, K., Demoustier-Champagne, S., Jonas, A.M., Nysten, B.: Atomic force microscopy investigation of the morphology and the biological activity of protein-modified surfaces for bio- and immunosensors. Anal. Chem. **79**, 6488–6495 (2007)
47. Razeghi, M.: Short-wavelength solar-blind detectors-status, prospects, and markets. Proc. IEEE **90**(6), 1006–1014 (2002)
48. Afzalian, A., Flandre, D.: Physical modeling and design of thin-film SOI lateral PIN photodiodes. IEEE Trans. Electron. Devices **52**(6), 1116 (2005)
49. Bulteel, O., Flandre, D.: Optimization of blue/UV sensors using PIN photodiodes in thin-film SOI technology. In: Proceedings of the 215th Electrochemical Society Meeting, San Francisco, CA, USA (2009)
50. OKI semiconductor website http://www2.okisemi.com
51. Kilchytska, V., et al.: Electrical characterization of true Silicon-On-Nothing MOSFETs fabricated by Si layer transfer over a pre-etched cavity. Solid-State Electron. **51**, 1238–1244 (2007)
52. Bulteel, O., Afzalian, A., Flandre, D.: Fully integrated blue/UV SOI CMOS photosensor for biomedical and environmental applications. In: Proceedings of TAISA, Lyon, France (2007). doi:10.1007/s10470-009-9402-y
53. Bulteel, O., et al.: Proceedings of eMBEC, Antwerp, Belgium (2008)
54. Karczemska, A., Sokolowska, A.: Materials for DNA sequencing chip. In: Proceedings of the 3rd International Conference on Novel Applications of Wide Bandgap Layers, Zakopane (2001). doi:10.1109/WBL.2001.946592
55. De Souza, M., Bulteel, O., Flandre, D., Pavanello, M.A.: Temperature influence on the behaviour of lateral thin-film SOI PIN photodiode in the blue and UV range. In: Proceedings of the Sixth Workshop of the Thematic Network on Silicon-on-Insulator Technology, Devices and Circuits, EuroSOI'10, Grenoble, France (2010)
56. Simon, I., Arndt, M.: Thermal and gas-sensing properties of a micromachined thermal conductivity sensor for the detection of hydrogen in automotive applications. Sens. Actuators A **98**, 104–108 (2002)
57. Hellmich, W., et al.: Field-effect gas sensitivity changes in metal oxides. Sens. Actuators B **43**, 132–139 (1997)

58. Sabaté, N., et al.: Mechanical characterization of thermal flow sensors membranes. Sens. Actuators A **125**, 260–266 (2005)
59. Noda, M., et al.: A new type of dielectric bolometer mode of detector pixel using ferroelectric thin film capacitors for infrared image sensor. Sens. Actuators **77**, 39–44 (1999)
60. Lee, S., Tanaka, T., Inoue, K.: Residual stress influence on the sensitivity of ultrasonic sensor having composite membrane structure. Sens. Actuators A **125**, 242–248 (2005)
61. Horrillo, M., Sayago, I., Arés, L., Rodriguo, J., Gutiérrez, J., Götz, A., Gràcia, I., Fonseca, L., Cané, C., Lora-Tamayo, E.: Detection of low $NO_2$ concentrations with low power micromachined tin oxide gas sensors. Sens. Actuators B **58**, 325–329 (1999)
62. Flandre, D., Adriaensen, S., Afzalian, A., Laconte, J., Levacq, D., Renaux, C., Vancaillie, L., Raskin, J.-P.: Intelligent SOI CMOS integrated circuits and sensors for heterogeneous environments and applications. In: IEEE Sensors, Orlando, FL, USA, vol. 28.2, pp. 1407–1412 (2002)
63. Rossi, C., Temple-Boyer, P., Estève, D.: Realization and performance of thin $SiO_2/SiN_x$ membrane for microheater applications. Sens. Actuators A **64**, 241–245 (1998)
64. Laconte, J., Iker, F., Jorez, S., André, N., Pardoen, T., Proost, J., Flandre, D., Raskin, J.-P.: Thin films stress extraction using micromachined structures and wafer curvature measurements. Microelectron. Eng. **76**, 219–226 (2004)
65. Laconte, J., Dupont, C., Flandre, D., Raskin, J-P.: SOI CMOS compatible low-power microheater optimization for the fabrication of smart gas sensors. IEEE Sens. J. **4**, 670–680 (2004)
66. Astié, S., Gué, A.M., Scheid, E., Guillemet, J.P.: Design of a low power $SnO_2$ gas sensor integrated on silicon oxynitride membrane. Sens. Actuators B: Chem. **67**, 84–88 (2000)
67. Briand, D., Krauss, A., van der Schoot, B., Weimar, U., Barsan, N., Göpel, W., de Rooij, N.F.: Design and fabrication of high-temperature micro-hotplates for drop-coated gas sensors. Sens. Actuators B: Chem. **68**, 223–233 (2000)
68. Jorez, S., Laconte, J., Cornet, A., Raskin, J.-P.: Low cost instrumentation for MEMS thermal characterization. Meas. Sci. Technol. **16**, 1833–1840 (2005)
69. Udrea, F., Gardner, J.W., Setiadi, D., Covington, J.A., Dogaru, T., Lu, C.C., Milne, W.I.: Design and simulations of SOI-CMOS micro-hotplate gas sensors. Sens. Actuators B **78**, 180–190 (2001)
70. Gardner, J.W., Pike, A., De Rooij, N.F., Koudelka-Hep, M., Clerc, P.A., Hierlemann, A., Goepel, W.: Integrated array sensor for detecting organic solvents. Sens. Actuators B **26**, 135–167 (1995)
71. Dibbern, U.: A substrate for thin-film gas sensor in microelectronic technology. Sens. Actuators B **2**, 63–67 (1990)
72. Demarne, V., Grisel, A.: An integrated low-power thin-film CO gas sensor on silicon. Sens. Actuators B **4**, 539–543 (1991)
73. Krebs, P., Grisel, A.: A low power integrated catalytic gas sensor. Sens. Actuators B **13**, 155–158 (1993)
74. Gall, M.: The Si planar pellistor array, a detection unit for combustible gases. Sens. Actuators B **16**, 260–264 (1993)
75. Zanini, M., Visser, J.H., Rimai, L., Soltis, R.E., Kovalchuk, A., Hoffman, D.W., Logothetis, E.M., Brewer, L., Bynum, O., Bonne, U., Richard, M.A.: Fabrication and properties of Si-based high-sensitivity microcalorimetric gas sensor. Sens. Actuators A **48**, 187–192 (1995)
76. Suehle, J., Cavicchi, R., Gaitan, M., Semancik, S.: Tin oxide gas sensor fabricated using CMOS micro-hotplates and in situ processing. IEEE Electron. Device Lett. **143**, 118–120 (1993)
77. Ivanov, P., Laconte, J., Raskin, J.-P., Stankova, M., Sotter, E., Llobet, E., Vilanova, X., Flandre, D., Correig, X.: SOI-CMOS compatible low-power gas sensors using sputtered and drop-coated metal-oxide active layers. J. Microsyst. Technol. **12**, 160–168 (2005)

78. Ivanov, P., Stankova, M., Llobet, E., Vilanova, X., Gracia, I., Cane, C., Correig, X.: Microhotplate sensor arrays based on sputtered and screen-printed metal oxide films for selective detection of volatile compounds. Sens. Trans. Mag. **36**, 16–23 (2003)
79. Korotchenkov, G., Brynzari, V., Dmitriev, S.: Electrical behaviour of $SnO_2$ thin films in humid atmosphere. Sens. Actuators B **54**, 197–201 (1999)
80. Barrettino, D., Graf, M., Zimmermann, M., Hagleitner, C., Hierlemann, A., Baltes, H.: A smart single-chip microhotplate-based gas sensor system in CMOS-technology. Analog Integr. Circuits Signal Process. **39**, 275–287 (2004)
81. Laconte, J., Rue. B., Flandre, D., Raskin, J.-P.: Fully CMOS-SOI compatible low-power directional flow sensor. In: Proceedings of the IEEE Sensors 2004 Conference, Vienna, Austria (2004)
82. Lim, H.C., et al.: Flexible membrane pressure sensor. Sens. Actuators A **119**, 332–335 (2005)
83. Singh, R., et al.: A silicon piezoresistive pressure sensor. In: Proceedings of the First IEEE International Workshop on Electronic Design, Test and Applications, vol. 1, pp. 181–184 (2002)
84. Jung, H.-M., Cho, S.-B., Lee, J.-H.: Design of smart piezoresistive pressure sensor. In: IEEE Proceedings of the Fifth Russian-Korean International Symposium on Science and Technology, pp. 202–205 (2001)
85. Pedersen, C., et al.: Combined differential and relative pressure sensor based on a double-bridged structure. In: Proceedings of IEEE Sensors, pp. 698–703 (2003)
86. Chau, M.-T., Dominguez, D., Bonvalot, B., Suski, J.: CMOS fully digital integrated pressure sensors. Sens. Actuators A **60**, 86–89 (1997)
87. Neumeister, J., Schuster, G., Von Münch, W.: A silicon pressure sensor using MOS ring oscillators. Sens. Actuators A **7**, 167–176 (1985)
88. Schörner, R., Poppinger, M., Eibl, J.: Silicon pressure sensor with frequency output. Sens. Actuators A **21**, 73–78 (1990)
89. Wang, Y., Zheng, X., Liu, L., Li, Z.: A novel structure of pressure sensors. IEEE Trans. Electron. Devices **38**, 1797–1802 (1991)
90. Gallon, C., et al.: Electrical analysis of mechanical stress induced by STI in short MOSFETs using externally applied stress. IEEE Trans. Electron. Devices **51**, 1254–1261 (2004)
91. Rinaldi, G., Stiharu, I., Packirisamy, M., Nerguizian, V., Landry, R., Raskin, J.-P.: Dynamic pressure as a measure of gas turbine engine (GTE) performance. Meas. Sci. Technol. (2010). doi:10.1088/0957-0233/21/4/045201
92. Wee, K.W., et al.: Novel electrical detection of label-free disease marker proteins using piezoresistive self-sensing micro-cantilevers. Biosens. Bioelectron. **20**, 1932–1938 (2005)
93. Lubecke, V.M., Barber, B., Chan, E., Lopez, D., Gross, M.E., Gammel, P.: Self-assembling MEMS variable and fixed RF inductors. IEEE Trans. Microwav. Theory Tech. **49**, 2093–2098 (2001)
94. Dahlmann, G.W., Yeatman, E.M., Young, P., Robertson, I.D., Lucyszyn, S.: Fabrication, RF characteristics and mechanical stability of self-assembled 3D microwave inductors. Sens. Actuators A: Phys. **97–98**, 215–220 (2002)
95. Zou, J., Liu, C., Trainor, D.R., Chen, J., Schutt-Ainé, J.E., Chapman, P.L.: Development of three-dimensional inductors using plastic deformation magnetic assembly (PDMA). IEEE Trans. Microwav. Theory Tech. **51**, 1067–1075 (2003)
96. Nguyen, H.D., Hah, D., Patterson, P.R., Chao, R., Piyawattanametha, W., Lau Erwin, K.: IEEE J. Microelectromech. Syst. **13**(3), 406–413 (2004)
97. Sasaki, M., Briand, D., Noell, W., de Rooij, N.F., Hane, K.: Three-dimensional SOI-MEMS constructed by buckled bridges and vertical comb drive actuator. IEEE J. Sel. Top. Quantum Electron. **10**, 455–461 (2004)
98. Freund, L.B., Suresh, S.: Thin film materials. Cambridge University Press, UK (2003)
99. Ohring, M.: Materials science of thin films, deposition and structure. Academic Press, USA (2002)

100. Raskin, J.-P., Iker, F., André, N., Olbrecht, B., Pardoen, T., Flandre, D.: Bulk and surface micromachined MEMS in thin film SOI technology. Electrochim. Acta **52**, 2850–2861 (2007)
101. André, N., Iker, F., Raskin, J.-P.: CMOS compatible 3D MEMS in SOI technology. In: Proceedings of the Third Workshop of the Thematic Network on Silicon on Insulator Technology, Devices and Circuits, EUROSOI'07, Leuven, Belgium, pp. 69–70 (2007)
102. Fan, Z., Chen, J., Bullen, D., Liu, C., Delcmyn, F.: Design and fabrication of artificial lateral line flow sensors. J. Micromech. Microeng. **12**, 655–661 (2002)
103. Chen, J., Liu, C.: Development and characterization of surface micromachined, out-of-plane hot-wire anemometer. IEEE J. Microelectromech. Syst. **12**, 979–988 (2003)
104. Iker, F., Andre, N., Pardoen, T., Raskin, J.-P.: Three-dimensional self-assembled sensors in thin film SOI technology. IEEE J. Microelectromech. Syst. **15**, 1687–1697 (2006)
105. Makinwa, K.A.A., Huijsing, J.H.: A smart wind sensor using thermal sigma-delta modulation techniques. Sens. Actuators A **97–98**, 15–20 (2002)
106. Kohl, F., et al.: Development of miniaturized semiconductor flow sensors. Measurement **33**, 109–119 (2003)
107. Fürjes, P., et al.: Thermal characterisation of a direction dependent flow sensor. Sens. Actuators A **115**, 417–423
108. Laconte, J., Flandre, D., Raskin, J.-P.: Micromachined Thin-Film Sensors for SOI-CMOS Co-Integration. Springer (2006). ISBN-10 0-387-28842-2
109. André, N., Rue, B., Renaux, C., Flandre, D., Raskin, J.-P.: 3D capacitive MEMS sensors co-integrated with SOI CMOS circuits. In: Proceedings of the Fourth Workshop of the Thematic Network on Silicon on Insulator Technology, Devices and Circuits, EUROSOI'08, Tyndall National Institute, Cork, Ireland, pp. 75–76 (2008)
110. Sobieski, S., André, N., Raskin, J.-P., Francis, L.A.: Temperature effect on Lorentz based magnetometer. Sens. Lett. **7**, 456–459 (2009)

# Floating-Body SOI Memory: The Scaling Tournament

M. Bawedin, S. Cristoloveanu, A. Hubert, K. H. Park and F. Martinez

**Abstract** In this paper, we present an overview of the typical device architectures of the single transistor capacitorless dynamic random access memory (1T-DRAM). This memory uses only one transistor and takes advantage of floating body effects in SOI and SOI-like devices. The principles of operation and key mechanisms for programming are described. The various approaches are compared in terms of architecture, performance and potential for aggressive scaling.

## 1 Introduction

As the storage capacitance must be kept constant, for the next generations of conventional 1T-1C (1 Transistor + 1 Capacitor) DRAM cells beyond the 22 nm technology node, the miniaturization of the bulky storage capacitor will become more and more difficult. In the future, several existing solutions like trench or stacked capacitors using high-k materials will remain expendable but at the expense of the lost of performances and cost rising. Several years ago [1–5] a new generation of DRAMs using only one transistor, called 1T-DRAM, was proposed as an alternative to the usual 1T-1C DRAM architecture. This memory cell uses the floating body of a single transistor to hold the information, i.e. to store the

---

M. Bawedin (✉) and F. Martinez
IES (UMR 5214), Université de Montpellier II, 34095, Montpellier, France
e-mail: maryline.bawedin@univ–montp2.fr

M. Bawedin, S. Cristoloveanu and K. H. Park
IMEP-LAHC (UMR 5130), Grenoble INP Minatec, 38016, Grenoble Cedex 1, France

A. Hubert
CEA-LETI, Minatec, 17 rue des Martyrs, 38054, Grenoble Cedex 9, France

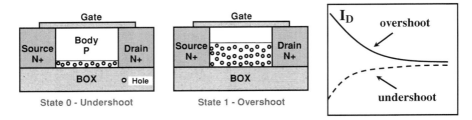

**Fig. 1** Schematic cross-section of a partially depleted (PD) SOI nMOSFET used as a 1T-DRAM memory cell and the corresponding current transient for 0 state (*undershoot*) and 1 state (*overshoot*) resulting from the lack or excess of majority carriers (*holes*), respectively

charge. As for 1T-1C DRAM cells, the crucial performance factors are a high retention time, sufficient sensing margin, low programming bias which enable low-power consumption, and further cell scalability. In that way, the floating body (Fig. 1) of the SOI or SOI-like transistors provides naturally an ideal storage environment to achieve competitive performances within a small storage volume. The 1T-DRAMs take advantage of parasitic floating-body effects in SOI transistors where the body is isolated. All these floating-body effects originate from the non-equilibrium of the body majority carriers.

In all 1T-DRAM variants, bit '1' (1-state) reflects the temporary generation of a majority carrier excess in the body which increases the potential and hence the drain current. Conversely, bit '0' (0-state) features a lower current due to the removal of majority carriers from the body. The majority carrier variation in the body can be sensed by measuring the current difference between the '0' and '1' states (Fig. 1).

These memory cells can be classified in different groups with respect to the way the excess of majority carriers is generated i.e. depending on the 1-state programming methods. In a nMOSFET based 1T-DRAM, the 1-state current level $I_1$ is triggered thanks to the excess hole charge generated into the body. This extra charge can be achieved by (1) impact ionization, (2) bipolar junction transistor (BJT) effect, (3) band-to-band (B2B) tunneling generation and (4) gate tunneling current. The programming of the 0-state current level $I_0$, i.e. the body hole charge removal, is usually achieved by forward biasing the drain- or source-body junction. Due to the lack of efficiency of the hole extraction during the 0-state programming, the current level $I_0$ remains of the same order of magnitude as $I_1$. As a consequence, the threshold voltage shift (due to the body hole charge variation) produces poor cell signal sensing margin, i.e. low $|I_1 - I_0|$ difference. To improve $|I_1 - I_0|$, dynamic coupling between front and back gates in fully depleted (FD) SOI or double-gate MOSFETs can be used.

The next two sections introduce the methods for programming and reading the '1' state (Sect. 2) and '0' state (Sect. 3). The physics principles of the various types of 1T-DRAM, their performance in terms of programming speed, bias and power consumption, and the technological solutions for further optimization and scaling are critically reviewed in Sects. 4 and 5. Finally, several candidates able to survive aggressive miniaturization and to meet the demand for higher speed and lower power consumption are presented. In this article, only n-channel MOSFETs, i.e. with P-type body, are considered, but the same issues can be applied to p-channel 1T-DRAM.

## 2 Programming 1-State

During the 1-state programming, the amount of majority carriers (holes) in the body is enlarged. This induces a dynamic decrease of the threshold voltage ($V_{TH} \rightarrow V_{TH} - \Delta V_{TH1}$) and an increase in drain current $I_1$. The threshold voltage shift $\Delta V_{TH1}$ (or $\Delta V_{TH0}$) results from the combined effects of the hole charge and the corresponding body potential variations. In the following, the body charging mechanisms are reviewed.

### 2.1 Impact Ionization (II)

A frequently used method for the 1-state programming consists in generating holes inside the body by impact ionization (Fig. 2a) [2, 3, 6–12]. A relatively high positive drain voltage $V_D$ is applied while the front interface is in inversion mode ($V_D > |V_{GF} - V_{TH}|$). The holes generated at the pinch-off region, close to the drain, move and accumulate into the body (Fig. 2a, inset).

During the programming, the front-gate voltage $V_{GF}$ remains unchanged (same as in hold and read conditions), while the drain voltage $V_D$ is pushed from a low value ($\sim$ mV) to a higher one ($\sim$ V). As the body/drain junction is reverse biased, when the holes are filling the body, the floating body potential increases above the steady-state level it had before the programming stage (Fig. 3a). For the 1-state reading (or hold), $V_D$ is switched back to its low level. As the excess hole charge remains inside the body, the potential shift $\Delta V_B$ achieved during the programming period is preserved, inducing a 'dynamic' $V_{TH}$ lowering, hence an increase in drain current. During the reading/holding period, as the body potential is higher than its steady-state value, the stored holes are gradually evacuated through the body-to-source/drain junctions.

Hence the body potential returns gradually ($\sim$ ms) downward to steady-state (Fig. 3a). This leads to a drain current overshoot or return to equilibrium with time (Fig. 2a). It is worth mentioning that the reading is non-destructive since the low drain voltage ($\sim$ mV) does not enhance significantly the 'parasitic' hole current leakage.

While the impact ionization condition allows fast 1-state writing speed and relatively large sensing window, hot-electron injection can induce an undesired shift in the threshold voltage $V_{TH}$ during cyclic 1T-DRAM operations. On the

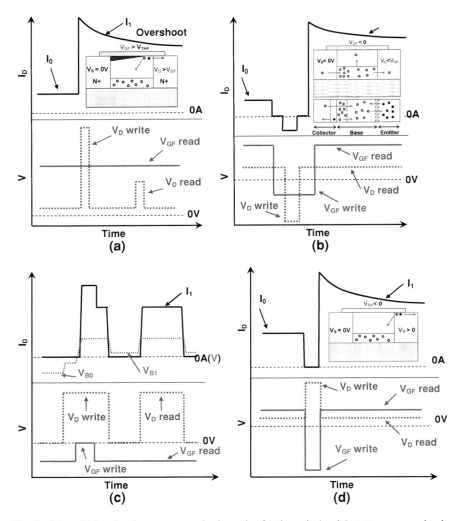

**Fig. 2** External bias signal sequences and schematics for the write/read 1-state programming by **a** impact ionization, **b** bipolar junction transistor (BJT) effect 1st method, **c** BJT effect 2nd method and **d** band-to-band tunnelling (B2B) generation. Plotted versus time are the front-gate voltage $V_{GF}$, drain voltage $V_D$, body potential $V_B$ and drain current $I_D$

other hand, the hot electrons injected into the gate oxide produce a premature degradation affecting the memory cell retention time. Furthermore, for the next 1T-DRAM generations to achieve a faster programming, the impact ionization rate will have to be enhanced. This unfortunately means higher $V_D$ and hence higher power consumption. Finally, higher voltages would require on-chip charge pump circuits which increase the area of peripheral circuits and affect the array efficiency.

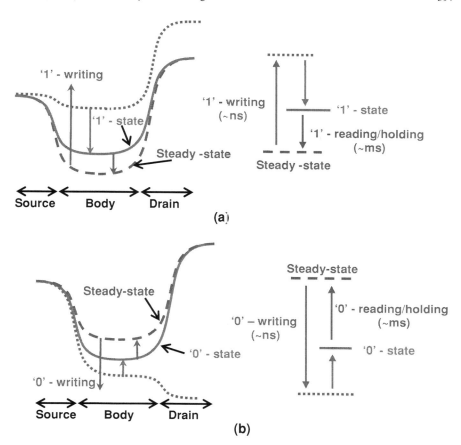

**Fig. 3** Front surface variation during the writing/reading of the **a** 1-state programmed by impact ionization and **b** 0-state programmed by the forward drain biasing method

## 2.2 Bipolar Junction Transistor (BJT)

The second method for 1-state programming takes advantage of the intrinsic BJT effect which can be activated in the floating body of SOI MOSFETs. In that case, the source (N+), body (P) and drain (N−) act as the emitter (or collector), base and collector (or emitter) of the BJT (Figs. 2b, c). To turn on the BJT effect, a hole current has to be generated in the base. Since the base is floating, the body potential increases during the programming. As for the previous method, BJT produces a threshold voltage lowering, hence a current increase. Two BJT programming methods are reviewed, their difference residing in the way the excess holes are created.

(a) Historically, the first BJT programming method [2, 3] was introduced at the same time as the impact ionization method. It consists in applying a negative $V_D$

pulse embedded into a negative gate $V_{GF}$ pulse (Fig. 2b). The programming starts by switching $V_{GF}$ to a negative value which has to be large enough ($\sim V$) to enable and keep an accumulation layer at the front interface. Then, $V_D$ is also pushed to a high negative value. At this stage, the body/drain and body/source junctions are forward and reverse biased, respectively. If $V_D$ is large enough, the energy of electrons flowing from the drain to the source enables the generation of electron/hole pairs by impact ionization at the body/source junction (Fig. 2b, inset). Before the reading (or holding) stage, the drain voltage is switched back to its reading low value ($\sim mV$) prior to the gate, in order to keep the stored hole charge in the body.

The main advantage of this method is the power consumption reduction due to the low effective current flowing between the drain and the source during the programming: (1) the BJT gain $\beta$ is low ($I_S = I_D \beta/(1 + \beta)$) and (2) the base hole current $I_h$ resulting from the impact ionization is small ($I_S = \beta I_h$).

(b) In the second BJT programming method [13–15], the bipolar effect is initiated with different bias conditions (Fig. 2c). To enable BJT, a $V_{GF}$ pulse is now embedded into a high positive drain voltage pulse. Notice that during the reading or holding, $V_{GF}$ is kept at a negative value lower than the front flat-band voltage $V_{FBF}$ ($V_G < V_{FBF} < 0$) in order to preserve the hole charge stored previously. At the programming onset, $V_D$ is increased while $V_{GF}$ is held below $V_{FBF}$: holes are generated by impact ionization in the body through the reverse biased body/drain junction. Nevertheless, as explained above, the resulting hole current $I_h$ is small and the BJT effect is not significant. To write the 1-state, the $V_D$ value is close to the breakdown voltage of the parasitic BJT but not enough to trigger on the bipolar current.

The next step consists in amplifying the hole generation. In order to do so, $V_{GF}$ is pushed to a higher value ($>V_{FBF}$) to lift up the body potential $V_B$ by dynamic gate coupling enough to increase the BJT current $I_h$ gain ($\beta$) and drive the latch condition, $\beta(M - 1) \approx 1$, where M is the impact-ionization multiplication factor. As the body potential is increased, the source/body junction becomes forward biased and the BJT turns on increasing the collector current $I_D$. Hence, the impact ionization and the resulting hole base current $I_h$ are drastically enhanced. As a result, electrons are injected into the body (base) from the source (emitter), due to the forward biased source/body diode and are collected by the drain (collector). The extra drain current enhances the impact ionization, which, in turn, drives the body to be more positive and enables a regenerative action to occur. Prior to reading (or holding), $V_{GF}$ is switched back to its negative reading value while $V_D$ is maintained high. Since $I_h$ is now large enough, the body potential (and current $I_D$) remains 'pinned' to its high level. The negative $V_{GF}$ ($<V_{FBF}$) stores the generated holes at the front interface. To perform the reading, a high positive $V_D$ pulse is applied. Consequently, the body potential $V_B$ increases by dynamic coupling and again turns on the BJT and drain current.

It is interesting to mention that the (drain) latch voltage decreases as the channel length is reduced. This dependence is due to a larger impact-ionization multiplication factor M and a higher gain $\beta$. The $(M - 1)$ value increases as the carrier-velocity saturation is more easily reached in short-channel devices

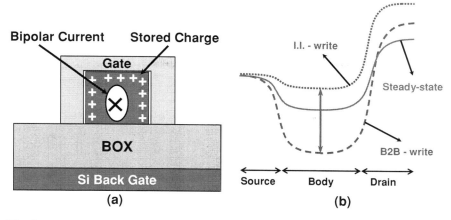

**Fig. 4** **a** Cross-section of a tri-gate 1TDRAM using the BJT effect programming technique [15] and **b** comparison between the impact ionization and the B2B tunnelling injection methods in terms of body potential variations during the 1-state writing [16]

providing a lower $V_D$ value at saturation. Furthermore, the higher $\beta$ value is due to the shorter base length ($L_{CH}$). Hence, under latch conditions, the value of $V_D$ is expected to be gradually lowered as $L_{CH}$ decreases.

The fact that the back gate is always grounded ($V_{GB} = 0$ V) stands as the main advantage of the second BJT method. It can be applied to a variety of advanced SOI technologies [15] such as partially depleted (PD), fully depleted (FD), double-gate, FinFET, or Tri-Gate (Fig. 4a).

Compared to the previous programming techniques, other improvements are the fast read and write operations ($\sim 2$ ns) [13]. It is worth noting that the impact ionization through the drain junction is not the only mechanism generating holes during the programming and reading. Indeed, the net potential drop at the gate edges can reach values up to 4 V [14]. Therefore, at the gate-to-drain (source) overlap region, holes can also be generated by band-to-band tunneling or impact ionization. This aspect is beneficial for improving the programming speed and stabilizing the 1-state drain current, but can deteriorate the retention time and the device reliability (see Sect. 4).

## 2.3 Band-to-Band (B2B) Tunneling

The holes generated by B2B arise from the gate-to-drain (source) overlap region (Fig. 2d, inset) [8, 16–20]. In order to produce the holes, the drain/oxide interface is in strong inversion; the band bending and the local electric field must be large enough to allow holes (electrons) tunneling from the conduction (valence) band into the valence (conduction) band. The holes are collected by the body and the electrons flow toward the drain contact. During the programming, a negative $V_{GF}$ pulse ($\leq V_{FBF}$) is applied while $V_D$ remains positive (Fig. 2d). At the programming onset, if a high negative gate bias is used ($V_{GF} \ll V_{FBF}$), the body potential decreases by dynamic gate coupling and becomes negative (Fig. 4b). As the holes cannot be supplied and accumulated instantly, the front interface is initially depleted. While the holes are gradually filling the body, the potential increases until it reaches equilibrium ($\sim 0$ V) when the accumulation layer is completed. Compared with the other programming methods, the holes are not able to escape through the source because the body potential is negative. Therefore, the storage and the body potential variation are more efficient. Obviously, if the programming time is too short, the resulting body potential increase and the programming efficiency are reduced. At the reading onset, $V_{GF}$ is increased up to or above $V_{THF}$ (Fig. 2d) and the holes stored at the front interface are pushed down toward the bottom interface.

If the net potential drop $|V_G| + |V_D|$ on the gate is increased, the electric field and hence the B2B generation are enhanced. As a result, the programming time can be readily shorted by increasing either $|V_G|$ or $V_D$. To reach a competitive programming speed compared with the II method, the total signal swing $|V_G| + |V_D|$ has to be increased by about 20% [16] but it remains similar to the BJT method. Moreover, the B2B method features a striking advantage which is low power consumption. Indeed, the B2B current is much lower than the drain current arising from impact ionization or BJT. Finally, the reliability of the memory cell, programmed with B2B, is expected to be superior to the impact ionization method.

## 2.4 Gate Tunneling Current

The use of the parasitic direct tunneling current through the gate oxide in a body contacted PD SOI pMOSFET was proposed and demonstrated by Guegan et al. [21] for 1T-DRAM application. In this specific device, a N+ body contact is left floating and the polysilicon gate covering the body contact is N+ doped in order to allow a strong injection of electrons (here the majority carriers) by tunneling from the polysilicon conduction band into the body (Fig. 5). The amount of electrons injected from the conduction band ECB into the body through the body contact determines the storage capability of the memory cell. Notice that as electron tunneling is more efficient, a pMOSFET was used instead of the conventional nMOSFET for the 1T-DRAM application. To program the 1-state, a negative gate bias is applied and the resulting ECB tunneling current induces the body potential decrease. As the n-type body potential becomes negative, a higher drain current can be observed. The main advantages of this method are: (i) the low bias and (ii) the power consumption reduction as compared with the 1-state programming by impact ionization and B2B tunneling (reduction by 6 and 2 orders of magnitude respectively).

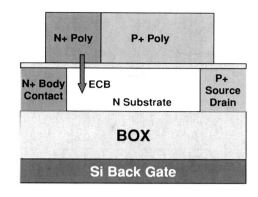

**Fig. 5** 1T-DRAM 1-state programming using the electron gate tunnelling current from the conduction band (ECB) of a N+ poly gate within a partially depleted p-MOSFET. The N+ body contact is left floating

## 3 Programming 0-State

To program the 0-state, a deficit of holes has to be created in the body. During the programming, the holes are removed and the body potential decreases. As a result, the effective threshold voltage is enhanced ($V_{TH} \rightarrow V_{TH} + \Delta V_{TH0}$) and the drain current lowered. It is obvious that the 0-state programming technique has to be compatible with that of the 1-state especially for embedded applications where the memory cell array configuration must be taken into account. Basically, two methods are available for removing the holes from the body.

### 3.1 Forward Biased Junction

In order to expel the holes (Figs. 3 and 6a), the front-gate voltage $V_{GF}$ is kept at the reading voltage ($\geq V_{THF}$) [2, 3, 7] (or pushed to a negative value [8, 10, 12]) while a negative voltage ($\sim V$) is applied on the drain contact. The body/drain and body/source junctions become forward and reverse biased, respectively. Consequently, the hole current injected from source inside the body is much lower than the one extracted through the drain junction: the holes are readily evacuated from the body.

This technique is reliable and fast since the time response of the forward body/drain junction to the bias switch is nearly instantaneous. Nevertheless, the resulting body potential variation is not really efficient and the 0-state current level usually keeps the same order of magnitude as the 1-state ($\sim \mu A$). This augments the power consumption during the reading and alters the retention time.

### 3.2 Capacitive Coupling

Let us assume that $V_{GF}$ is below the front flat-band voltage $V_{FBF}$ and a hole accumulation layer stands at the front interface. A fast removal can be performed

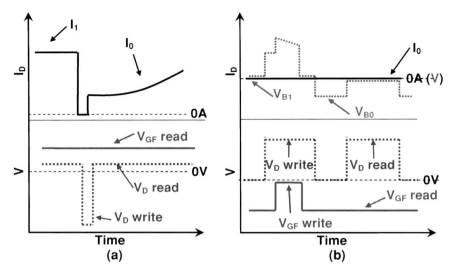

**Fig. 6** External bias signal sequences for the write/read '0'-state operations by **a** 'forward bias junction' and **b** capacitive coupling

by applying a front-gate pulse higher than $V_{FBF}$ (Fig. 6b) [13–15]. When the gate voltage increases, the body potential follows by capacitive coupling. As the potential increases, the front interface becomes depleted and the body/drain and/or body/source junctions become 'forward biased'. Therefore, the holes are evacuated through the junctions. For reading, the gate is switched back to a negative value and the body potential decreases by capacitive coupling to a negative value. Since the holes cannot be accumulated instantly, the body potential is pinned to its negative value as well. The return to equilibrium during the reading or holding depends on hole charging mainly via the junction leakage current $I_{LEAK}$. Leakage is detrimental because it makes the 0-state drain current increase, altering the retention time.

In Fig. 6b, the capacitive coupling method is applied to produce a body potential $V_B$ decrease which prevents the BJT activation during the reading.

During the 0-state programming, the drain voltage keeps a high positive value for compatibility with the whole BJT 2nd method. This means that a positive small source voltage pulse has to be applied in order to avoid that the BJT turns on and charges the body. On the other hand, the source voltage cannot be too high in order to allow the holes to leave the body.

Despite the signal complexity increase, an additional drain level can be used to write 0 (3 instead of 2 different $V_D$ bias levels). As previously, the front-gate voltage is pulsed up to raise the body potential while $V_D$ is kept below the breakdown voltage of the BJT. Hence, the holes are readily evacuated through the body/source junction without latch activation [22, 23].

In FD SOI or double-gate 1T-DRAM, the capacitive coupling method can also be used to cut-off an inversion channel standing at one interface by applying a negative voltage pulse on the opposite gate [17, 18]. This method, currently used in the MSDRAM (see Sect. 5.3), allows achieving very fast programming time and ultra low 0-state drain current level (1 nA/μm). Consequently, the drain current sensing ratio ($I_1/I_0$) and the retention time are greatly improved.

The main difficulty resides in combining all these 0 and 1-state programming methods to achieve simultaneously the best compatibility with a specific technology (PD SOI, FD SOI, double-gate, FinFET...) and high performances, i.e. low programming voltage, high retention time and high signal sense margin (maximum $\Delta V_{TH}$). The retention time is defined by the minimum current difference ($I_1 - I_0$) detectable by the sensing circuit. The recently developed current sense amplifiers are able to detect a current difference down to 3 μA. All together, the performance improvement implies a reduction in power consumption.

When a memory cell is embedded into an array, the entire bias configuration and the disturb conditions have to be taken into account. During the programming of a cell standing in a word line, the word (row) and bit (column) line biases are also applied to other cells belonging to the same row and column. Moreover, the number of signal lines has to be limited in order to reduce the complexity of the control circuits. To be reliable, a single memory cell should have a stable retention time of a few seconds at room temperature to secure a 'disturb free' memory array.

## 4 Performance Improvement and Scaling

The partially depleted (PD) SOI 1T-DRAM (Fig. 7a) allows reducing the memory cell size compared to usual 1T-1C DRAMs. The next question concerns the scalability and reliability of the 1T-DRAM itself as well as its compatibility with more advanced SOI technologies (e.g. FD, FinFET).

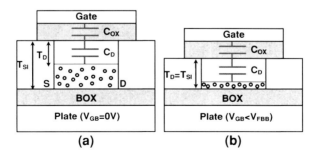

**Fig. 7** Cross-section of SOI nMOSFETs: **a** partially depleted (PD) with a hole quasi-neutral zone and **b** fully-depleted (FD) with a back accumulation layer. $T_D$ and $T_{Si}$ are the depletion layer and silicon film thicknesses, respectively. $C_{OX}$ and $C_D$ are the front-gate oxide and depletion layer capacitances; $V_{GB}$ and $V_{FBB}$ are the back-gate and back flat-band voltages

As the effective channel length $L_{CH}$ decreases, the improvement of the retention time and current signal margin ($|I_1 - I_0|$ or $\Delta V_{TH}$) remains the critical target and also the 1T-DRAM weakness. When $L_{CH}$ shrinks, the hole storage volume is reduced and the maximum electric field is enhanced. This is beneficial for improving the 1-state writing speed, i.e. to reduce the programming voltage and/or time. However, the 0-state retention time can be noticeably altered by parasitic hole generation. If the current sensing margin can be increased, the retention time is improved because it takes longer for the memory cell to recover its steady state and reach the minimum acceptable $|I_1 - I_0|$ difference.

On the other hand, short-channel effects (SCE) can reduce further the storage volume and the threshold variation $\Delta V_{TH}$ during the 1-state programming. Another drawback of the SCE which deteriorates the 1-state programming is the drain-induced barrier lowering (DIBL). Indeed, when a high drain voltage is applied to induce the impact ionization, the body/source barrier is lowered. Hence, the generated holes can escape more easily through the source and the storage efficiency is deteriorated. Notice that for this reason, the DIBL effect is less relevant for B2B generation technique as the body potential is negative at the programming onset (Fig. 4b).

In PD devices, the channel doping can be increased in order to suppress the SCE and enlarge $\Delta V_{TH}$ (= $\Delta V_{TH0} + \Delta V_{TH1}$) (Fig. 8a) [7, 10]. Hence, the junction depletion spreading effect is reduced while a higher current sense margin is achieved. It corresponds to increased effective storage volume and current difference $|I_1 - I_0|$, leading to longer retention time. On the other hand, when the body doping increases, the effective potential barrier at the source/body junction is enhanced reducing the hole leakage into the source for a more efficient hole

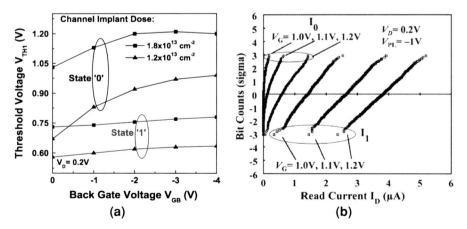

**Fig. 8 a** Threshold voltage versus back-gate (or field plate) bias $V_{GB}$ in 0 and 1-states of a fully-depleted 1T-DRAM for two different channel dopings (adapted from [7]). **b** Statistical distribution of the memory cell 0-state $I_0$ and 1-state $I_1$ read currents of 1,000 transistors in a 96 kb memory array plotted for several front-gate voltages $V_{GF}$ (from [7])

storage. Unfortunately, a too high body doping augments the junction leakage current $I_{LEAK}$ during the reading and induces random dopant fluctuation (RDF) effects [7, 24]. While $I_{LEAK}$ degrades the retention time, the additional RDF effect produces random changes in $\Delta V_{TH}$ and fluctuations in read current (Fig. 8b). Since the read current also varies with time and only its statistical distribution is known, the comparison with a reference cell current to identify the 0 and 1-states is problematic and requires special sensing circuit design [25–27]. Notice that, when the impact ionization programming technique is used, if the body doping decreases, the maximum electric field at the body/drain junction is lowered. Then, higher $V_D$ and/or longer programming time are required to achieve the same performance. This means that it becomes difficult for a PD 1T-DRAM to be scaled aggressively [6, 19].

The RDF effect and the junction leakage current are attenuated if the channel doping decreases. This is beneficial for the memory reliability and stability as well as for retention time improvement. However, when the doping is lowered, $\Delta V_{TH}$ and therefore the current sense margin decrease. It is not clear if by optimizing other technological parameters one can simultaneously take advantage of a low doping and keep a decent current sensing margin.

In PD SOI MOSFETs, the threshold variation $\Delta V_{TH}$ can be approximated by $\Delta V_B(C_D/C_{OX})$ [7], where the body potential variation $\Delta V_B$ is proportional to the change in hole density. This model remains valid for FD SOI MOSFET if the back channel is in accumulation mode i.e. behaves like PD devices. To increase the depletion layer capacitance $C_D$ and hence $\Delta V_{TH}$, the depletion layer thickness $T_D$ has to become thinner (Fig. 7a). This means that the channel doping in PD MOSFETs should be increased. A more efficient approach is to use FD devices where a thin depletion layer can be achieved by decreasing the silicon film thickness $T_{Si}$ ($\approx T_D$) while the back interface remains accumulated by applying a negative voltage $V_{GB}$ on the back gate (Fig. 7b). It can be observed that, as $V_{GB}$ becomes more negative, the 1-state threshold shift $\Delta V_{TH1}$ is nearly constant whereas the 0-state $\Delta V_{TH0}$ increases and finally saturates when the back interface enters in strong accumulation regime ($V_{GB} < -2$ V, Fig. 8a). Remark that if the back interface is depleted, then $C_D$ is connected in series with the buried oxide capacitance, hence the resulting $C_D$ is much lower and $\Delta V_{TH}$ is drastically reduced.

In FD devices (Fig. 7b), the channel doping can be decreased to the 'intrinsic' value while keeping an excellent current sense margin as long as a sufficiently negative $V_{GB}$ is applied. Since $I_{LEAK}$ and the RDF are reduced, the retention time is enhanced. As the silicon film thickness decreases, $I_{LEAK}$ reduction also originates from the smaller effective junction area ($I_{LEAK} \propto T_{Si}$). When a $P^+$ doped back gate or a $P^+$ field plate is used with FD devices (Figs. 7b and 9a, [10, 12]), the maximum $\Delta V_{TH}$ (i.e. the maximum sensing margin) can be further improved by increasing the plate doping [7]. It is also obvious that the optimal $|V_{GB}|$ can be lowered to a less negative bias if a thinner BOX is used. The amplification factor ($C_D/C_{OX}$) of $\Delta V_{TH}$ can be enlarged with thicker front-gate oxide $T_{OX}$. Finally, thinner films are suitable for scaling [28].

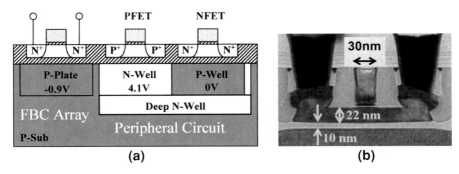

**Fig. 9** Cross-section of **a** well structure implanted for compatibility purpose between the floating body cell (FBC) memory array and the peripheral circuits (from [10]) and **b** very thin silicon film 1T-DRAM using a 45 nm logic technology (TEM picture from [11])

These considerations naturally lead to adopt FD SOI as the solution to overcome the short-channel effect and improve the retention time and the reliability of the 1T-DRAM by reducing the silicon film thickness $T_{Si}$ and the channel doping. However, in fully depleted SOI, the use of a common back-gate voltage $V_{GB}$ different from 0 V induces compatibility problems with the peripheral circuits and requires special implantations under the back gate (Fig. 9a). On the other hand, a common $V_{GB}$ limits the possible combinations of the different 0 and 1-state programming methods. To the best of our knowledge, only the second generation BJT method (Fig. 2c) allows keeping a common grounded $V_{GB}$ for all the memory cells embedded in the same array while enabling the capacitive coupling method which is the most efficient 0-state programming technique without disturb issues. This is made possible because the front inversion channel is not required during the reading and programming periods (Fig. 4a). Regarding the dependence on $T_{Si}$, there are two opposing mechanisms, the body coefficient ($C_D/C_{OX}$) and the body storage volume capacity. The body coefficient increases as $T_{Si}$ decreases while the storage capacity is reduced. Due to these competing mechanisms, an optimum point was found between 50 and 80 nm [29, 30].

In FD SOI, as the back-gate oxide decreases, the electric field at the gate-to-drain overlap region is enhanced, deteriorating the retention time of 0-state by enabling parasitic B2B tunneling generation due to the negative $V_{GB}$. However, it is demonstrated [12] that the FD SOI 1T-DRAM can be scaled down while keeping the electric field strength and the electrical characteristics constant. In this case, the operation bias and the thicknesses of the front oxide, back-gate oxide and silicon film should also be reduced. In order to keep the electric field constant, special technological 'improvers' can also be used, for example lightly doped drain LDD [10] or gate non-overlap regions [14]. These techniques tend to reduce the parasitic B2B tunneling generation and consequently extend the retention time. Regarding the back-gate voltage, if it is not sufficiently high, the holes cannot be kept in the body efficiently, so the $V_{TH}$ shift and, consequently, the current sensing margin is reduced. For too large negative back-gate bias, in addition to hole leakage increase arising from parasitic B2B tunneling at the back gate-to-drain overlap, the hole accumulation at the back-gate interface will remain even at the 0-state hence, the current sensing margin drops down to zero.

All previous considerations highlight that the ultimate technological step to be considered for the 1T-DRAMs is their implementation within double-gate (DG) devices [6, 9, 13, 17, 19, 31, 32]. Indeed, with FD devices, in order to improve memory cell performances and achieve low bias operation, it is necessary to use thinner silicon film and BOX. In addition, as a negative $V_{GB}$ is required to improve the hole storage, separated wells have to be implanted below the back gates in order to ensure that the other surrounding circuits work properly [10]. Finally, in double-gate devices, the back gate of each transistor can be controlled individually, which enables to get rid of additional technological steps and take full advantage of the different programming methods [9, 13, 17].

In double-gate devices, as the thin body efficiently 'collapses' the electric field lines from the drain, a thicker gate oxide can be used avoiding drain-induced barrier lowering (DIBL) while keeping a moderate off-state leakage. By combining a thin silicon film thickness and thick gate oxide, double gate DG 1T-DRAM has the capability to convert the small body potential gains $\Delta V_B$ into larger $\Delta V_{TH}$ shift. With vertical structures (FinFET with separated gates), one gate can easily be manufactured with a thicker oxide.

## 5 Are We Ready for the 22 nm Tournament?

We have evoked the advantages of the different programming techniques and the scaling effects on the SOI 1T-DRAM performance. We now review several combinations of the programming methods and technological processes which highlight the most promising candidates to scaling. In the following, variants of 1T-DRAMs using FD, PD and double-gate devices are briefly described. The performance in terms of programming bias, retention time, and current sense margin are provided quantitatively, for various programming methods. We will also introduce more exotic and hopefully promising memory cell architectures, demonstrated so far only by numerical simulations.

### 5.1 Toshiba Approach

In the first example (Fig. 9a, [10, 12, 24]), the feasibility of a FD SOI 1T-DRAM cell was experimentally demonstrated for 90 nm technology node in a 128 Mb memory array. Numerical simulations were performed for the devices toward the 32 nm technology. The impact ionization and the forward bias techniques are used respectively for the 1 and 0-state programming. To improve the retention time and the current sensing margin, the CMOS process includes LDD, a relatively high channel doping ($3 \times 10^{17}$ cm$^{-3}$) and P-doped field plate. In addition, Co salicide and Cu wirings are used to reduce the parasitic source and drain resistances which usually degrade the current signal. The gate length, the front-gate oxide, buried oxide and the silicon film thicknesses were 145, 6, 25 and 43 nm, respectively. The 1 (0)-state programming voltages were +1.5 V (−2.3 V) for the front gate and +2.2 V (−1.5 V) for the drain. The plate bias $V_{GB}$ is fixed at −2.5 V. A $\Delta V_{TH}$ of 420 mV and a retention time of 70 ms were achieved for a temperature of 85°C.

## 5.2 Intel Approach

A stand-alone FD SOI 1T-DRAM memory cell (Fig. 9b, [11]) was fabricated with a 45 nm logic technology. Simulations were used to demonstrate the viability of the memory for 16 nm technology node. No details were provided on the programming methods and bias levels. Nevertheless, since (i) the worst-case disturb condition occurs at high drain voltage and lowers the retention time due to residual B2B tunneling generation and (ii) the 0-state current level has the same order of magnitude as the 1-state, it is reasonable to assume that the impact ionization and the forward bias programming methods were applied. To improve the retention time, low doped source/drain implants were used. A high-k dielectric was implemented. The gate length, width, and the thicknesses of the BOX and Si film were 55, 65, 10 and 22 nm, respectively. The thinner BOX allowed reducing significantly the back-gate voltage.

Since the silicon film is very thin, the channel can be undoped. This allows suppressing the RDF effects and lowering significantly the junction leakage current. A $\Delta V_{TH}$ of 400 mV can be achieved with a $V_{GB}$ of −2 V. The maximum retention time at the worst disturb condition is 25 ms for a temperature of 85°C. Notice that, if the worst-case disturb condition is not taken into account, the hold retention time can reach about 100 s at 85°C and in an array configuration up to 25 ms at 85°C. It is explained that the main physics phenomenon responsible for the retention time degradation is the SRH recombination/generation.

## 5.3 Z-RAM

PD SOI 1T-DRAMs were fabricated with 100 nm design rules and tested in the memory cell array condition [13–15]. The second generation BJT effect and the capacitive coupling techniques were respectively used to program the 1 and 0-states (Fig. 2c and 6b). To improve the retention time, a non-overlap structure was implemented. It allows decreasing the maximum electric field and hence reducing the leakage current during the holding and reading operations. Indeed, the maximum B2B tunneling generation peak usually occurs at the gate-to-drain and -source overlap regions. Moreover, the lower electric field drastically decreases the SRH generation rate by four orders of magnitude compared with conventional devices. Notice that a high channel doping ($10^{18}$ cm$^{-3}$) is needed in order to increase the impact ionization generation during the 1-state programming. This allows reducing the programming time and drain voltage.

The gate length, front-gate oxide and Si film thicknesses were 55, 5 and 80 nm, respectively. The gate voltage ($-1$ V) and drain voltage (3 V) were the same for 1 and 0-state programming, whereas a source voltage of 0.5 V was used only for the 0-state programming. The back gate is kept grounded. For this particular programming technique, a $\Delta V_{TH}$ larger than 500 mV and an excellent $I_1/I_0$ current ratio ($10^4$–$10^6$) are achieved. Thanks to the underlap regions, a retention time of 70 ms could be obtained at 85°C.

These programming modes, previously proposed by Okhonin et al. [13], can be applied to advanced technologies like MuGFET devices (Fig. 4a). For a fin width of 11 nm and a gate length of 50 nm, a retention time of 1 ms at 125°C was reported [15]. Recently [33], it was demonstrated with the second Z-RAM generation in 45 nm CMOS SOI technology, that the effect of the source/drain asymmetry can produce a significant bias reduction (by 15%) and retention improvement (3 times higher).

## 5.4 MSDRAM

The Meta-Stable DRAM (MSDRAM) memory cell physics principles are based on the MSD hysteresis effect illustrated in Fig. 10a [34]. Note the very wide memory window and the current ratio $I_1/I_0$ which exceeds six orders of magnitude. The MDRAM [17] is the first 1T-DRAM taking full advantage of double-gate operation which allows combining the low consumption and superior reliability (thanks to the B2B tunneling for 1-state programming) with the efficiency of the 0-state programming by dynamic gate coupling. This preliminary demonstration used large area FD MOSFETs fabricated with unoptimized technology, which explains the disproportionate bias levels. It is clear that specific source and drain architectures [18] allow enhancing the retention time while reducing the programming voltage and/or time.

The MSDRAM scalability was demonstrated by 2D simulations of a 50 nm long channel memory cell with DG configuration and ultra thin BOX (Fig. 10b). The gate and drain voltages used for programming, reading, and holding have been restricted to less than 2 V, without affecting the MSD effect. The performance is remarkable: $I_1/I_0 = 10^3$ at the reading onset, 14 s retention time for $I_1/I_0 = 10$, and 5 ns programming time. Recent measurements confirm that the MSD memory effect is maintained in small MOSFETs with gate area of 0.1 µm$^2$ (0.35 × 0.35 µm). However, the sensing margin needs to be further improved by appropriate technological solutions [18, 29].

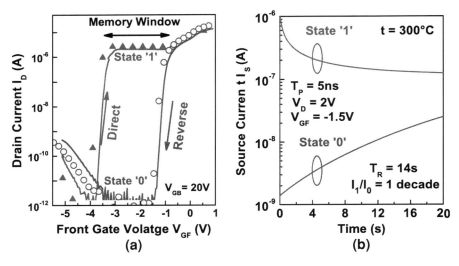

**Fig. 10 a** Measured drain current $I_D$ versus decreasing (*reverse scan*) and increasing (*direct scan*) front-gate bias $V_{GF}$. MSD effect is observed for direct scan where $I_D$ saturates. The applied back-gate $V_{GB}$, source $V_S$, and drain $V_D$ voltages are 30, 0 and 0.1 V, respectively. This n-MOSFET had 400 nm thick BOX, 6 nm thick gate oxide, 80 nm thick Si film and $5 \times 10^{16}$ cm$^{-3}$ doping. The gate length and width are respectively 1.5 and 20 μm. **b** Simulated short-channel DG-MSDRAM: drain current $I_D$ versus time during the '0'-state and '1'-state reading cycle. $V_{GB}$ and $V_D$ are 0.25 and 0.1 V and the silicon film thickness and the channel length are 40 and 50 nm respectively. The front and back-gate oxide thicknesses are 3 and 6 nm and the body doping is $10^{16}$ cm$^{-3}$. A retention time $T_R$ of 14 s with $I_1/I_0 = 10$ is achieved at 30°C. During the 1-state writing by B2B tunnelling, $V_D$ and $V_{GF}$ are equal to 2 and $-1.5$ V with a programming time $T_P$ equal to 5 ns

## 5.5 Vertical Channel 1T-DRAM

In order to reach the ultimate integration with an effective feature size of 4F$^2$, a vertical version of the double gate 1T-DRAM (Fig. 11a) has been developed [35–37]. The first reported variant proposed a gate-all-around MOSFET architecture with a vertical channel called the Surrounding Gate Vertical Channel (SGVC) cell (Fig. 11b, [35, 36]). The main advantage is the channel length which is not an area limiting factor anymore. The SGVC cell can be fabricated on bulk Si substrates and uses a common source structure, which makes the 4F$^2$ structure possible and allows achieving superior scalability. The memory operation has been investigated by 2D numerical simulation and SGVC cells were successfully fabricated. The majority carriers can be generated either by impact ionization or by B2B tunneling. Notice that the experimental demonstration was performed on an array of memory cells with relaxed size (6F$^2$) for practical purpose. Considering a sensing margin of about 3 μA (here 40 μA/μm), the effective retention time at room temperature is 4 ms. As the retention time required for embedded eDRAM application is a few ms at 85°C, the SGVC cell has still to be improved since the retention time will be

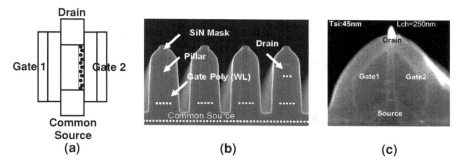

**Fig. 11 a** Vertical independent double gate 1T-DRAM with common source (**b**) called the Surrounding Gate Vertical Channel (SGVC) cell [35]. **c** Similar concept tested for scalable DRAM down to the 22-nm technology node [37]

strongly deteriorated with increasing temperature. Another attempt was made with the same type of structure (Fig. 11c, [37]). Similar performances were achieved with a retention time slightly improved (10 ms at room temperature for a sensing margin of 3 µA/µm).

## 5.6 The Thyristor-RAM (T-RAM)

The T-RAM aims to compete with standard 6T-SRAMs and 1T-1C embedded eDRAM. An SOI thyristor provides a positive regenerative feedback which results in very large sensing margin [38, 39]. The difference is that the four-transistor CMOS used as bistable latching circuitry for SRAM is replaced by the PNP–NPN bipolar latch of a single thyristor device, which reduces the cell area by a factor of four, enables high-density and provides low-power operation. Despite the density of integration is not as high as for Z-RAMs, the development of 32 and 22 nm versions of the SOI T-RAM was announced for low-power cache applications. The Thyristor-RAM technology has been successfully implemented on Bulk and SOI CMOS and its manufacturability was demonstrated within an 18 Mb synchronous SRAM memory chip with excellent yield and reliability.

## 5.7 Engineered Body 1T-DRAM

In order to improve the retention and sensing margin performances, several exotic architectures and material combinations were also tested. First, we introduce briefly a non exhaustive list of engineered body techniques which were used to improve the hole storage efficiency and were demonstrated via extensive 2D numerical simulations.

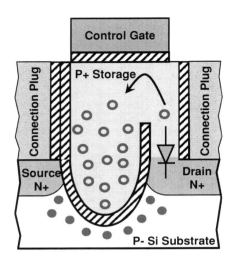

**Fig. 12** The floating junction gate (FJG) cell

(a) The floating junction gate (FJG) cell is a $4.5F^2$ capacitorless DRAM cell with a 'U-shaped' floating gate (FG) connected to the drain through a gated p–n diode inside the body (Fig. 12). The U-shaped FG is used to improve the storage volume as well as to extend the channel length and hence reduce SCE like DIBL and punchthrough. To program the FJG cell, the FG is electrically charged or discharged by the current flowing through the gated $p^+$ (top region of the FG)–$n^+$ (drain) diode.

The 1-state is achieved by injecting positive charges into the FG which produce a relatively high potential increase in this region. This FG potential shift induces in turns by 'dynamic' coupling an increase of the electron channel density, i.e. of the drain current below the U-shaped FG. As the p+/n+ junction is reversed biased, a high positive drain and negative MOS-gate voltages have to be applied. Hence, the reverse hole current is enhanced thanks to B2B tunneling in the gated diode. 0-state is programmed by removing holes from the FG by forward biasing of the p–n junction diode. The hole current flows from the FG into the drain and pulls down the FG potential. According to the simulation results, the retention time can reach 6 s at room temperature thanks to the low leakage diode current density. The sensing margin is about 300 µA/µm at the reading onset.

(b) A-RAM Concept is an original architecture for 1T-DRAM proposed in [40, 41]. A-RAM enables the coexistence of electron and hole layers even in SOI transistors with ultrathin body, by suppressing the supercoupling effect [42]. The supercoupling prevents hole and electron layers facing each other when the silicon film thickness becomes lower than 10 nm. The A-RAM physically isolates holes and electrons in two semi-bodies which respectively serve for majority carrier storage and electron current sense (Fig. 13). The A-RAM isolation offers the possibility to maintain electron and hole layers facing each other, on each side of the dielectric layer MOX, even in ultrathin SOI films that are necessary for CMOS scaling. To write 1 (Fig. 13c), the excess of holes is generated by impact ionization or band-to-band

Floating-Body SOI Memory: The Scaling Tournament 413

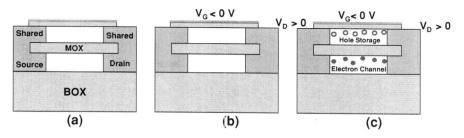

**Fig. 13 a** A-RAM cell schematics. Channel length is compatible with 45 nm node and beyond. MOX is typically 4 nm thick. **b** A-RAM reading bit '0' and **c** reading bit '1'

tunneling in the upper storage semi-body. These extra holes induce a dynamic increase of the upper semi-body potential. Hence, by 'electrostatic' coupling an electron channel (i.e. a drain current) can be sensed in the lower semi-body. By contrast, if no holes are stored, the electron current in the sense semi-body is zero (bit '0', Fig. 13b).

Different operation modes can be considered depending on the body thickness. A-RAM is compatible with single-gate (SG-SOI), double-gate (DG-SOI) and FinFETs devices. This concept has been validated so far by numerical simulations. The proof-of-concept still needs to be verified at the experimental level (processing and measurements). Although similar programming principles are used for the FJG cell (see Sect. (a) and a comparable retention time is achieved, the A-RAM, architecture seems to be more realistic for practical application in terms of process complexity and scaling. Remark that the A-RAM structure and operation remind the HRAM which was proposed previously in GaN/AlGaN heterostructures [43].

(c) The single-transistor quantum well (QW) 1T-DRAM [44] uses an engineered body integrating within the Si film a thin layer of a material with a different band gap (here SiGe). This layer serves as storage well for holes (Fig. 14a). The QW also gives the possibility to modulate the spatial hole distribution within the device (option not available with the conventional 1T-DRAMs). It was demonstrated theoretically an improvement in the current sensing margin and scalability characteristics. Compared with 1T-DRAM, this QW memory has the ability to store the holes closer to the front gate inducing enhanced $V_{TH}$ shift and retention time. As the QW devices are more scalable thanks to the introduction of the extra 'storage space', the effect of the volume reduction with the channel length is lessened.

(d) A convex channel 1T-DRAM structure (Fig. 14b) using the BJT programming technique was proposed to improve the retention time [45]. The holes are stored beneath a raised gate oxide which may be filled by a smaller bandgap material (e.g. SiGe). As the holes stored during the 1-state programming reduce the body/source (drain) potential barrier, they easily diffuse through these junctions filling the SiGe region. The convex channel architecture provides a physical well for more effective storage of holes. Moreover, if a smaller bandgap material is

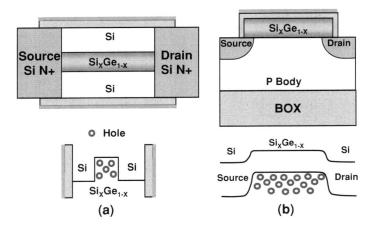

**Fig. 14** Engineered body 1T-DRAM with **a** body Si/SiGe/Si stack [44] and **b** convex channel structure filled by SiGe [45]

used in the convex channel region, a deeper potential well is formed improving further the sensing margin and retention time. Similar concepts have also been investigated in vertical and planar [18] double gate devices.

## 5.8 Engineered Source/Drain 1T-DRAM

If using a hetero-layer within the Si-body would improve the storage efficiency, another way to increase the retention time can be achieved by reducing the current leakage through the source (drain)-body junction.

(a) The band-gap engineered source and drain floating body cell (BESD-FBC) [46] was proposed to induce a deeper potential well in the body within a fully depleted SOI MOSFET. To increase the body/source (drain) potential barrier, a higher energy band offset was created with silicon–carbon (SiC) source and drain regions (Fig. 15a). As a deeper potential well is formed due to valence band offset ($\Delta E_v$), more holes can be stored in the body. Impact ionization was performed for writing 1 and forward biased drain–body p–n junction current was used for writing 0. The BESD-FBC has been demonstrated using 2D simulations and compared with conventional SOI 1T-DRAM. Due to $\Delta E_v$, the sensing margin would reach 100 µA/µm, which is about three times larger than usual 1T-DRAM (36.5 µA/µm). It was shown that the hole leakage at source can be two or three orders of magnitude lower during the 1-state programming.

(b) The dopant segregated Schottky barrier (DSSB) [47] technique was implemented within the source/drain region of a FinFET (Fig. 15b). This device also uses a specific material gate stack which allows combining the 1T-DRAM with the non-volatile functionality of a Flash memory (see Sect. 5.10).

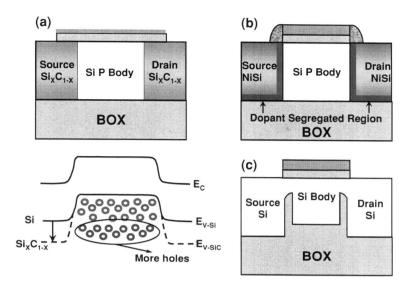

**Fig. 15** a The band-gap engineered source and drain floating body cell (BESD-FBC) [46]. b The dopant segregated Schottky barrier (DSSB) [47]. c Silicon-On-Insulator with Block Oxide process for 1T-DRAM application [48]

This particular engineered DSSB S/D aims to speed up the Flash memory programming operation while attenuating effectively the short-channel effects within the FinFET device. At the same time, as SCE are significantly reduced, the body volume available for the hole storage is increased together with the 1T-DRAM retention time and sensing margin. To take advantage from the Schottky S/D barriers in a capacitorless 1T-DRAM, a dopant segregation technique with a partial silicidation process at the S/D was used. It allows the hole current flowing into the floating body what would be impossible with 'pure' metallic Schottky barriers. By considering a minimum sensing margin of 3 µA (for the device considered, 100 µA/µm), the DSSB retention time at room temperature was about 70 ms. Notice that, in order to improve the 1T-DRAM performances, a comparable architecture with a specific Si–SiO$_2$ process was developed [48]: The fully (partially) depleted SOI with self aligned block oxide (bFD(PD)SOI, Fig. 15c). The simulation results showed that an improvement of 87% for the sensing current margin (at the reading onset) is expected as compared to equivalent SOI devices (same body thickness with raised S/D).

## 5.9 Optically Assisted 1T-DRAM

Even if the exploration of photonics for the use in computers has increased in the last decades, electronics still remains superior to optics in many ways. Totally optical computers are not a reality yet, but computers which combine together electronics and optics, called electro-optic hybrids, are under development. On the other hand, the optical interconnections have attracted interest as candidates for next-generation interconnections in advanced ICs. If optical interconnections replace the electrical ones, they could also be applied to conventional memory technologies. Nevertheless optically operated memory in DRAM applications has not been studied yet. For the first time, 1T-DRAM programming assisted by optically generated hole was proposed and demonstrated experimentally in [49]. The 1-state programming method is based on the BJT operation triggered by a source of light (halogen). Thanks to this method, the gate voltage pulse which was used to pull up the body potential and switch on the bipolar latch is not required anymore. During all the operations (read, hold, program), the gate voltage keeps a constant negative value. The maximum sensing margin at the reading onset is 54 µA (here 900 µA/µm with a fin width of 60 nm) allowing a data retention time over a few seconds at room temperature. Although the worst disturbance for the 0-state holding, arising from the light pulse programming, has not been investigated yet, the proposed programming method could be considered as a promising candidate for electro-optic hybrids in memory applications.

## 5.10 The Unified RAM (URAM)

The ideal memory device or the so called 'universal memory' would satisfy three requirements: high speed, high density and non-volatility (Fig. 16a). Unfortunately, such a memory has not been developed yet and these three different targets have been 'magnified' separately (SRAM, DRAM, Flash). Hence, if a single transistor can integrate different memory functions, a paradigm shift from 'scaling' to 'multifunction' would drive the evolution of the silicon technology

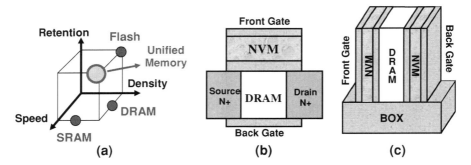

**Fig. 16** **a** Performances targeted for the ideal or the so-called 'Universal Memory'. Schematics of the unified memory combining the non-volatile (NVM) functionality of the ONO flash and the volatile functionality of the 1T-DRAM in **b** a separated double gate planar MOFET with one ONO gate and **c** a FinFET with separated gates and two ONO layers for multi-bit application

(more Moore). The URAM aims to combine the non-volatile memory (NVM) and DRAM functionalities in a single transistor.

Up to now, non-volatile memories like Flash and 1T-1C DRAM could not be integrated on a single chip without additional process steps and high cost. On the other hand, NVM needs high reliability (good cycling and retention at elevated temperatures) combined with low voltages and high program speed. Therefore, it is increasingly difficult to integrate conventional NVM and DRAM on a single chip while maintaining these severe requirements and low costs.

The URAM concept consists in implementing, a charge trapping layer for NVM applications and a PD or FD SOI device to insure the 1T-DRAM effect (Fig. 16b). Indeed, this new concept allows combining both volatile and non-volatile functions in a single memory transistor, leading to reduced processing costs. The NVM core region can be composed by an oxide/nitride/oxide (ONO) stack; electrons being injected and trapped in the silicon nitride layer like in a standard SONOS memory. The programming and erasing steps can be achieved using either Fowler–Nordheim or hot carrier injection through a tunnel oxide above the channel of the transistor. On the other hand, the floating body is used as a storage area for the 1T-DRAM function. The URAM concept is attractive and preliminary results were published by Korean groups [50–53]. Excess holes are usually generated by impact ionization for the 1T-DRAM 1-state programming and are accumulated in the lowest potential area of the floating body. The main method to write a '0' is to sweep out the excess holes of the floating body by forward biasing the body/drain junction as usual in 1T-DRAM.

The combination of a high density non-volatile memory with floating-body 1T-DRAMs, integrated on the same cell structure, is a promising architecture in terms of density increase and cost per bit decrease. However, it is difficult to avoid interference between programming voltages of 1T-DRAM and NVM. Thus, the disturbance between the two memory functions (especially in programming and erasing modes) has to be further investigated [54–59]. Moreover, the threshold voltage shift resulting from each memory function (volatile and non-volatile) must be de-correlated. In particular, optimized reading procedures will have to be developed overall if a multi-bit functionality (number of states >2) is targeted (Fig. 16c).

# 6 Conclusion

The floating-body SOI memory is not an intellectual exercise anymore; it could compete in the DRAM market although many improvements have still to be achieved. One of the serious concerns is the retention time deterioration at high temperature (>85°C). There are several 1T-DRAM options, the principles of which have been discussed and compared. Among the basic mechanisms, we prioritize B2B tunneling for 1-state programming and capacitive coupling for 0-state as in MSDRAM. In terms of CMOS scaling, double-gate transistors are best suited devices and vertical configuration seems to be promising candidate to reach the minimum cell feature size.

**Acknowledgments** Part of this work has been supported by the European Union programs EUROSOI+ and NANOSIL.

# References

1. Wann, H.J., Hu, C.: A capacitorless DRAM cell on SOI substrate. In: Electron Devices Meeting IEDM '93 Technical Digest International, pp. 635–638 (1993)
2. Okhonin, S., Nagoga, M., Sallese, J.M., Fazan, P.: A SOI capacitor-less 1T-DRAM concept. In: SOI Conference 2001 IEEE International, pp. 153–154 (2001)
3. Okhonin, S., Nagoga, M., Sallese, J.M., Fazan, P.: A capacitor-less 1T-DRAM cell. Electron Device Lett. IEEE **23**, 85–87 (2002)
4. Hu, C., King, T.-J., Hu, C.: A capacitorless double-gate DRAM cell. Electron Device Lett. IEEE **23**, 345–347 (2002)
5. Kuo, C., King, T.J., Hu, C.: A capacitorless double-gate DRAM cell design for high density applications. In: Electron Devices Meeting IEDM '02 Digest International, pp. 843–846 (2002)
6. Kuo, C., King, T.J., Hu, C.: A capacitorless double gate DRAM technology for sub-100-nm embedded and stand-alone memory applications. Electron Devices IEEE Trans. **50**, 2408–2416 (2003)
7. Shino, T., Ohsawa, T., Higashi, T., Fujita, K., Kusunoki, N., Minami, Y., Morikado, M., Nakajima, H., Inoh, K., Hamamoto, T., Nitayama, A.: Operation voltage dependence of memory cell characteristics in fully depleted floating-body cell. Electron Devices IEEE Trans. **52**, 2220–2226 (2005)
8. Okhonin, S., Fazan, P., Jones, M.E.: Zero capacitor embedded memory technology for system on chip. In: Memory Technology, Design, and Testing MTDT 2005 IEEE International Workshop, pp. xxi–xxv (2005)
9. Ban, I., Avci, U.E., Shah, U., Barns, C.E., Kencke, D.L., Chang, P.: Floating body cell with independently-controlled double gates for high density memory. In: Electron Devices Meeting IEDM '06 International, pp. 1–4 (2006)
10. Hamamoto, T., Minami, Y., Shino, T., Kusunoki, N., Nakajima, H., Morikado, M., Yamada, T., Inoh, K., Sakamoto, A., Higashi, T., Fujita, K., Hatsuda, K., Ohsawa, T., Nitayama, A.: A floating-body cell fully compatible with 90-nm CMOS technology node for a 128-Mb SOI DRAM and its scalability. Electron Devices IEEE Trans. **54**, 563–571 (2007)
11. Avci, U.E., Ban, I., Kencke, D.L., Chang, P.L.D.: Floating body cell (fbc) memory for 16-nm technology with low variation on thin silicon and 10-nm box. In: SOI Conference 2008 IEEE International, pp. 29–30 (2008)
12. Hamamoto, T., Ohsawa, T.: Overview and future challenges of floating body RAM (FBRAM) technology for 32 nm technology node and beyond. In: Solid-State Device Research Conference ESSDERC 2008 38th European, pp. 25–29 (2008)
13. Okhonin, S., Nagoga, M., Carman, E., Beffa, R., Faraoni, E.: New generation of Z-RAM. In: Electron Devices Meeting IEDM 2007 IEEE International, pp. 925–928 (2007)
14. Song, K.W., Jeong, H., Lee, J.W., Hong, S.I., Tak, N.K., Kim, Y.T., Choi, Y.L., Joo, H.S., Kim, S.H., Song, H.J., Oh, Y.C., Kim, W.S., Lee, Y.T., Oh, K., Kim, C.: 55 nm capacitor-less 1T-DRAM cell transistor with non-overlap structure. In: Electron Devices Meeting IEDM 2008 IEEE International, pp. 1–4 (2008)
15. Okhonin, S., Nagoga, M., Lee, C.W., Colinge, J.P., Afzalian, A., Yan, R., Dehdashti Akhavan, N., Xiong, W., Sverdlov, V., Selberherr, S., Mazure, C.: Ultra-scaled Z-RAM cell. In: SOI Conference IEEE International, pp. 157–158 (2008)

16. Yoshida, E., Tanaka, T.: A capacitorless 1T-DRAM technology using gate-induced drain-leakage (GIDL) current for low-power and high-speed embedded memory. Electron Devices IEEE Trans. **53**, 692–697 (2006)
17. Bawedin, M., Cristoloveanu, S., Flandre, D.: A capacitorless 1T-DRAM on SOI based on dynamic coupling and double-gate operation. Electron Device Lett. IEEE **29**, 795–798 (2008)
18. Bawedin, M., Cristoloveanu, S., Flandre, D., Renaux, C., Crahay, A.: Double-gate floating body memory device. Patent WO2009087125 (A1), 2008
19. Tanaka, T., Yoshida, E., Miyashita, T.: Scalability study on a capacitorless 1T-DRAM: from single-gate PD-SOI to double-gate finDRAM. In: Electron Devices Meeting 2004 IEDM Technical Digest IEEE International, pp. 919–922 (2004)
20. Puget, S., Bossu, G., Fenouiller-Beranger, C., Perreau, P., Masson, P., Mazoyer, P., Lorenzini, P., Portal, J.M., Bouchakour, R., Skotnicki, T.: FD-SOI floating body cell eDRAM using gate-induced drain-leakage (GIDL) write current for high speed and low power applications. In: Memory Workshop IMW '09 IEEE International, pp. 1–2 (2009)
21. Guegan, G., Touret, P., Molas, G., Raynaud, C., Pretet, J.: A novel capacitor-less 1T-DRAM on partially depleted SOI pMOSFET based on direct-tunneling current in the partial n+ poly gate. In: Proceedings of Ext. Abs. Solid State Devices Mater (SSDM) (2010)
22. Fossum, J.G., Lu, Z., Trivedi, V.P.: New insights on 'capacitorless' floating-body DRAM cells. Electron Device Lett. IEEE **28**, 513–516 (2007)
23. Zhou, Z., Fossum, J.G., Lu, Z.: Physical insights on BJT-based 1T-DRAM cells. Electron Device Lett. IEEE **30**, 565–567 (2009)
24. Furuhashi, H., Shino, T., Ohsawa, T., Matsuoka, F., Higashi, T., Minami, Y., Nakajima, H., Fujita, K., Fukuda, R., Hamamoto, T., Nitayama, A.: Scaling scenario of floating body cell (FBC) suppressing Vth variation due to random dopant fluctuation. In: SOI Conference 2008 IEEE International, pp. 33–34 (2008)
25. Blagojevic, M., Kayal, M., Pastre, M., Harik, L., Declercq, M.J., Okhonin, S., Fazan, P.C.: Capacitorless 1T-DRAM sensing scheme with automatic reference generation. Solid-State Circuits IEEE **41**, 1463–1470 (2006)
26. Ohsawa, T., Fujita, K., Hatsuda, K., Higashi, T., Shino, T., Minami, Y., Nakajima, H., Morikado, M., Inoh, K., Hamamoto, T., Watanabe, S., Fujii, S., Furuyama, T.: Design of a 128-Mb SOI DRAM using the floating body cell (FBC). Solid-State Circuits IEEE **41**, 135–145 (2006)
27. Ohsawa, T., Fukuda, R., Higashi, T., Fujita, K., Matsuoka, F., Shino, T., Furuhashi, H., Minami, Y., Nakajima, H., Hamamoto, T., Watanabe, Y., Nitayama, A., Furuyama, T.: Autonomous refresh of floating body cell (FBC). In: Electron Devices Meeting IEDM 2008 IEEE International, pp. 1–4 (2008)
28. Cristoloveanu, S.: Electrical Characterization of Silicon-on-Insulator Materials and Devices. Kluwer Academic, Boston (1995)
29. Hubert, A., Bawedin, M., Cristoloveanu, S., Ernst, T.: Dimensional effects and scalability of meta-stable dip (MSD) memory effect for 1T-DRAM SOI MOSFETs. Solid-State Electron. **53**, 1280–1286 (2009)
30. Ertosun, M.G., Saraswat, K.C.: Investigation of capacitorless double-gate single-transistor DRAM: with and without quantum well. Electron Devices IEEE Trans. **57**, 608–613 (2010)
31. Bossu, G., Puget, S., Masson, P., Portal, J.M., Bouchakour, R., Mazoyer, P., Skotnicki, T.: Independent double gate—potential for non-volatile memories. In: Silicon Nanoelectronics Workshop SNW 2008 IEEE, pp. 1–2 (2008)
32. Puget, S., Bossu, G., Mazoyer, P., Portal, J.M., Masson, P., Bouchakour, R., Skotnicki, T.: On the potentiality of planar independent double gate for capacitorless eDRAM. In: Silicon Nanoelectronics Workshop SNW 2008 IEEE, pp. 1–2 (2008)
33. Mohapatra, N.R., vanBentum, R., Pruefer, E., Maszara, W.P., Caillat, C., Chalupa, Z., Johnson, Z., Fisch, D.: Effect of source/drain asymmetry on the performance of Z-RAM® devices. In: SOI Conference 2009 IEEE International, pp. 1–2 (2009)
34. Bawedin, M., Cristoloveanu, S., Yun, J.G., Flandre, D.: A new memory effect (MSD) in fully depleted SOI MOSFETs. Solid-State Electron. **49**, 1547–1555 (2005)

35. Jeong, H., Song, K.W., Park, I.H., Kim, T.H., Lee, Y.S., Kim, S.G., Seo, J., Cho, K., Lee, K.G., Shin, H., Lee, J.D., Park, B.G.: A new capacitorless 1T-DRAM cell: surrounding gate MOSFET with vertical channel (SGVC cell). Nanotechnol. IEEE Trans. **6**, 352–357 (2007)
36. Chung, H.K., Jeong, H., Lee, Y.S., Song, J.Y., Kim, J.P., Kim, S.W., Park, J.H., Lee, J.D., Shin, H., Park, B.G.: A capacitor-less 1T-DRAM cell with vertical surrounding gates using gate-induced drain-leakage (GIDL) current. In: Silicon Nanoelectronics Workshop SNW 2008 IEEE, pp. 1–2 (2008)
37. Ertosun, M.G., Cho, H., Kapur, P., Saraswat, K.C.: A nanoscale vertical double-gate single-transistor capacitorless DRAM. Electron Device Lett. IEEE **29**, 615–617 (2008)
38. Nemati, F., Plummer, J.D.: A novel thyristor-based SRAM cell (T-RAM) for high-speed, low-voltage, giga-scale memories. In: Electron Devices Meeting, 1999. IEDM Technical Digest. International, pp. 283–286 (1999)
39. Cho, H.J., Nemati, F., Roy, R., Gupta, R., Yang, K., Ershov, M., Banna, S., Tarabbia, M., Sailing, C., Hayes, D., Mittal, A., Robins, S.: A novel capacitor-less DRAM cell using thin capacitively-coupled thyristor (TCCT). In: Electron Devices Meeting 2005 IEDM Technical Digest IEEE International, pp. 311–314 (2005)
40. Rodriguez, N., Cristoloveanu, S., Gamiz, F.: A-RAM: novel capacitor-less DRAM memory. In: SOI Conference 2009 IEEE International, pp. 1–2 (2009)
41. Rodriguez, N., Gamiz, F., Cristoloveanu, S.: A-RAM memory cell: concept and operation. Electron Device Lett. IEEE **31**, 972–974 (2010)
42. Eminente, S., Cristoloveanu, S., Clerc, R., Ohata, A., Ghibaudo, G.: Ultra-thin fully-depleted SOI MOSFETs: special charge properties and coupling effects. Solid-State Electron. **51**, 239–244 (2007)
43. Bawedin, M., Uren, M.J., Udrea, F.: DRAM concept based on the hole gas transient effect in a AlGaN/GaN HEMT. Solid-State Electron. **54**, 616–620 (2010)
44. Ertosun, M.G., Kapur, P., Saraswat, K.C.: A highly scalable capacitorless double gate quantum well single transistor DRAM: 1T-QW DRAM. Electron Device Lett. IEEE **29**, 1405–1407 (2008)
45. Cho, M.H., Shin, C., Liu, T.J.K.: Convex channel design for improved capacitorless DRAM retention time. In: Simulation of Semiconductor Processes and Devices SISPAD '09 International Conference, pp. 1–4 (2009)
46. Poren, T., Ru, H., Dake, W.: Performance improvement of capacitorless dynamic random access memory cell with band-gap engineered source and drain. Jpn J. Appl. Phys. **49**, 04DD02 (2010)
47. Choi, S.J., Han, J.W., Kim, S., Kim, H., Jang, M.G., Yang, J.H., Kim, J.S., Kim, H.K., Lee, G.S., Oh, J.S., Song, M.H., Park, Y.C., Kim, J.W., Choi, Y.K.: High speed flash memory and 1T-DRAM on dopant segregated Schottky barrier (DSSB) finFET SONOs device for multi-functional SoC applications. In: Electron Devices Meeting IEDM 2008 IEEE International, pp. 1–4 (2008)
48. Tseng, Y.M., Lin, J.T., Eng, Y.C., Kang, S.S., Tseng, H.J., Tsai, Y.C., Jheng, B.T., Lin, B.H.: A new process for self-aligned silicon-on-insulator with block oxide and its memory application for 1T-DRAM. In: Solid-State and Integrated-Circuit Technology ICSICT 2008 9th International Conference, pp. 1154–1157 (2008)
49. Moon, D.I., Choi, S.J., Han, J.W., Choi, Y.K.: An optically assisted program method for capacitorless 1T-DRAM. Electron Devices IEEE Trans. **57**, 1714–1718 (2010)
50. Han, J.W., Ryu, S.W., Kim, C., Kim, S., Im, M., Choi, S.J., Kim, J.S., Kim, K.H., Lee, G.S., Oh, J.S., Song, M.H., Park, Y.C., Kim, J.W., Choi, Y.K.: A unified-RAM (URAM) cell for multi-functioning capacitorless DRAM and NVM. In: Electron Devices Meeting IEDM 2007 IEEE International, pp. 929–932 (2007)
51. Bae, D.I., Gu, B., Ryu, S.W., Choi, Y.K.: Multiple data storage of URAM (unified-RAM) with multi dual cell (MDC) method. In: Silicon Nanoelectronics Workshop SNW 2008 IEEE, pp. 1–2 (2008)
52. Han, J.W., Ryu, S.W., Kim, S., Kim, C.J., Ahn, J.H., Choi, S.J., Kim, J.S., Kim, K.H., Lee, G.S., Oh, J.S., Song, M.H., Park, Y.C., Kim, J.W., Choi, Y.K.: A bulk finFET unified-RAM

(URAM) cell for multifunctioning NVM and capacitorless 1T-DRAM. Electron Device Lett. IEEE **29**, 632–634 (2008)
53. Han, J.W., Ryu, S.W., Kim, S., Kim, C.J., Ahn, J.H., Choi, S.J., Choi, K.J., Cho, B.J., Kim, J.S., Kim, K.H., Lee, G.S., Oh, J.S., Song, M.H., Park, Y.C., Kim, J.W., Choi, Y.K.: Band offset finFET-based URAM (unified-RAM) built on SiC for multi-functioning NVM and capacitorless 1T-DRAM. In: VLSI Technology 2008 Symposium, pp. 200–201 (2008)
54. Han, J.W., Ryu, S.W., Choi, S.J., Choi, Y.K.: Gate-induced drain-leakage (GIDL) programming method for soft-programming-free operation in unified ram (URAM). Electron Device Lett. IEEE **30**, 189–191 (2009)
55. Han, S.W., Kim, C.J., Choi, S.J., Kim, D.H., Moon, D.I., Choi, Y.K.: Gate-to-source/drain nonoverlap device for soft-program immune unified RAM (URAM). Electron Device Lett. IEEE **30**, 544–546 (2009)
56. Han, J.W., Kim, C.J., Choi, S.J., Kim, D.H., Moon, D.I., Choi, Y.K.: Gate-to-source/drain nonoverlap device for soft-program immune unified RAM (URAM). IEEE Electron Device Lett. **30**, 544–546 (2009)
57. Choi, S.J., Han, J.W., Kim, C.J., Kim, S., Choi, Y.K.: Improvement of the sensing window on a capacitorless 1T-DRAM of a finFET-based unified RAM. IEEE Trans. Electron Devices **56**, 3228–3231 (2009)
58. Park, K.H., Jeong, M.K., Kim, Y.M., Han, K.R., Kwon, H.I., Kong, S.H., Lee, J.H.: Novel capacitorless double-gate 1T-DRAM cell having nonvolatile memory function. In: Silicon Nanoelectronics Workshop SNW 2008 IEEE, pp. 1–2 (2008)
59. Park, K.H., Park, C.M., Kong, S.H., Lee, J.H.: Novel double-gate 1T-DRAM cell using nonvolatile memory functionality for high-performance and highly scalable embedded DRAMs. Electron Devices IEEE Trans. **57**, 614–619 (2010)

# Part V
# Afterword

# A Selection of SOI Puzzles and Tentative Answers

S. Cristoloveanu, M. Bawedin, K.-I. Na, W. Van Den Daele, K.-H. Park,
L. Pham-Nguyen, J. Wan, K. Tachi, S.-J. Chang, I. Ionica, A. Diab,
Y.-H. Bae, J. A. Chroboczek, A. Ohata, C. Fenouillet-Beranger,
T. Ernst, E. Augendre, C. Le Royer, A. Zaslavsky and H. Iwai

**Abstract** Recent research on advanced SOI materials and devices has delivered rich and informative data, enabling further progress in science and technology. However, some of the results still look intriguing, likely to open new space for investigation and developments. In this chapter, we have selected a variety of multi-angle problems which may stimulate dedicated SOI research. When available, experimental arguments and scenarios are proposed.

## 1 Introduction

As bulk CMOS is confronted with multiple challenges in terms of scaling, power and performance, SOI technology is gaining momentum. The roadmap shows that

---

S. Cristoloveanu (✉), M. Bawedin, K.-I. Na, W. Van Den Daele, K.-H. Park,
L. Pham-Nguyen, J. Wan, K. Tachi, S.-J. Chang, I. Ionica, A. Diab, Y.-H. Bae,
J. A. Chroboczek, A. Ohata and A. Zaslavsky
IMEP-LAHC (UMR 5130), Grenoble INP Minatec, BP257, 38016 Grenoble Cedex 1, France
e-mail: sorin@enserg.fr

K. Tachi, C. Fenouillet-Beranger, T. Ernst, E. Augendre and C. Le Royer
CEA-LETI, Minatec, 17 rue des Martyrs, 38054 Grenoble Cedex 9, France

Y.-H. Bae
Uiduk University, Gangdong, Gyeoju, Korea

C. Fenouillet-Beranger
STMicroelectronics, 850 rue Jean Monnet, 38926 Crolles Cedex, France

A. Zaslavsky
Division of Engineering, Brown University, Providence, RI 02912, USA

H. Iwai
Frontier Research Center, Tokyo Institute of Technology, Yokohama 226-8502, Japan

SOI is situated at the crossing of the two Moore Avenues: one is going vertically, the other laterally. The vertical avenue is measured in nanometers, for the ultimate transistor feature size, and in billions for their number in a System-on-a-Chip. SOI will certainly take us beyond, simply because the device miniaturization is far easier if transistors are fabricated 'on-insulator'. The SOI world will feature denser, smaller, and faster circuits. As for the lateral avenue, it departs from the traditional miniaturization-driven scaling in order to meet the needs for enriched circuit functionality. SOI is compatible with the co-integration of heterogeneous technologies and multi-functional devices.

Converting dreams into reality requires dedicated work in material and device science, technology and circuit design. Although SOI MOSFETs already exhibit high performance, our goal is to explore innovative physics-based and technological solutions for further improvement. This task is not easy because SOI is a complicated structure with nanometer-thick stacked layers and special mechanisms governing its electrostatics and transport properties. For example, gate inter-coupling and floating-body effects can be detrimental or advantageous depending on the device architecture, designer skills and envisioned applications. The SOI field of research is vast, with many questions that are still open.

In this paper, we take a non-conventional approach. Instead of demonstrating how we succeeded in solving a particular SOI dilemma, we will present a selection of problems that today remain without a convincing solution. We will describe these problems in terms of systematic measurements and, when necessary, numerical simulations. More or less mature arguments will be offered that may help to complete the understanding of the SOI-related mechanisms.

## 2 How can the Density of Traps be Measured in Ultrathin SOI Wafers?

Pseudo-MOSFET (or $\Psi$-MOSFET, Fig. 1a) is an undisputable technique for the *in situ* material-level characterization [1]. Recent progress shows that the method can be enriched by adding magnetic field and noise measurements.

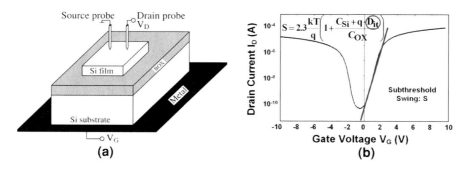

**Fig. 1** a Pseudo-MOSFET configuration and b typical current–voltage characteristics

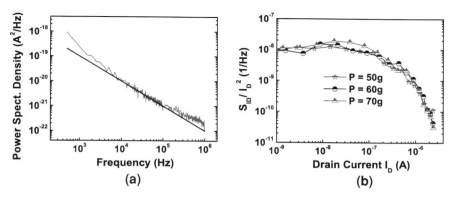

**Fig. 2 a** Power spectral density versus frequency in pseudo-MOSFET showing 1/f noise. **b** Normalized noise versus drain current for three values of probe pressure P

Pseudo-MOSFET current–voltage characteristics are shown in Fig. 1b. The density of back interface traps $D_{it}$ is usually deduced from the subthreshold slope. Typical values for good quality SOI wafers, mandatory for high performance CMOS circuits, are $D_{it} = 10^{11}$ cm$^{-2}$ eV$^{-1}$ (detection limit) or less.

Noise is well known to yield an accurate evaluation of interface trap densities in MOSFETs. The question is whether the noise in pseudo-MOSFETs can be used in order to improve the detection limit of $D_{it}$.

Figure 2 shows, for the first time, noise measurements using the $\Psi$-MOSFET set-up with pressure probes. The noise follows the celebrated $1/f$ law (Fig. 2a) and it generally arises from carrier trapping-detrapping (carrier number fluctuations) rather than mobility fluctuations [2]:

$$\frac{S_{Id}(f)}{I_d^2} = \frac{g_m^2}{I_d^2} \frac{\lambda kTq^2 N_{it}^2}{WLC_{ox}^2 f^\gamma} \tag{1}$$

where $S_{ID}$ is the noise spectral density of the drain current $I_D$, W·L is the device area, $g_m$ is the transconductance, $C_{ox}$ is the oxide capacitance, kT is the temperature in energy units, $\lambda$ is a characteristic tunneling distance, and $N_{it}$ is the density of slow traps (also known as border traps) located in the oxide next to the interface. The densities of border traps and interface states $D_{it}$ are inter-related by $D_{it} = \lambda N_{it}$ [2].

The model of carrier number fluctuations is confirmed by the noise dependence on drain current (Fig. 2b): a marked plateau in weak inversion, followed by a decrease in strong inversion. Two additional observations can be invoked:

- The $(g_m/I_D)^2$ curve parallels the variation of the spectral noise density as predicted by Eq. 1.
- The noise is nearly independent of the probe pressure, hence it is not generated by the contact resistance. (A higher pressure is known to lower the contact series resistance [1].)

The problem comes when $D_{it}$ is extracted from these noise curves using the equation above. We obtain a trap density 1–2 orders of magnitude higher than the value indicated by the subthreshold slope.

In Ψ-MOSFETs, the channel width and length are not well-defined. The aspect ratio W/L was determined from comparison with both four-point probe measurements and numerical simulations: W/L = 0.75 [1]. For $D_{it}$ extraction we had to guess the device area (S = W·L). Since the distance between probes was 1 mm, we assumed L = 1 mm and S = 0.75 mm². This simple calculation is most likely naive. Indeed, the back gate induces an inversion layer over the entire sample, which is much larger than the area defined by the source-drain separation. Minority carriers are trapped everywhere on the inverted area, impacting the noise level. In other words, the effective surface for drain current flow and noise are not equivalent. As further evidence, when the probe separation is increased, the noise level does not scale as $L^2$. Why?

The role played by the surface area in the excess noise observed in Ψ-MOSFETs requires further investigation. But the fact that the noise is higher than expected also points on the possible role of the top wafer surface which may act as a source of additional noise. This brings us to the next question.

## 3 Can the Channel-to-surface Coupling in Ψ-MOSFETs be Used as a Detection Platform?

It has been demonstrated that the Ψ-MOSFET characteristics in ultrathin SOI films depend on the quality of the top surface. Comparing wafers with passivated and bare (unpassivated) surfaces, we find a significant threshold voltage shift (on the order of volts) and mobility difference (by tens of %) [3]. The coupling between the channel and the opposite interface is well documented in SOI MOSFETs, but

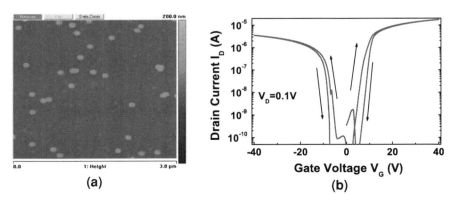

**Fig. 3** **a** Image of the film surface (30 nm Si on top of 145 nm buried oxide) with deposited gold nanospheres and **b** Ψ-MOSFET current–voltage characteristics showing a strong hysteresis due to particle charging. Slow $V_G$ scan with 0.2 V steps and 0.2 s delay time

often ignored in Ψ-MOSFETs. The idea is to take advantage of this coupling for sensing applications.

Gold nanoparticles, which are promising for detection and medical inspection, were deposited on the surface of SOI wafers (Fig. 3a). We used a two-step procedure for particle attachment: first, the SOI surface was treated with an amino-propyl-tri-ethoxy-silane (APTES) solution and, secondly, colloidal gold particles of 5 nm or 50 nm diameter were dispersed on the surface. According to the direction of the gate-bias scan, the nanoparticles get charged or discharged, leading to a strong hysteresis in the Ψ-MOSFET current–voltage characteristics, as shown in Fig. 3b.

Increasing the gate voltage $V_G$ from zero to a positive value creates an electron inversion channel and the gold particles become gradually charged with electrons. This trapped negative surface charge induces an electric field that lowers the body potential and the channel density. This is why reversing the $V_G$ direction from a large positive value to zero produces an $I_D(V_G)$ curve with a clear increase in threshold voltage. A similar hysteresis is observed for $V_G$ scan between zero and a negative bias. Here, the threshold voltage of the accumulation P-channel (i.e., the flat-band voltage) is increased by the net positive charge trapped in the gold nanoparticles.

Figure 3b shows that the hysteresis amplitude reaches 3 V in 30 nm thick SOI film. In thinner films the effect is even larger. Knowing the capability of sense amplifiers (in the µA and mV range), shifts of several volts are a luxury. But we still need to examine the charging mechanism and its efficiency. A rough estimate indicates that a $\Delta V_G \sim 1$ V shift corresponds to a charging of $\Delta Q = C_{ox} \cdot \Delta V_G \approx 2 \times 10^{11}$ electrons/cm$^2$ at the interface where the electrons flow. In our case, it is the opposite interface that is actually charged, hence the value above should be modified by the interface coupling coefficient which depends on the capacitances of depleted film, buried oxide (BOX) and surface traps [4]. The multiplication coefficient converting $\Delta Q$ at the interface to $\Delta Q$ at 30 nm above the interface is roughly $(1 + C_{it}/C_{si})$. Since the density of gold particles is $4 \times 10^8$ cm$^{-2}$, we infer that each 50 nm diameter sphere traps about $10^3$ electrons.

We now can formulate appropriate questions related to Au particles:

- Is this analysis quantitatively correct?
- What is the maximum charge the balls can accommodate?
- For ultimate sensitivity, should the golden balls be larger or more numerous?
- Is the method sensitive enough for detection of DNA and other molecules attached to the nanoparticles?

## 4 What is the Origin of Parasitic Conduction in P-channel GeOI MOSFETs?

Germanium-on-insulator (GeOI) devices are attractive mainly because the hole mobility is high, leading to excellent performance of P-channel MOSFETs [5–8]. GeOI can be fabricated by Smart-Cut starting from bulk-Ge or epi-Ge wafers [7]. GeOI can also be synthesized by the 'Ge condensation' technique, where the Si atoms of an SOI layer with SiGe cap are replaced by Ge atoms [9]. Increasing the Ge concentration from 10% to 100% causes a remarkable improvement in hole mobility whereas the electron mobility is dramatically degraded [9]. This 'mobility balance' is still a puzzle; ad-hoc technological solutions are envisioned for co-integrating P-channel MOSFETs on GeOI and N-channel MOSFETs on SOI.

Figure 4a shows $I_D(V_G)$ characteristics of an undoped GeOI pMOSFET, where two undesirable features can be observed: a positive threshold voltage and a parasitic channel that carries a large leakage current. The leakage is suppressed when the back interface is driven in accumulation ($V_{BG} = +60$ V) which is the sign of a conducting back channel for $V_{BG} = 0$ V. In other words, the threshold voltage of the back interface is also shifted towards positive values of $V_{BG}$. Intentional N-type doping is an efficient way to disable the back-channel leakage since the threshold voltages at the front and back interfaces are both increased (Fig. 4a). However, body doping is not a preferred solution for fully depleted (FD) transistors because the hole mobility is strongly degraded by Coulomb scattering.

The origin of the parasitic back channel can be explained by a mysterious P-type contamination of Ge film. The experimental behaviour of Fig. 4a is reproduced by numerical simulations if very large ($N_A = 3 \times 10^{17}$ cm$^{-3}$) body doping is assumed. For $V_{BG} = 0$ V, the transistor would be only partially depleted: the undepleted P-doped region connects the source and drain, enabling the leakage current to flow (Fig. 4b). For $V_{BG} = +60$ V, the transistor becomes fully depleted, suppressing the leakage path.

Atomistic simulations indicate that hydrogen passivation in Ge may be responsible for parasitic P-type doping [10]. At the experimental level, it is known that the threshold voltage shift at low temperature depends on the Fermi level variation and yields a precise doping value. Our recent measurements (Fig. 5)

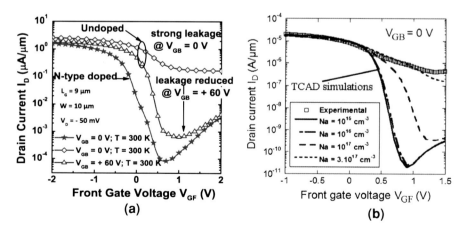

Fig. 4 a Drain current versus front-gate bias in two GeOI pMOSFETs. The severe leakage observed in the undoped transistor is reduced by applying $V_{BG} = +60$ V. The intentionally doped transistor features well behaved characteristics with no excessive leakage. b Simulated curves for various P-type parasitic doping levels

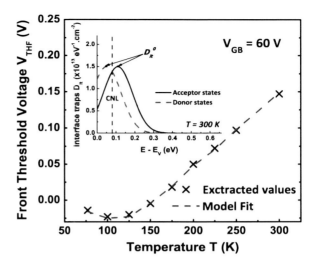

**Fig. 5** Front-channel threshold voltage versus temperature in 9 μm long GeOI pMOSET. The asymmetric double $D_{it}$ distribution (*inset*) reproduces the low temperature behavior as well as the experimental characteristics of Fig. 4a

demonstrated that the doping level is very low ($4 \times 10^{15}$ cm$^{-3}$) in GeOI MOSFETs [11]. This means that the leakage current cannot be attributed to large body doping (Fig. 4b).

An alternative scenario assumes the co-existence of donor and acceptor interface states, both related to unpassivated dangling bonds and located near the valence band edge [12]. The two Gaussian distributions illustrated in the inset of Fig. 5 enabled us to reproduce the $V_{th}(T)$ behaviour (Fig. 5) as well as the measured $I_D(V_G)$ characteristics (Fig. 4a). The acceptor states are responsible for a negative charge large enough to bring the back interface into weak/strong inversion regime even for $V_{BG} = 0$ V. This negative charge also explains the positive threshold voltage in P-MOSFETs and the poor performance of N-MOSFETs, where the inversion layer formation is delayed and the electron mobility degraded.

Thus our results indicate that parasitic conduction is not caused by doping contamination of the Ge channel. We can explain the data by invoking interface traps with a particular energy distribution. But perhaps there exists yet a third explanation...

## 5 Are there Strong Competing Effects of Strain and Neutral Defects in Short Channel MOSFETs?

Strain is an efficient booster of the carrier mobility. A popular solution is to transfer strain from the Contact Etch Stop Layer (CESL) into the body [13]. The strain intensity depends on the transistor geometry and dimensions. CESL-engineered strain can be tuned to maximize performance of short transistors.

Figure 6 shows the effect of compressive strain on the hole mobility. The mobility gain reaches an impressive maximum (+80%) for 100 nm long P-MOSFETs. Then, for even shorter devices, the mobility decays.

The mobility variation with channel length indicates the strain localization at the channel extremities. Detailed mechanical simulations confirm the presence of narrow ($\sim$ 50 nm) 'stressed pockets' at the corner between source/drain regions and offset spacer [14]. In a long MOSFET ($\sim$ 1 µm), most of the channel region remains unstressed. By contrast, in sub-100 nm long devices, the strain is strong and uniformly distributed along the channel.

An intriguing feature in Fig. 6 is the superior mobility improvement at the back channel. It is clear that the strain effect is transmitted from CESL through the ultrathin film. However, is there a logical reason why the strain increases in the vertical direction, as we go from the top to the bottom interface?

The next question concerns the severe degradation in hole mobility for devices shorter than 100 nm. Among several competing mechanisms, we can cite:

1. Stress relaxation: disagrees with numerical simulations.
2. Series resistance: our mobility extraction method eliminates these effects.
3. Semi-ballistic transport, which would give $\mu \sim L$, is unlikely to occur above 30–50 nm.
4. Additional scattering mechanisms in short devices, presumably related to neutral defects [15].

Neutral defects, generated during the source/drain implantation or the gate stack processing [16], are concentrated near the source/drain junctions. Long devices are free from their influence whereas, in short transistors, the defective 'edge' regions overlap leading to an increased density of defects. Also possible is the *Remote Coulomb Scattering* due to lateral depletion in the source/drain regions, next to the metallurgical junction [17].

Mobility measurements at low temperature bring additional information, as shown in Fig. 7. The mobility is expected to increase steadily from 300 K to 77 K as a consequence of attenuated phonon scattering. This variation is indeed observed for long channels (Fig. 7) and also for short or long back channels [17]. However, the front mobility in short MOSFETs shows an unusual saturation. The benefit of strain, visible at 300 K, tends to disappear at 77 K.

The mobility dependence on temperature and gate length reveals the adverse effects of strain, neutral defects and Coulomb scattering rate, which all are highly inhomogeneous along the channel. A recent analytical model [17], combining these ingredients with the Matthiessen rule, was able to reproduce the experimental results of Figs. 6 and 7. But with enough adjustable parameters, any experiment can be matched... Is Matthiessen's rule able to sort out all possible contributions?

Two arguments can be invoked to explain why the back-channel mobility is higher and exhibits a more predictable evolution with temperature. First, the channel is located at a Si–SiO$_2$ interface and does not suffer from scattering on the Coulomb centers located in the high-K top dielectric. Second, there may be less

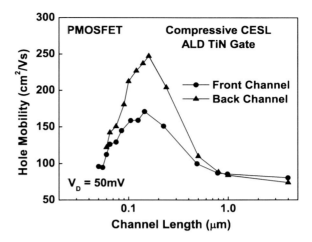

**Fig. 6** Hole mobility versus channel length in SOI MOSFETs with compressive CESL stressor. 8 nm thick SOI film

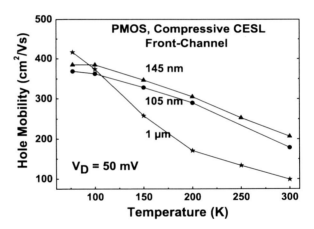

**Fig. 7** Hole mobility versus temperature in long and short SOI MOSFETs with compressive CESL

impact from the neutral defects. However, it is unclear whether one can reasonably assume a vertical distribution of defects when the Si film is less than 10 nm thick.

## 6 What is the Origin of Hysteretic Effects in Multiple-Gate MOSFETs?

The floating body of SOI transistors can be manipulated such as to trigger strong hysteresis effects, which are useful for memory applications: meta-stable dip (MSD), bipolar transistor latch, etc. [18]. The principle of the capacitorless

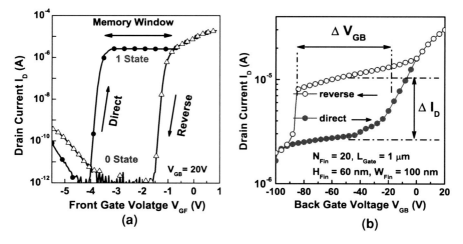

**Fig. 8 a** MSD hysteresis induced by scanning the front gate of a fully depleted MOSFET. **b** MSD-like hysteresis in Triple-Gate FinFET induced by scanning the back gate

single-transistor dynamic memory (1T-DRAM) is to modify the amount of majority carriers stored in the body for achieving two distinct values of drain current.

Figure 8a illustrates the MSD hysteresis effect used in MSDRAMs [19]. The back gate $V_{GB}$ is biased in moderate inversion and the front gate $V_{GF}$ is scanned from depletion to strong accumulation and vice versa. When $V_{GF}$ is switched from 0 V to $-3$ V, the body goes into deep depletion and the drain current is suppressed. This 0-state is maintained as long as the front accumulation channel is not formed. The 1-state is programmed by increasing $V_{GF}$ to $-6$ V, at which point band-to-band tunneling at the drain supplies enough holes to complete the accumulation channel. Switching $V_{GF}$ to $-3$ V results in a large electron current at the back channel simply because the transistor is in equilibrium mode.

We have tried to reproduce the MSD hysteresis in Triple-Gate SOI FinFETs. The front-gate voltage is biased in inversion, while the back-gate voltage is swept back and forth from positive to negative values, i.e., the body/BOX interface varies between depletion and accumulation. As shown in Fig. 8b, the hysteresis on $I_D(V_{GF})$ curves is strong enough for memory applications. Surprisingly, the sense of hysteresis is the reverse of what we observe in the MSD effect of Fig. 8a. Is this contrast due to the 3D configuration of FinFETs?

When the back-gate voltage is swept from negative to positive value (direct sweep), the majority carrier generation is efficient enough to build the back accumulation layer 'instantly'. The device is at equilibrium and the current flows in the top and lateral channels. The current variation simply reflects the gradual decrease of the front threshold voltage with back-gate bias.

During the reverse sweep, the back channel moves from depletion into accumulation regime. However, the back accumulation layer cannot be formed immediately, as the hole generation is too slow. One could expect the body to reach deep depletion, cutting off the front inversion channel. Nevertheless, this prediction is not confirmed in Fig. 8b, where the 'reverse' current remains surprisingly high, larger than the 'direct' current.

Why is the MSD effect upside-down in FinFETs? Presumably, the body does not experience deep depletion because the electrostatics is dominated by the front gate, which covers three sides of the fin. This explains why the front current is not switched off. Since the back interface is depleted (not accumulated as for the 'direct' scan), the front threshold voltage is comparatively low and the current is higher. The two current levels merge for back-gate voltage below −80 V, when enough holes become available and accumulate at the BOX interface. However, the way that holes are generated when the back-gate bias is scanned remains puzzling: band-to-band tunnelling near the very thick BOX or front-gate leakage current induced by back-gate coupling?

Two experimental facts bring light on the origin of this specific current behaviour. First, bringing light actually suppresses the hysteresis effect because photo-generated holes promptly complete the accumulation charge. Second, in very narrow fins, the MSD hysteresis tends to disappear due to the neutralization of the back-gate effect which is blocked by the lateral gates. Will we see 1T-DRAMs based on this reciprocal MSD effect?

Memory effects can be enhanced by functionalizing the BOX. Our measurements were performed on Triple-Gate FinFETs with oxide-nitride-oxide (ONO) BOX [20]. The $Si_3N_4$ layer is used to store the charge for volatile or non-volatile memorization. Hysteresis effects occur when the back-gate voltage $V_{GB}$ is sufficiently high to inject/remove electrons into/from the nitride film (Fig. 9).

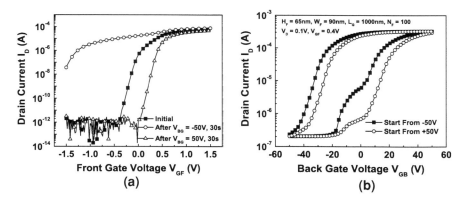

Fig. 9 Hysteresis in Triple-Gate FinFETs with ONO BOX. a Current versus front-gate bias measured at zero back-gate voltage. The hysteresis is due to the pre-charging of the BOX. b Current versus back-gate bias measured with the front-gate voltage in moderate inversion. The hysteresis results from the dynamic change of the charge trapped in the BOX

The combination of coupling and trapping effects is illustrated in Fig. 9a. $I_D(V_{GF})$ curves were measured at $V_{GB} = 0$, after long electron trapping ($V_{GB} = +50$ V for 30 s) or detrapping ($V_{GB} = -50$ V for 30 s). The hysteresis denotes a shift in front-channel threshold voltage corresponding to the amount of charge trapped in the $Si_3N_4$ layer. Excess electrons in the BOX decrease the back surface potential. By electrostatic coupling, the front-channel threshold voltage increases and shifts the curves to the right. The opposite effect is observed after electron detrapping.

Unlike flash memories, where the carrier flow and trapping occur at the same interface, in ONO FinFETs it is the remote trapping in the BOX which induces the front-gate threshold voltage variation. The advantage is that the carrier flow no longer disturbs the stored charge.

Dynamic memory effects are shown in Fig. 9b. The drain current was measured as a function of back-gate bias, back and forth, after an initial hold time of 30 s. When $V_{GB}$ is initially negative, holes are trapped in the ONO, hence the front-channel threshold voltage is low and the current is high. As $V_{GB}$ increases to positive values, electrons are dynamically trapped and increase the threshold voltage. As a consequence, the 'reverse' curve shows much lower current. Starting the measurements from positive to negative $V_{GB}$ leads to similar conclusions, except that the curves are shifted to the right.

The current hysteresis (memory window) is amplified when the initial starting voltage is higher because the amount of detrapped electrons (or trapped holes) increases. The memory window decreases in narrow fins where the fringing fields from the lateral gates tend to prevent Fowler–Nordheim tunneling by blocking the ONO potential.

An interesting feature in Fig. 9b is the hump observed for 'reverse' scans. We speculate that the hump reflects the contribution of the back channel current, which is no longer masked by the front channel when its threshold voltage is large. A more important question refers to the applicability of this ONO hysteresis effect for developing 'unified' memories combining volatile and non-volatile functionalities.

## 7 Why is the Mobility Degraded in Small Nanowires?

Gate-all-around (GAA) nanowire transistors (NWTs) are most promising candidates for future CMOS due to reduced short-channel effects. Simulation results indicate that the gate lengths in these devices can be scaled down to as small a dimension as the NW diameter itself [21].

3D-stacked Si and SiGe NWTs with high-K/metal gate stacks were fabricated at LETI (Fig. 10a). Circular shape NWTs were formed by annealing in hydrogen ambient [22]. All NWTs exhibit near-ideal subthreshold swing ($\sim 60$ mV/dec) and very low DIBL (7 mV/V) for L = 100 nm gate length. On-currents at $V_{DD} = 1.2$ V are equal to 13.3 mA/μm and 7.5 mA/μm in N- and P-channel

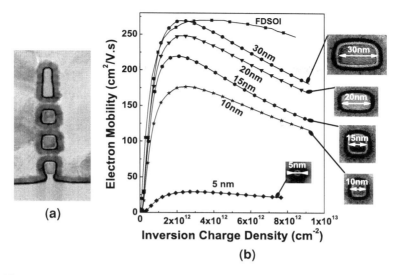

**Fig. 10** **a** Vertically stacked 3D nanowire transistors with high-K/metal gate and **b** effective electron mobility versus inversion charge for NWT with different cross-sections

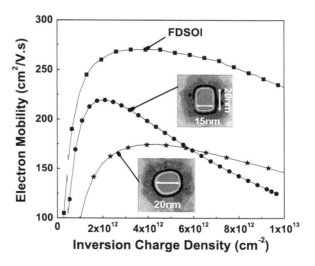

**Fig. 11** Effective mobility versus inversion charge in nanowire transistors with rectangular and circular shape

NWTs, respectively. These extremely high values are due to the 3D configuration of the 3-level vertically-stacked NWTs and to the normalization by the channel width as seen from the top.

In all NWTs, the effective mobility drops as the width shrinks below 10 nm [22]. In particular, very poor mobility is observed in 5 nm circular NWT (Fig. 10b). Is this degradation related to subband splitting and carrier/phonon confinement?

The comparison between rectangular and circular NWTs with comparable size is shown in Fig. 11. Rectangular NWTs exhibit lower mobility than FD planar MOSFETs which is partly due to the contribution of (110) sidewalls. A circular shape leads to marked mobility degradation at low inversion charge. In this region, the mobility is limited by Coulomb scattering involving oxide charges, high-K dipoles and interface traps. Why is this type of scattering more pronounced in circular NWTs?

Charge pumping measurements indicate a threefold increase in the density of traps despite the beneficial effect of H-anneal [22, 23]. Is it due to the continuously varying surface orientation in circular NWTs?

Figure 11 also shows that in very strong inversion, where surface roughness scattering prevails, the mobility is higher for the circular NWTs. We believe that the surface roughness has been improved during hydrogen annealing.

A more generic debate is open on the applications of nanometer-size NWTs. The combination of mesoscopic properties and technological limitations leads to concerns about variability issues. Can the extreme sensitivity of nanowires to fluctuations and surface charges be turned into a practical advantage? Can the NWTs be used for memory and sensor applications?

## 8 How can the Tunneling Current be Discriminated in Tunnel-FETs?

The tunneling field-effect transistor (TFET) [24] is of interest due to its complete CMOS process compatibility. The TFET structure is similar to that of MOSFET, except that the dopant types in source and drain are different. This gated PIN diode is operated in reverse mode. The current is induced by band-to-band tunneling (BTBT), which makes it theoretically possible to reach extremely low OFF currents as well as an ultra-low subthreshold slope (S) below the ideal MOSFET value of 60 mV/decade at room temperature [25, 26]. In order to enhance the ON current, TFETs with multiple-gate structure [27] and lower bandgap semiconductors, such as Ge and SiGe, have been reported [28, 29].

One difficulty is to simultaneously achieve high $I_{ON}$, low S, and low leakage. In symmetrical TFETs, the leakage current under opposite gate bias is large, because interband tunneling occurs at either the source-channel or the drain-channel junction depending on the sign of the gate bias. This problem was solved in our TFETs by inserting an intrinsic regions $L_{IN}$ separating the drain contact from the channel (Fig. 12a). Since the BTBT rate is directly determined by the maximum electric field $E_{max}$, it can be largely reduced at the drain side due to the increase of $L_{IN}$. At the source side $E_{max}$ does not change and the tunneling rate in ON state is constant.

Figure 12b shows the experimental results for NTFETs with different $L_{IN}$. In agreement with the simulation, the leakage current can be effectively suppressed by increasing $L_{IN}$ to 50 nm. It should be noted that the threshold voltage and S are

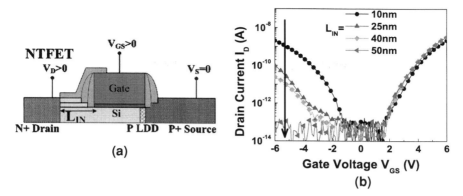

**Fig. 12 a** Configuration of asymmetric N-channel TFET. The gate is positively biased and the BTBT occurs at the P$^+$ source junction. **b** Experimental current versus gate voltage characteristics for variable length of the intrinsic region (L = 400 nm, $V_D$ = 1 V)

independent of gate length. This is due to the fact the tunneling current, governed by the maximum field at the tunneling junction, is unaffected by the carrier transport in the channel.

A challenging aspect is to assess the precise origin of $I_{ON}$. In general, the tunneling current is smaller than theoretically predicted by simple models and not clearly distinguishable from other possible current sources. Our low-temperature measurements indicate that the temperature variation of $I_{ON}$ follows the Kane tunneling model [30], where temperature-induced bandgap changes determine the tunneling rate. But, is there any other mechanism likely to contribute to $I_{ON}$? Thinner gate oxides do improve the subthreshold characteristics and ON current. Nevertheless, the subthreshold swing in our devices is never below 60 mV/decade at 300 K. Why?

Digging further, we performed noise measurements. In very small area MOSFETs, the typical 1/$f$ noise, due to carrier trapping and detrapping, turns into 1/$f^2$ RTS noise (Random Telegraph Signal). In TFETs with *large* area, RTS noise is always observed, suggesting that only a discrete number of traps are active. This tends to confirm that the dominant current mechanism is tunneling. Indeed, the tunneling rate is only affected by the trapping process at the Si–SiO$_2$ interface just above the tunneling junction which is very narrow (around 10 nm). Is this argument convincing enough? How can we enhance the tunneling current?

# 9 Conclusions

In this paper, we have raised a number of questions related to state-of-the-art SOI materials and devices. Each topic was illustrated with measurements and/or simulations. Tentative explanations have been proposed, even if they still are at a preliminary stage. These issues have been selected to develop curiosity and stimulate further research. But, the list of questions is far from being exhausted.

**Acknowledgments** We would like to thank our supporting organizations (Eurosoi+, Nanosil, WCU, etc.) and our SOI colleagues.

# References

1. Cristoloveanu, S., Li, S.S.: Electrical Characterization of Silicon-On-Insulator Materials and Devices. Kluwer Academic, Boston (1995)
2. Ghibaudo, G.: Low-frequency noise and fluctuations in advanced CMOS devices. In: Proceedings of SPIE, vol. 5113, pp. 16–28. Presented at Noise in Devices and Circuits, Santa Fe, USA (2003)
3. Hamaide, G., Allibert, F., Hovel, H., Cristoloveanu, S.: Impact of free-surface passivation on silicon on insulator buried interface properties by pseudo transistor characterization. J. Appl. Phys. **101**, 114513 1–6 (2007)
4. Rodriguez, N., Cristoloveanu, S., Gamiz, F.: Revisited pseudo-MOSFET models for the characterization of ultrathin SOI wafers. IEEE Trans. Electron Devices **56**, 1507–1515 (2009)
5. Shang, H., Frank, M., Gusev, E., et al.: Germanium channel MOSFETs: opportunities and challenges. IBM J. Res. Dev. **50**(4–5), 377–386 (2006)
6. Mitard, J., De Jaeger, B., Leys, F.E., et al.: Record ION/IOFF performance for 65 nm Ge pMOSFET and novel Si passivation scheme for improved EOT scalability. Tech. Dig. IEDM art. no. 4796837 (2008)
7. Akatsu, T., Deguet, C., Sanchez, L., et al.: Germanium-on-insulator (GeOI) substrates: a novel engineered substrate for future high performance devices. Mater. Sci. Semiconductor Process. **9**(4–5 SPEC. ISS), 444–448 (2006)
8. Romanjek, K., Hutin, L., Le Royer, C., et al.: High performance 70 nm gate length germanium-on-insulator pMOSFET with high-k/metal gate. Solid State Electron. **53**(7), 723–729 (2009)
9. Nguyen, Q.T., Damlencourt, J.F., Vincent, B., Clavelier, L., Morand, Y., Gentil, P., Cristoloveanu, S.: High quality germanium-on-insulator wafers with excellent hole mobility. Solid State Electron. **51**, 1172–1179 (2007)
10. Weber, J.R., Janotti, A., Rinke, P., et al.: Dangling bond defects and hydrogen passivation in germanium. Appl. Phys. Lett. **91**(14), 142101 (2007)
11. Van Den Daele, W., Augendre, E., Le Royer, C., Damlencourt, J.F., Grandchamp, B., Cristoloveanu, S.: Low temperature characterization and modeling of advanced GeOI pMOSFETs: mobility mechanisms and origin of the parasitic conduction. Solid State Electron. **54**, 205–212 (2009)
12. Dimoulas, A., Tspias, T.: Germanium surfaces and interfaces. Microelectronic Eng. **86**(7–9) 1577–1581 (2009)
13. Eneman, G., Verheyen, P., Keersgieter, A.D., Jurczak, M., De Meyer, F.: Scalability of stress induced by contact-etch-stop layers: a simulation study. IEEE Trans. Electron Devices **54**, 1446–1453 (2007)
14. Payet, F., Boeuf, F., Ortolland, C., Skotnicki, T.: Nonuniform mobility enhancement techniques and their impact on device performance. IEEE Trans. Electron Devices **55**, 1050–1057 (2008)
15. Cros, A., Romanjek, K., Fleury, D., Harrison, S., Cerruti, R., Coronel, P., Dumont, B., et al.: Unexpected mobility degradation for very short devices: a new challenge for CMOS scaling. Proc. IEDM art. no. 4154291 439–442 (2006)

16. Pham-Nguyen, L., Fenouillet-Beranger, C., Vandooren, A., Skotnicki, T., Ghibaudo, G., Cristoloveanu, S.: In situ comparison of Si/High-K and Si/SiO$_2$ channels properties in SOI MOSFETs. IEEE Electron Device Lett. **30**, 1075–1077 (2009)
17. Pham-Nguyen, L., Fenouillet-Beranger, C., Ghibaudo, G., Skotnicki, T., Cristoloveanu, S.: Mobility enhancement by CESL strain in short-channel ultrathin SOI MOSFETs. Solid State Electron. **54**, 123–130 (2010)
18. Bawedin, M., Cristoloveanu, S., Flandre. D.: Innovating SOI memory devices based on floating-body effects. Solid State Electron. **51**, 1252–1262 (2007)
19. Bawedin, M., Cristoloveanu, S., Flandre, D.: A capacitor-less 1T-DRAM on SOI based on double gate operation. IEEE Electron Device Lett. **29**, 795–798 (2008)
20. Chang, S.J., Na, K.I., Bawedin, M., Bae, Y.H., Park, K.H., Lee, J.H., Xiong, W., Cristoloveanu, S.: Investigation of hysteresis memory effects in SOI FinFETs with ONO buried insulator. 2010 IEEE International SOI Conference, San Diego, CA, USA, 2010
21. Colinge, J.P.: Multiple-gate SOI MOSFETs. Solid State Electron. **48**, 897–905 (2004)
22. Tachi, K., Casse, M., Jang, D., Dupré, C., Hubert, A., Vulliet, N., Maffini-Alvaro, V., Vizioz, C., Carabasse, C., Delaye, V., Hartmann, J.M., Ghibaudo, G., Iwai, H., Cristoloveanu, S., Faynot, O., Ernst, T.: Relationship between mobility and high-k interface properties in advanced Si and SiGe nanowires. IEDM Tech. Dig. art. no. 5424360 12.7.1–12.7.4 (2009)
23. Casse, M., Tachi, K., Thiele, S., Ernst, T.: Spectroscopic charge pumping in Si nanowire transistors with a high-kappa/metal gate. Appl. Phys. Lett. **96**, 3–123506 (2010)
24. Reddick, W.M., Amaratunga, G.A.J.: Silicon surface tunnel transistor. Appl. Phys. Lett. **67**, 494–496 (1995)
25. Choi, Y., Park, B.G., Lee, J.D., Liu, T.K.: Tunneling field-effect transistors (TFETs) with subthreshold swing (SS) less than 60 mV/dec. IEEE Electron Device Lett. **28**, 743–745 (2007)
26. Mayer, F., Le Royer, C., Damlencourt, J.F, Romanjek, K., Andrieu, F., Tabone, C., Previtali, B., Deleonibus, S.: Impact of SOI, Si$_{1-x}$Ge$_x$OI and GeOI substrates on CMOS compatible tunnel FET performance. IEDM Tech. Dig. art. no. 4796641 163–166 (2008)
27. Leonelli, D., Vandooren, A., Rooyackers, R., Verhulst, A.S., De Gendt, S., Heyns, M.M., Groeseneken, G.: Performance enhancement in multi gate tunneling field effect transistor by scaling the fin-width japanese. J. Appl. Phy. **49**(4 Part 2), art. no. 04DC10 (2010)
28. Krishnamohan, T., Kim, D., Raghunathan, S., Saraswat, K.: Double-gate strained-Ge heterostructure tunneling FET (TFET) with record high drive currents and ≪ 60 mV/dec subthreshold slope. IEDM Tech. Dig. art. no. 4796839 947–949 (2008)
29. Kazazis, D., Jannaty, P., Zaslavsky, A., Le Royer, C., Tabone, C., Clavelier, L., Cristoloveanu, S.: Tunneling field-effect transistor with epitaxial junction in thin germanium-on-insulator. Appl. Phys. Lett. **94**(26), art. no. 263508 (2009)
30. Kane, E.O.: Theory of tunneling. J. Appl. Phys. **32**(1), 83–91 (1961)

# Index

**μ**
μ-Czochralski approach, 69

**2**
2D interconnection routing, 68
2-D microcantilevers, 327
2G strained Si MOSFET, 128

**3**
3D integration, 68
3D integration of Nanowires, 132
3-D MEMS, 377
3-D microbeams, 380
3-D microcantilevers, 377
3D structure, 142
3D Volmer-Weber type growth, 14

**6**
6T-SRAM, 145

**A**
Accumulation-mode pi-gate transistor, 180
Acoustic phonon confinement, 172
Acoustic phonon-dominated scattering, 309
Ambipolar transport, 224
Anodization process, 51
Anti-reflection coating, 363

**B**
Back bias, 164
Ballistic mobility, 314
Ballistic mobility, 242
Ballistic transport, 227, 230
Ballisticity, 125, 128
Beyond-CMOS devices, 123
Biaxially strained devices, 283
Biaxially strained triple gate devices, 278
Bio particles, 319
Biochip, 351
Bioliquid cell, 351
Biosensors, 347
Bolometers, 383
Boltzmann transport equation, 219
Bovine serum albumin, 352

**C**
Capacitance plateau, 325
Capacitor-less 1T-DRAM, 134
Carbon nanotube field-effect transistors, 216
Carbon nanotubes, 216
Centroid of the inversion charge, 158
Channel doping, 156
Charge centroid, 329, 332
CMOS circuits, 111
CMOS ICs, 1
CMOS technologies, 92
CMOS technology, 4
CMOS-MEMS, 357
CNT structures, 133
Coaxially-gated CNTFET, 221
Compliance approach, 48
Confinement effective mass, 170
Contact etch-stop layer, 165
Coulomb, 283
    blockade, 253
    blockade oscillations, 254
    blockade phenomena, 252
Coulomb scattering, 238

**C** (*cont.*)
Coupled MOS-SET, 257
CPW line, 106, 113
CPW transmission line, 106
Cut-off frequencies, 105
Cut-off frequency, 99

**D**
Deep Reactive Ion Etching, 356
Density of state, 202
Deoxidation, 60
DG mode, 332
DG-FinDRAMs, 134
DG-MOSFET, 229
DGSOI, 174
DGSOI devices, 178
DIBL, 190
Digital substrate noise, 110, 111
Discretized Hamiltonian depends, 178
Dopant fluctuations, 169
Double etch scheme, 147
Double gate transistors, 417
Double-gate MOSFET, 128
Drain Induced Barrier Lowering, 158
DRAM, 327
DRIE, 356
Drift-diffusion mobility, 315
Dynamic performance, 161
Dynamic Threshold MOS, 271

**E**
Effective mobility, 328
Effective oxide thickness, 25, 287
Electron conduction band tunneling, 221
Electron-phonon interaction, 239
Electrostatic Integrity, 191
Electrostatics, 156, 162
Epitaxial lateral overgrowth, 21
Equivalent oxide thickness, 324
EVB tunneling, 295
Excimer laser crystallization, 69
Extrinsic diffusivity coefficient, 10
Extrinsic transconductance, 98

**F**
Fast Coupled Mode Space, 203
FD SOI MOSFET, 324
FD-SOI devices, 311
FD-SOI MOSFET, 269
FDSOI transistor, 159
Fermi Golden Rule, 182

Ferro-electric FET, 202
Film thickness, 159, 163
FinFET, 19, 96, 99, 134, 299
  tri-gate, 274
FinFlash, 135
Finite element analysis simulations, 367
FinTFTs, 69
Flash lamp annealing, 10
Floating body cells, 134
Floating body effects, 256
Front-gate split C-V measurements, 325
Full potential of SOI technology, 169
Fully Depleted, 268
Fully-depleted SOI MOSFET, 323

**G**
GAA transistors, 308
GAA-Gate All Around, 128
Gate induced floating body effect, 273
Gate workfunction, 157, 158
Gate-to-channel split, 308
Gate-to-channel tunneling current, 333
GeOI photodetectors, 18
GeOI substrates, 11
Germanium on Quartz, 22
Germanium on sapphire, 22
Germanium-on-insulator, 1
GHz range, 92
Global effective mobility, 319
Global straining technique, 303
GR fluctuations, 288
Graded-Channel SOI MOSFET, 270
Graphene, 194
Ground-plane, 154

**H**
Hafnium dioxide, 6
Harmonic distortion, 113
High resistivity silicon, 92
High-k, 311
High-k gate dielectric, 92
High-k gate stack, 311
High-k layer, 304
Hole mobility, 178

**I**
I-MOS, 202
Inelastic scattering, 246
Interlayer dielectrics, 68
Intra-subband acoustic phonons, 219
Inversion-layer centroid, 331

Index 445

Inversion-mode pi-gate MOSFET, 189
Ion beam induced epitaxial crystallisation, 8
Irreversible plastic, 49

**J**
Junctionless gated resistor, 186
Junctionless pi-gate device, 189
Junctionless transistor, 186

**K**
Kubo-Greenwood formula, 170

**L**
Lab-on-chip, 351
Lateral crystallisation proceeds, 20
Lattice mismatches, 47
Lectron backscattered diffraction, 76
Lift-off process, 72
Lightly doped drain, 95
Lim-Fossum model, 326
Line edge roughness, 147
Line width roughness, 147
Linear kink effect, 273
Longitudinal confinement, 207
Lorentzian fluctuations
    Back-Gate-Induced, 287
    Linear Kink Effect, 287
Low Schottky barrier contacts, 96
Low-field mobility mobility, 247
Low-temperature direct wafer bonding, 32
LPCVD, 72
LPCVD-deposition, 108
LPE germanium, 18

**M**
Mathiessen rule, 318
Matthiessen-rule-like expression, 314
Meier-Wingreen formalism, 240
Metal-induced lateral crystallization, 68
Metallic SET, 257
Metal-Oxide-Semiconductor
    Field-Effect-Transistors, 287
Metal-Oxide-Semiconductor Single
    Electron Transistor, 251
*Metamorphic growth*, 48
Micro-Electro-Mechanical Systems, 356
Microfluidic biochips, 345
Microheater, 371
Midgap interface state density, 5
MOCVD, 10

Model of Gibson and Ashby, 50
Modulated Barrier Resonant
    Tunneling effect, 202
MOS-SET, 257
Multi-bridge-channel MOSFET, 128
Multi-channel thin film MOSFET, 123
Multi-gate MOSFET, 126
Multiple Gate Field Effect Transistors, 277
Multi-Users Multi-Chips, 357

**N**
Nano-biosensors, 361
Nanochip, 344
Nano-interdigitated arrays, 358
Nanoribbons, 70, 76, 195
Nanotubes
    "armchair-type", 218
    "zigzag-type", 218
Nanowire, 251, 252, 343
Nanowire FET, 238
Nanowire field effect transistors, 251
Nanowire-nFET, 124
NEMS, 377
nIDA capacitance sensors, 358
nIDA-gate, 359
nIDA-gate MOSFET, 358
Nitrogen radical activated samples, 35
Noise
    Generation-Recombination, 288
    McWhorter, 291
    Nyquist, 288
Non volatile SRAM, 136
Non-equilibrium Green's function, 217
Non-translational invariant movement, 207
Nuclear Reaction Analysis, 8
Null transverse field condition, 157
NW FET, 343, 345
NW sensors, 344

**O**
Orientation effects, 175
Oswald Ripening Mechanism, 40

**P**
Paramorphic approach, 48
Partially Depleted, 268
PECVD, 17
Performance, 173
PH-limited, 247
Phonon scattering, 193
Phonon scattering, 172, 173

**P** (cont.)
Phonon scattering rate, 173
Piezoresistor, 385
PIN diode, 363
Plasma Enhanced Chemical Vapor Deposition, 365
Polysilicon microheater, 369
Porosification, 52, 56
Porous silicon oxidation, 49
Protein BSA molecules, 344
PS heterostructures, 54
*Pseudomorphic* growth, 48
Pseudo-MOSFET, 82
Pseudosubstrate, 49, 60

**Q**
QG-Quadruple Gate, 128
QM "darkspace", 328
QM effect, 326, 327
QM effective gate dielectric thickening, 327
QM modes, 326
Quantum confinement, 157
Quantum effects, 170
Quantum-mechanical effects, 324

**R**
Radio frequency, 92
Random Dopant Fluctuation, 163
Reactive Ion Etching, 360
Remote Coulomb scattering, 311, 314
Remote-charge scattering, 237
Resonant tunneling effect, 209
Resonant tunneling -FET, 202
Reversible elastic, 49
RF applications, 112
RF FinFET, 101
RF performance, 99, 100
RMS roughness, 24
Roughness RMS, 242

**S**
Scalability, 162
SCE, 97
Schottky barrier, 202
Schottky barrier MOSFET, 97, 98
Schottky barrier CNTFET, 222
Schottky contacts, 126
Schottky-like source and drain contacts, 216
Screen–printing, 367
Self-aligned tungsten gate process, 6

Self-assembled nanoporous materials, 69
Self-consistent Born approximation, 5
Self-consistent three-dimensional Non-Equilibrium Green Function quantum simulator, 202
Semiconductor nanowires, 69
Series resistance, 159
SG mode, 332
SG MOSFET, 105
Short channel effect, 141
Short channel effects, 158, 169, 253, 307
Short time anneal, 4, 38
Short-Channel Effect, 190
Short-channel effects, 99, 134, 191, 323
Silicon nanowire, 25, 138
Silicon SET, 239, 257, 245
Silicon-on-Nothing, 364
Silicon-on-sapphire, 22
Single dopant, 251, 261
Single dopant effects, 252
Single electron effects, 255
Single gate SOI, 127
Single gate transistor, 134
Single particle Monte Carlo method, 170
SiOC hard mask, 147
Six-band k•p model, 170
Small slope switches, 132
Smart-cut process, 23
Smart-Cut SOI, 85
SOI and Germanium on Insulator substrates, 38
SOI fully depleted, 308
SOI Gate-all-Around, 211
SOI MOSFET, 262
SOI nanowire transistor, 344
SOI NW FET, 344
SOI PIN photodetector, 363
SOI RF-MEMS, 357
SOI-CMOS-MEMS, 356
SOI-nanowires transistors, 345
Solid-phase crystallization, 67, 68
SONOS, 136
Source/drain engineering, 133
S-parameters, 102
Spin-on-dopant, 10
Spreading Resistance analysis, 13
SR scattering, 174
SRAM, 123, 146, 150
SR-limited, 238, 247
SR-limited mobility, 243
Stability diagram, 257
Static noise margin, 163
Sticking/unsticking, 52
Strained MOSFET, 96

Index                                                                                    447

Strained silicon on insulator, 123
Strained-SOI, 166
Stress memorization technique, 166
Subband modulation, 170
Subband modulation effect, 174
Subband spatial fluctuations, 248
Subthreshold slope, 196, 272
Super lateral growth, 69
Super-cell, 30
Supersaturation, 70
Surface roughness, 109, 172
Surface-roughness scattering, 238
Suspended-Gate FET, 202
System-on-a-chip, 357
System-on-chip, 91, 92
Systems-on-chip, 110

T
TCAD simulations, 320
TeraHertz sensors, 356
Thermal coefficients of expansion, 11
Thermally induced decomposition, 75
Threshold voltage, 143, 155
Thru-Line-Reflect method, 107
Top-down approach, 343
Transconductance, 198, 216, 267
Transferred germanium layer, 44
Transferred layer, 13
Transport anisotropy, 177
Transport boosters, 164
Trigate MOSFET, 196
Tunnel barriers, 253
Tunnel FET, 133
Tunnel-FET, 202

U
Ultrashallow junctions, 187
Ultra-thin body, 96
Ultrathin buried oxide, 155
Ultra-thin FDSOI, 125
Ultrathin SOI, 170, 178
Ultra-thin SOI devices, 172
Uncooled infrared, 356
Uncooled IR sensors, 356
Underlap geometry, 356
Uniaxially strained devices, 252, 259
UTB SOI, 283
UTB SOI MOSFET, 324, 336, 327

V
Vapor-liquid-solid catalytic
        growth, 67, 69
Variability, 163
VLS growth, 79
Voltage Controlled Oscillators, 371
Voltage-Doping Transformation, 190
Volume inversion, 170

W
Wavefunction confinement, 180
Wheatstone bridge, 374
Wheatstone structure, 387
Wigner current, 231
Wigner Monte Carlo
        simulation, 217
Wigner-Boltzmann equation, 220